WITHDRAWN
WRIGHT STATE UNIVERSITY LIBRARIES

Mathematics and Visualization

Series Editors

Gerald Farin
Hans-Christian Hege
David Hoffman
Christopher R. Johnson
Konrad Polthier
Martin Rumpf

Lars Linsen
Hans Hagen
Bernd Hamann

Editors

Visualization in Medicine and Life Sciences

With 127 Figures, 35 in Color

Lars Linsen
School of Engineering and Science
Jacobs University Bremen
P.O. Box 750561
28725 Bremen, Germany
E-mail: l.linsen@jacobs-university.de

Hans Hagen
Technische Universität Kaiserslautern
67653 Kaiserslautern, Germany
E-mail: hagen@informatik.uni-kl.de

Bernd Hamann
Department of Computer Science
University of California
One Shields Avenue
Davis, CA 95616-8562, U.S.A
E-mail: hamann@cs.ucdavis.edu

Library of Congress Control Number: 2007935102

Mathematics Subject Classification: 68-06, 68U05

ISBN-13 978-3-540-72629-6 Springer Berlin Heidelberg New York

This work is subject to copyright. All rights are reserved, whether the whole or part of the material is concerned, specifically the rights of translation, reprinting, reuse of illustrations, recitation, broadcasting, reproduction on microfilm or in any other way, and storage in data banks. Duplication of this publication or parts thereof is permitted only under the provisions of the German Copyright Law of September 9, 1965, in its current version, and permission for use must always be obtained from Springer. Violations are liable for prosecution under the German Copyright Law.

Springer is a part of Springer Science+Business Media
springer.com
© Springer-Verlag Berlin Heidelberg 2008

The use of general descriptive names, registered names, trademarks, etc. in this publication does not imply, even in the absence of a specific statement, that such names are exempt from the relevant protective laws and regulations and therefore free for general use.

Typesetting by the authors and SPi using a Springer LATEX macro package

Cover design: *design & production* GmbH, Heidelberg

Printed on acid-free paper   SPIN: 12066520   46/SPi/3100   5 4 3 2 1 0

Preface

Visualization technology has become a crucial component of medical and biomedical data processing and analysis. This technology complements traditional image processing methods as it allows scientists and practicing medical doctors to visually interact with large, high-resolution three-dimensional image data. Further, an ever increasing number of new data acquisition methods is being used in medicine and the life sciences, in particular in genomics and proteomics. The book contains papers discussing some of the latest data processing and visualization techniques and systems for effective analysis of diverse, large, complex, and multi-source data.

Internationally leading experts in the area of data visualization came together for a workshop dedicated to visualization in medicine and life sciences, held on the island of Rügen, Germany, in July 2006. About 40 participants presented state-of-the-art research on this topic. Research and survey papers were solicited and carefully refereed, resulting in this collection.

The research topics covered by the papers in this book deal with these themes:

- Segmentation and Feature Detection
- Surface Extraction
- Volume Visualization
- Graph and Network Visualization
- Visual Data Exploration
- Multivariate and Multidimensional Data Visualization
- Large Data Visualization

The workshop was supported, in part, by the Deutsche Forschungsgemeinschaft (DFG).

Bremen, Germany *Lars Linsen*
Kaiserslautern, Germany *Hans Hagen*
Davis, California, U.S.A. *Bernd Hamann*

June 2007

Contents

Part I Surface Extraction Methods from Medical Imaging Data

Towards Automatic Generation of 3D Models of Biological Objects Based on Serial Sections
Vincent Jasper Dercksen, Cornelia Brüß, Detlev Stalling, Sabine Gubatz, Udo Seiffert, and Hans-Christian Hege 3

A Topological Approach to Quantitation of Rheumatoid Arthritis
Hamish Carr, John Ryan, Maria Joyce, Oliver Fitzgerald, Douglas Veale, Robin Gibney, and Patrick Brennan 27

3D Visualization of Vasculature: An Overview
Bernhard Preim and Steffen Oeltze 39

3D Surface Reconstruction from Endoscopic Videos
Arie Kaufman and Jianning Wang 61

Part II Geometry Processing in Medical Applications

A Framework for the Visualization of Cross Sectional Data in Biomedical Research
Enrico Kienel, Marek Vančo, Guido Brunnett, Thomas Kowalski, Roland Clauß, and Wolfgang Knabe 77

Towards a Virtual Echocardiographic Tutoring System
Gerd Reis, Bernd Lappé, Sascha Köhn, Christopher Weber, Martin Bertram, and Hans Hagen 99

Supporting Depth and Motion Perception in Medical Volume Data
Jennis Meyer-Spradow, Timo Ropinski, and Klaus Hinrichs 121

Part III Visualization of Multi-channel Medical Imaging Data

Multimodal Image Registration for Efficient Multi-resolution Visualization
Joerg Meyer 137

A User-friendly Tool for Semi-automated Segmentation and Surface Extraction from Color Volume Data Using Geometric Feature-space Operations
Tetyana Ivanovska and Lars Linsen 153

Part IV Vector and Tensor Visualization in Medical Applications

Global Illumination of White Matter Fibers from DT-MRI Data
David C. Banks and Carl-Fredrik Westin 173

Direct Glyph-based Visualization of Diffusion MR Data Using Deformed Spheres
Martin Domin, Sönke Langner, Norbert Hosten, and Lars Linsen 185

Visual Analysis of Bioelectric Fields
Xavier Tricoche, Rob MacLeod, and Chris R. Johnson 205

MRI-based Visualisation of Orbital Fat Deformation During Eye Motion
Charl P. Botha, Thijs de Graaf, Sander Schutte, Ronald Root, Piotr Wielopolski, Frans C.T. van der Helm, Huibert J. Simonsz, and Frits H. Post 221

Part V Visualizing Molecular Structures

Visual Analysis of Biomolecular Surfaces
Vijay Natarajan, Patrice Koehl, Yusu Wang, and Bernd Hamann 237

BioBrowser – Visualization of and Access to Macro-Molecular Structures
Lars Offen and Dieter Fellner 257

Visualization of Barrier Tree Sequences Revisited
Christian Heine, Gerik Scheuermann, Christoph Flamm, Ivo L. Hofacker, and Peter F. Stadler 275

Part VI Visualizing Gene Expression Data

Interactive Visualization of Gene Regulatory Networks with Associated Gene Expression Time Series Data
Michel A. Westenberg, Sacha A. F. T. van Hijum, Andrzej T. Lulko, Oscar P. Kuipers, and Jos B. T. M. Roerdink 293

Segmenting Gene Expression Patterns of Early-stage Drosophila Embryos
Min-Yu Huang, Oliver Rübel, Gunther H. Weber, Cris L. Luengo Hendriks, Mark D. Biggin, Hans Hagen, and Bernd Hamann 313

Color Plates 329

Part I

Surface Extraction Methods from Medical Imaging Data

Part 1

Towards Automatic Generation of 3D Models of Biological Objects Based on Serial Sections

Vincent Jasper Dercksen[1], Cornelia Brüß[2], Detlev Stalling[3], Sabine Gubatz[2], Udo Seiffert[2], and Hans-Christian Hege[1]

[1] Zuse Institute Berlin, Germany {dercksen,hege}@zib.de
[2] Leibniz Institute of Plant Genetics and Crop Plant Research, Gatersleben, Germany {bruess,gubatz,seiffert}@ipk-gatersleben.de
[3] Mercury Computer Systems Inc., Berlin, Germany dstallin@mc.com

Summary. We present a set of coherent methods for the nearly automatic creation of 3D geometric models from large stacks of images of histological sections. Three-dimensional surface models facilitate the visual analysis of 3D anatomy. They also form a basis for standardized anatomical atlases that allow researchers to integrate, accumulate and associate heterogeneous experimental information, like functional or gene-expression data, with spatial or even spatio-temporal reference. Models are created by performing the following steps: image stitching, slice alignment, elastic registration, image segmentation and surface reconstruction. The proposed methods are to a large extent automatic and robust against inevitably occurring imaging artifacts. The option of interactive control at most stages of the modeling process complements automatic methods.

Key words: Geometry reconstruction, surface representations, registration, segmentation, neural nets

1 Introduction

Three-dimensional models help scientists to gain a better understanding of complex biomedical objects. Models provide fundamental assistance for the analysis of anatomy, structure, function and development. They can for example support phenotyping studies [J.T06] to answer questions about the relation between genotype and phenotype. Another major aim is to establish anatomical atlases. Atlases enable researchers to integrate (spatial) data obtained by different experiments on different individuals into one common framework. A multitude of structural and functional properties can then jointly be visualized and analyzed, revealing new relationships. 4D atlases can provide insight into temporal development and spatio-temporal relationships.

We intend to construct high-resolution 4D atlases of developing organisms, that allow the integration of experimental data, e.g. gene expression patterns.

Such atlases can support the investigation of morphogenesis and gene expression during development. Their creation requires population-based averages of 3D anatomical models representing different developmental stages. By interpolating these averages the development over time can be visualized.

In this work we focus on the mostly automatic creation of individual 3D models as an essential step towards a standard atlas. Individual models can be created from stacks of 2D images of histological sections or from true 3D image data. With histological sections higher resolutions are possible than for example with 3D Nuclear Magnetic Resonance (NMR) imaging. Furthermore, fluorescent dye penetration problems, which can occur when imaging (thick) plant tissue with for example (3D) Confocal Laser Scanning Microscopy (CLSM), can be avoided. During data acquisition often multiple images per section have to be created to achieve the desired resolution. These sub-images have to be stitched in an initial mosaicing step. The 3D model construction then continues with the alignment of the image stack to restore the 3D coherence. In the following segmentation step, the structures of interest need to be identified, delimited and labeled. The model is completed by creating a polygonal surface, marking the object boundary and separating the structures it consists of. This sequence of nontrivial processing steps is here called the *geometry reconstruction pipeline* (see Fig. 1).

With our flexible general-purpose 3D visualization system [SWH05], three highly resolved 3D models of grains have recently been created basically manually [S. 07] (see Fig. 2(a)). Experience gained during the highly interactive modeling however clearly showed the need for the automation and facilitation of the many repetitive, time-consuming, and work-intensive steps.

When creating such detailed models, the size of the data sets is a complicating factor. Due to the high resolution of the images and the size of the subject, the data sets frequently do not fit into the main memory of common workstations and must therefore be processed out-of-core. Another problem is that due to the cutting and handling, histological sections are susceptible to imaging artifacts, like cracks, contrast differences and pollution. Very robust processing methods that are able to deal with such imperfect data are therefore required.

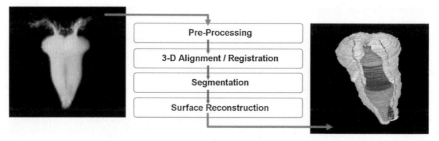

Fig. 1. From physical grains to 3D models: processing steps of the geometry reconstruction pipeline.

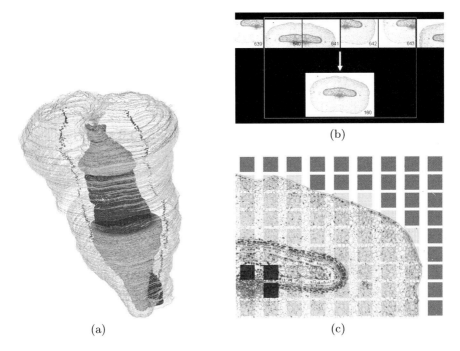

Fig. 2. (a) Example of a 3D surface model of a grain, created by a basically manual modeling procedure [S. 07]. (b) Exemplary cross-section image (no. 160) assembled from 4 sub-images (nos. 640 - 643). (c) A regular grid of rectangular cutouts is used to find the correct position of this sub-image with respect to the previous one (not shown). The optimal position is searched for each cutout. Here three patches have been successfully matched with an adequate translation vector and sufficient correlation values (darkened squares), the result is considered reliable and the search for the remaining patterns is skipped. The medium gray cutouts are not considered because of too little texture.

Early work for digitally representing and managing anatomical knowledge in graphical 3D models was concerned with human anatomy [K. 92, STA96, J.F99], later with model organisms, e.g. mouse [R. 03, M. 97, DRJ01] or fly [PH04]. Since the beginning, neuroscience has been a dominant application area, resulting in the generation of various brain atlases, e.g. [J.C97, RZH99, PH04, R. 05, A. 04, J.P05].

The problem of 3D model construction based on 2D section images has a long history, going back to the 1970's [LW72]. Though much work has been done since then [SWM97, U. 05, Ju05], the proper alignment of the sections still remains a difficult problem, especially when there is no 3D reference object available.

In plant biology, NMR imaging has been used to visualize developing barley caryopses, including the distribution of chemical compounds [Gli06]. Direct volume rendering visualizations of rice kernels [Y. 01] and soybeans [KOS02] based on 2D serial sections have also been reported. A relatively new technique for 3D imaging is Optical Projection Tomography (OPT) [K. 06]. Haseloff [Has03] has created three-dimensional models of *Arabidopsis* meristem cells based on CLSM data. So far, no large scale 3D anatomical, geometrical models of plant seeds, based on serial sections have been created using automatic methods.

Here we propose a coherent set of methods to reconstruct 3D surface models from high-resolution serial sections with minimal user interaction. This set consists of: 1) a robust automatic stitching algorithm; 2) a fast automatic rigid alignment algorithm included in an interactive environment, which provides effective visual feedback and allows for easy interactive correction-making; 3) an elastic registration method, which compensates for non-linear deformations; 4) automatic labelling achieved by an artificial neural network-based segmentation; and 5) an algorithm for the simultaneous triangulation and simplification of non-manifold surfaces from out-of-core label data. All methods are able to process out-of-core data and special attention was paid to the robustness of all automatic algorithms.

We applied these methods exemplarily for the reconstruction of plant seeds. However, all methods (except segmentation) can also be applied to data sets of different objects, and/or to data obtained with different imaging techniques, e.g., different histological staining or true 3D data. In the latter case one would skip the first part of the pipeline and start with the segmentation. To obtain a good segmentation of complex objects, it is imperative to incorporate expert knowledge. The presented segmentation method uses such knowledge and is therefore problem-specific. The neural network approach is however generally applicable. Each part of the pipeline will be described in detail in the next section.

2 Geometry Reconstruction Pipeline

2.1 Image Pre-Processing

The input for the reconstruction pipeline is a set of sub-images, each representing a part of a histological section (see Fig. 2(b)). In an automatic pre-processing step the sub-images are stitched to form a single image for each section. Szeliski [Sze05] provides an overview of image stitching. The set of n sub-images constituting a particular sectional image is known in advance. Neighboring sub-images normally share a common region of overlap, which forms the basis for the computation of their correct relative positions. For n sub-images there are $n(n-1)/2$ possible combinations to check for overlap. In our case, we know that successively photographed images share a common region which reduces the number of pairs of sub-images to investigate to $n-1$ plus one additional check of the last to the first sub-image for consistency and reliability.

The translation vector $T_{I_2} = (x_{I_2}, y_{I_2})$ which optimally positions image I_2 with respect to I_1 is implicitly determined via the translation vector T_P for a small cutout P of I_2. The latter is found by using the windowed *Normalized Cross-Correlation* (NCC) [GW02]:

$$\mathrm{NCC}(x,y) = \frac{\sum_{i=0}^{I-1}\sum_{j=0}^{J-1} \left(I_1(i+x, j+y) - E(I_1(x,y,I,J))\right)\left(P(i,j) - E(P)\right)}{\mathrm{std}(I_1(x,y,I,J))\,\mathrm{std}(P)} \quad (1)$$

where I and J mark the size of P, while E and *std* represent the expectation value and the standard deviation of the overlapping cutouts in both gray-value sub-images, respectively. The maximum of the NCC corresponds to the translation vector resulting in the best match between P and I_1. To be considered reliable, the maximum NCC value must exceed an empirically defined threshold. The search process is accelerated by using a multi-scale approach. Using a cutout P instead of an entire image I_2 ensures comparable correlation values, as the NCC is computed from a constant number $(I*J)$ of values (the insertion positions are chosen such that $I_1(i+x, j+y)$ is always defined). The size of P should be chosen such that it is smaller than the region of overlap and large enough to obtain statistically significant NCC values.

For reliability purposes, we compute the NCC for multiple cutouts of I_2. An incorrect T, e.g. caused by a match of a cutout containing a moving particle, is effectively excluded by searching for an agreeing set of patches/translation vectors (see Fig. 2(c)). In rare cases an image cannot be reliably assembled. Such cases are automatically detected and reported. Appropriate user actions – such as an additional calculation trial of a translation vector with a manually defined cutout or a visual check in case of arguable correlation values – are provided.

2.2 Alignment

In the alignment step, the three-dimensional coherence of the images in the stack, lost by sectioning the object, is restored. The input for this step is an ordered stack of stitched section images. This step is divided into two registration steps: a rigid alignment restores the global position and orientation of each slice; the subsequent elastic registration step compensates for any non-linear deformations caused by, e.g., cutting and handling of the histological sections. The latter is required to obtain a better match between neighboring slices, resulting in a smoother final surface model. Maintz [MV98] and Zitova [ZF03] provide surveys of the image registration problem.

The rigid alignment of the image stack is approached as a series of pairwise 2D registration problems, which can be described as follows: for each slice $R^{(\nu)} : \Omega \subset \mathbb{R}^2 \to \mathbb{R}$, $\nu \in \{2, \ldots, M\}$, find a transformation $\varphi^{(\nu)}$, resulting in a slice $R'^{(\nu)} = R^{(\nu)} \circ \varphi^{(\nu)}$, such that each pair of transformed consecutive slices $(R'^{(\nu-1)}, R'^{(\nu)})$ matches best. Allowed are translations, rotations and flipping. In the pair-wise alignment only one slice is transformed at a time, while the other one serves as a reference and remains fixed. An advantage of this approach is that it allows the alignment of a large, out-of-core image stack, as only two slices need to be in memory at any time. The image domain Ω is discretized, resulting in $\overline{\Omega}$, a grid consisting of the voxel centers of the image. The image range is also discretized to integer values between 0 and $2^g - 1$. Rounding is however deferred to the end of the computations, intermediate results are stored as floating point numbers. Values at non-grid points are defined by bi-linear interpolation.

We implemented an automatic voxel-based registration method, which seeks to maximize a quality function Q, with respect to a transformation $\varphi^{(\nu)}$. Because the images have comparable gray-values, we use the sum of squared differences (SSD) metric to base our quality function on:

$$Q = 1 - \sqrt{\frac{\sum_{x \in \overline{\Omega}'} \left(R'^{(\nu-1)}(x) - R'^{(\nu)}(x)\right)^2}{|\overline{\Omega}'| \cdot (2^g - 1)^2}} \qquad (2)$$

where $\overline{\Omega}' = \overline{\Omega} \cap \varphi^{(\nu-1)}(\Omega) \cap \varphi^{(\nu)}(\Omega)$ is the part of the discrete grid $\overline{\Omega}$ within the region of overlap of both transformed images. The size of $\overline{\Omega}'$ and thus, the number of grid points which values contribute to the sum in Eq. (2), is denoted by $|\overline{\Omega}'|$. To avoid that the optimum is reached when the images do not overlap, Q is multiplied with a penalty term $\sqrt{f/0.5}$, when the overlap $f = |\overline{\Omega}'|/|\overline{\Omega}|$ is less than 50%. We use a multi-resolution approach to increase the capture range of the optimizing method and for better computational performance. The optimization algorithm consists of 3 stages and can be summarized as follows:

Stage	Level	Search strategy
1	Max	Extensive search
2	Max-1...1	Neighborhood search
3	0	Best direction search

The user may interactively provide a reasonable starting point for the search. At the coarsest level, an extensive search is performed, which looks for the maximum in a large neighborhood. It first searches for the best translation in a grid $(dx, dy) = (dx + 2kh, dy + 2kh)$, $-K \leq k \leq K$ by evaluating the quality function for each vector. It then searches for an angle $dr = dr + n\Delta\theta$, $0 \leq n\Delta\theta \leq 360$ in the same way. Both searches are repeated until no improvement is found. The translation stepsize h is chosen to be r_l units, where r_l is the voxel size at resolution level l. As default values we use $K = 12$ and $\Delta\theta = 10$. The found optimum is used as a starting point for refinement within decreasingly smaller neighborhoods in the next stage. The neighborhood search [SHH97] first looks for the best direction in parameter space in terms of quality increase by trying the six possibilities $dx \pm h, dy \pm h, dr \pm \Delta\theta$, i.e., one step forward and back in each parameter dimension. It then moves one step in the direction with the largest quality increase. These two steps are repeated until no improvement is found. The translation and rotation step-sizes, h and $\Delta\theta$, are chosen to be r_l units (resp. degrees). Thus, the search neighborhood becomes smaller with each level. At the highest resolution level, a best direction search takes the solution to its (local) optimum. It does this by first searching for the best direction as above and then move into this direction until the quality does not increase anymore. These steps are repeated until no improvement can be found. The algorithm runs very quickly due to the multi-resolution pyramid and an efficient implementation, which uses the GPU for the frequently required evaluation of the quality function.

If the automatic alignment does not lead to satisfactory results, the corrections can be made interactively, using the effective graphical interface (see Figure 3(a)). The main window displays the alignment status of the current pair of slices. A simple but very effective visualization is created by blending both images, but with the gray values of the reference image inverted. Thus, the display value $V(x)$ equals $\frac{1}{2}(2^g-1-R'^{(\nu-1)}(x)+R'^{(\nu)})$, resulting in medium gray for a perfect match (see Fig. 4). This also works for color images, in which case all color channels of the reference image are inverted. In addition to the main alignment window, two cross-sectional planes, orthogonal to the image

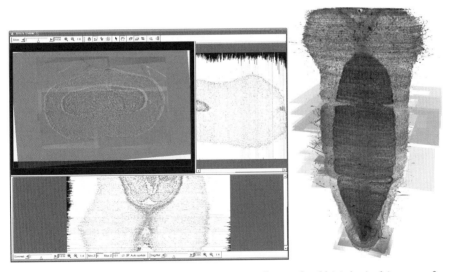

Fig. 3. Left: User interface showing alignment of a stack of histological images of a grain. Right: Direct volume rendering of the same rigidly aligned stack (minimum intensity projection). Clearly visible are the imaging artifacts, e.g. pollution and contrast differences. Data dimensions: 1818 slices of 3873x2465 voxels (∼17 GB). The used volume rendering technique is similar to [LHJ99]. It allows the rendering of large out-of-core data-sets at interactive rates by incremental and viewpoint-dependent loading of data from a pre-processed hierarchical data structure.

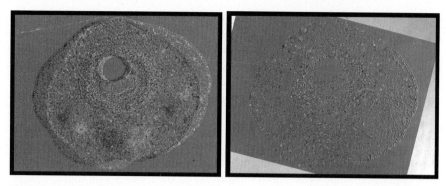

Fig. 4. Visualization of the alignment result using an inverted view before (left) and after (right) alignment.

plane, provide information about the current alignment status in the third dimension. Small alignment errors may propagate and deteriorate towards the end of the stack. The orthogonal views reveal such errors and support their correction. Experience showed that this enables a substantial improvement of the alignment result. Creating such an orthogonal cross-section may take very long when data is stored out-of-core, because an arbitrarily oriented line needs to be extracted from each image on disk. We showed that by rearranging the data into z-order, the number of required disk accesses can be reduced, leading to a significant speed-up [DPH05]. Figure 3(b) shows a volume rendering of an entire rigidly aligned stack.

Several methods have been presented in the past for the elastic registration of an image stack. Montgomery [MR94], for example, applies a local Laplacian smoothing on corresponding points on neighboring contours, which have to be (manually) delineated in each image. Ju [T. 06] deforms each slice by a weighted average of a set of warp functions mapping the slice to each of its neighboring slices within a certain distance. The warp function he uses is however very limiting in the deformations it can represent. Our method for the correction of the non-linear deformations of the slices is an extension to the variational approach as described in [S. 04]. The basic idea is to deform each image $R^{(\nu)} : \Omega \subset \mathbb{R}^2 \to \mathbb{R}$, $\nu \in \{1, \ldots, M\}$, in the stack so that it matches its neighboring slices in some optimal way. The deformations are described by transformations $\varphi^{(\nu)} = \mathbb{I} - u^{(\nu)}$, one for each image ν, which are based on the displacement fields

$$u(x) := \left(u^{(1)}(x), \ldots, u^{(M)}(x) \right), \quad u^{(\nu)} : \mathbb{R}^2 \to \mathbb{R}^2. \tag{3}$$

The first and the last slice are not transformed: $\varphi^{(1)} = \varphi^{(M)} = \mathbb{I}$. The optimal deformation is defined as the minimum of an energy function $E[R, u]$, which consists of a distance measure $D[R, u]$ and a regularizer (or smoother) $S[u]$:

$$E = D[R, u] + \alpha \cdot S[u]. \tag{4}$$

The regularizer S is based on the elastic potential

$$S[u] = \sum_{\nu=1}^{M} \int_\Omega \frac{\mu}{4} \sum_{j,k=1}^{2} \left(\partial_{x_j} u_k^{(\nu)} + \partial_{x_k} u_j^{(\nu)} \right)^2 + \frac{\lambda}{2} \left(\text{div } u^{(\nu)} \right)^2 dx. \qquad (5)$$

The scalars λ and μ are material parameters. The distance measure used in [S. 04] is the sum of squared differences (SSD)

$$d^{(\nu)}[R, u] = \int_\Omega \left(R'^{(\nu)} - R'^{(\nu-1)} \right)^2 dx \qquad (6)$$

extended for image series:

$$D[R, u] = \sum_{\nu=2}^{M} d^{(\nu)}[R, u]. \qquad (7)$$

To make the method more robust against distorted slices, we extend the distance measure to a weighted average of the SSD of the current slice with the slices within a distance $d \geq 1$:

$$D[R, u] = \sum_{\nu=2}^{M-1} \sum_{i=max(-d,1-\nu)}^{min(d,M-\nu)} w_d(\nu, i) \int_\Omega \left(R'^{(\nu)} - R'^{(\nu+i)} \right)^2 dx \qquad (8)$$

Note that for $w = \frac{1}{2}$ and $d = 1$ this results in the original distance measure of Eq. (7). The distance measure serves also as an error measure. The weights are positive, i.e. $w_d(\nu, i) \geq 0$, and they are normalized for each ν,

$$\sum_{i=max(-d,1-\nu)}^{-1} w_d(\nu, i) = \sum_{i=1}^{min(d,M-\nu)} w_d(\nu, i) = \frac{1}{2} \qquad (9)$$

in order to balance the influence of the slices above and below the current slice. Obvious choices for the weighting function w are a normalized constant or $1/i^2$, so the weight factor decreases with an increasing distance between the slices. A slice-specific function $w_d(i, \nu)$ which – automatically or by user intervention – penalizes images involved in high error values may also be chosen.

In order to minimize $E[R; u]$, a stationary point u of its Gâteau derivative is computed: $dE[R; u] = 0$. The resulting system of non-linear partial differential equations,

$$\sum_{\nu=1}^{M} \left(f^{(\nu)} - \alpha(\mu \Delta u^{(\nu)} + (\lambda + \mu) \nabla \text{div } u^{(\nu)}) \right) = 0 \qquad (10)$$

the Navier-Lamé equations (NLE) for serially connected slices, describes the elastic deformation of an object subject to a force f, where f is the derivative of the chosen distance function. We are using the efficient solution scheme for this problem described in [FM99] and [Mod04]. In addition, a multi-resolution approach is used to achieve faster convergence and for a larger region of capture [S. 04]. For implementation details, see [FM99, Mod04, S. 04]. As computations are performed for each slice independently, requiring only data within a small window, the process can be parallelized and out-of-core data can easily be registered. An example of an elastically registered image stack is shown in Figures 5 and 6

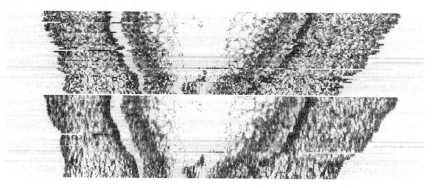

Fig. 5. Rigidly aligned image stack before (top) and after (bottom) elastic registration. A multi-resolution pyramid with four levels was used to register this stack. At the lowest two resolution levels, a window width $d = 2$ was used with Gaussian weighting. A value of $d = 1$ was used for the highest two resolution levels.

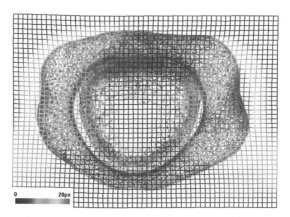

Fig. 6. Visualization of the elastic deformation of one transformed image from the stack in Figure 5. A regular grid is overlaid onto the image and transformed according to the computed deformation field. The grid is colored by deformation vector magnitude.

2.3 Image Segmentation

During segmentation, abstract conceptional knowledge is imposed on images by classifying 2D and/or 3D image regions. The biologically relevant materials, e.g., tissues or organs, must be specified by an expert and then identified and assigned in all images of the aligned stack. For the models described above [S. 07] (see Fig. 2(a)) this crucial task was done interactively for all section images, a very laborious and subjective task. For the generation of elaborate models and a 4D atlas automation of the segmentation task is inevitable. This is however a difficult task, mainly because criteria to base the automatic discrimination of materials on are difficult to establish. In addition, tissue characteristics are often not constant, not even within a single data set, due to changes during development.

To overcome these problems, we chose an approach in which histological expert knowledge for the discrimination between materials is incorporated implicitly. To attempt to automatically perform such human-like perception-based recognition, artificial neural networks (ANN) [SVM96, dCZ04, For04] have proved to be very powerful and adaptive solutions. Due to its trainability and generalization ability, sample based learning without the necessity to explicitly specify pattern (object) models can easily be implemented. The choice of both borderline-learning and prototype-based neural methods turned out to be very advantageous. In contrast to, for example, often applied support vector machines, these techniques offer good solutions also for classification tasks where more than just a few classes are to be distinguished.

Our approach can be summarized as follows: 1) from the image stack, the biologist selects a sub-stack with images showing a high degree of similarity; 2) the biologist interactively segments a representative set of images from this sub-stack; 3) feature vectors are computed from these exemplary images; 4) the neural networks are trained; 5) in the recall phase, all pixels of the remaining images of the sub-stack are classified; 6) the final segmentation is determined in a post-processing step. These steps are repeated until the entire stack has been segmented. Two essential parts of this approach, the feature vector extraction and the segmentation using the neural network, are described in more detail below.

Feature Extraction

A feature vector containing particular properties of the image pixels has to be defined in such a way, that the selected features are significant to distinguish between different materials of the image. In doing so, each single feature considered separately cannot discriminate between all materials. However, once combined to a vector, a proper classification becomes feasible. Depending on the complexity of the image material and the applied classification system the vector length may vary significantly.

In the work presented here about 1000 features from the following list were used [GW02, SHB99]:

- Mainly statistical features based on gray-level histograms (standard features like mean, maximum possibility, entropy, energy, 2nd to 7th central moments for different neighborhood sizes and for the color channels red, green, blue according to the RGB color model and the value channel of the HSV model),
- Color features (derived from RGB and HSV color model),
- Geometry and symmetry features (such as Cartesian and polar coordinates, respectively according to the centroid of the image, absolute angle to the second main component axis – as an obvious symmetry axis of the cross-section, see Fig. 7 a),
- Several features representing applications of Gaussian filters (with several variances) and Laws-Mask filters.

By the built-in feature rating of SRNG a specific subset of 170 features was selected as a starting point.

Segmentation using Artificial Neural Networks

Artificial neural networks generate their own internal model by learning from examples whereas in conventional statistical methods knowledge is made explicit in the form of equations or equation systems. This learning is achieved by a set of learning rules controlled by a number of parameters which adapt the synaptic weights of the network in response to the presented examples.

Supervised trained networks are used since a classification of the above mentioned features is required. In general, a supervised trained network receives pairwise vectors containing the features of a pixel and the corresponding class (e.g. 1-of-n coded) for all pixels of an image. Due to the ability of ANNs to generalize, only a tiny fraction of all pixels of an image or of several selected images has to be used to train the network. In order to improve the speed and thereby facilitate larger models, a parallel implementation is used [CS04, Sei04].

In this work two different supervised trained ANNs are applied. At first instance the well established *Multiple Layer Perceptron (MLP)* [MP69, RHW86, Lip87] is used, which has undoubtedly proven its ability to solve complex and high-dimensional classification problems. Alternatively, a prototype-based system is applied, the *Supervised Relevance Neural Gas* (SRNG). It is an initialization-tolerant version of Kohonen's well-established Learning Vector Quantization (LVQ) classifier based on the minimization of a misclassification cost function [HSV05].

With these two classifiers we aimed at the superposition of the strengths of the borderline-learning MLP – which is representing large contiguous materials well – against receptive fields of neural prototypes generated by the SRNG – as an adequate means to represent even small isolated material patches. Exemplary results of both variants are shown in Fig. 7(e)-(h). See [BSS06] for more detailed information. In a post-processing step small misclassifications

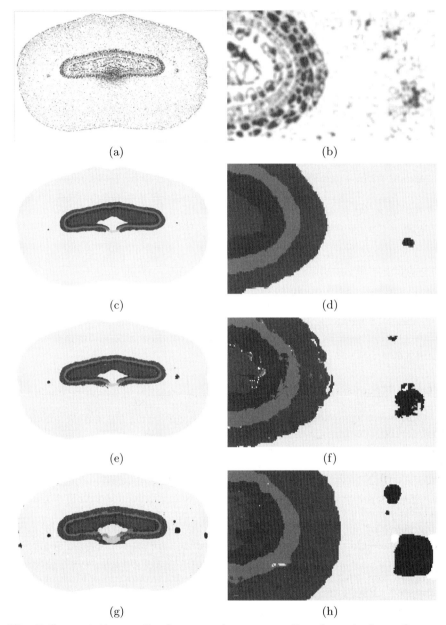

Fig. 7. Segmentation results of an exemplary cross-section of a grain chosen from an automatically registered and automatically segmented image stack with 11 materials. Left column: a) original color image of a cross-section, c) reference segmentation manually created by a biologist, e) and g) results of an automatic segmentation with ANN, MLP and SRNG respectively (in case of MLP a 4-layer architecture was chosen with 120, 80, 36, 11 neurons, in case of SRNG 10 prototypes per material were used; both classifiers are based on an initial feature vector of 170 features). Right column: panels b), d), f) and h) depict cut-outs corresponding to a), c), e) and g).

are eliminated using structural knowledge, e.g. size, neighborhood relations and the distance to the image center of gravity of all material patches. The final material borders are found using a morphological watershed [GW02]. Using accepted material patches as initial catchment basins, all misclassified regions in between are "flooded" and assigned to the surrounding materials. An exemplary segmentation result is shown in Figure 9.

2.4 Surface Reconstruction and Simplification

The last step in the geometry reconstruction pipeline is the extraction of a polygonal surface model representing the material boundaries from the segmented voxel data. Although the labeled image volume can already be considered as a 3D model, a surface representation is more suitable for several reasons. First, a smooth surface resembles the physical object much better than the blocky voxel representation. Second, polygonal surfaces can be rendered more efficiently using modern graphics hardware. Last, they can easily be simplified, allowing models at different levels of detail to be created: high-resolution models for accurate computations, smaller ones for fast interactive visualization. The surface reconstruction algorithm must be able to create non-manifold surfaces, because more than two materials may meet at any point. An example of a reconstructed surface model of a grain is shown in Figure 2(a).

Surface Reconstruction

A well-known approach to compute surfaces separating the voxels belonging to different structures is the standard marching-cubes algorithm [LC87]. This method however cannot produce non-manifold surfaces. Therefore, we use a generalized reconstruction algorithm capable of handling multiple labels at once [H.-97]. Like standard marching-cubes the generalized technique works by processing each hexahedral cell of a labeled voxel volume individually. While in standard marching-cubes each of the eight corners of a cell is classified as either lying below or above a user-defined threshold, here each corner may belong to one of up to eight different labels. The algorithm generates a triangular surface separating all corners assigned to different labels. For simple, non-ambiguous configurations with up to three different labels this surface is read from a pre-computed table. For more complex configurations, the surface is computed on-the-fly using a subdivision approach. First, the cell is subdivided into 6^3 subcells. Each subcell is assigned one particular label. In order to do so, a binary probability field is constructed for each label with values 1 or 0 at the corners, depending on whether a corner is assigned to that label or not. The probability fields are tri-linearly interpolated at a point inside a subcell and the subcell is assigned to the label with the highest probability. Afterwards the boundaries of the subcells are extracted and the resulting surface is simplified such that at most one vertex per edge, one vertex per face, and one vertex

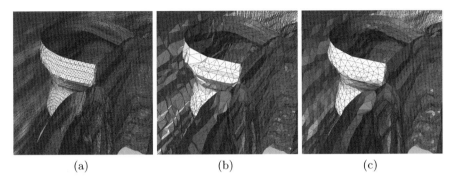

Fig. 8. Out-of-core surface reconstruction, simplification and remeshing. (a) Reconstruction of weighted label volume. (b) Simplified surface using our method (reduction 90% compared to (a)). (c) Simplified surface after remeshing step.

inside the cell is retained. Although the overall procedure involves several non-trivial processing steps, it turns out to be quite fast while producing well-shaped, high-quality results.

A problem with the described surface reconstruction method as well as with the standard marching-cubes method applied to binary labels is, that the resulting triangles are always oriented perpendicular to the principal axes or to the diagonals of a cell. Thus the final surface looks quite coarse and jagged. The situation can be improved by applying a smoothing filter to binary labels as described in [Whi00]. In case of multiple labels and non-manifold surfaces a similar technique can be applied. For each label a weight factor is computed. These weights can then be used to shift the vertices along the edges or faces of a cell, so that a smooth surface is obtained (see Fig. 8(a)). When computing the weights, special care has to be taken not to modify the original labeling [Wes03]. Otherwise especially very thin structures could get lost.

Surface Simplification

Applying the described surface reconstruction method for the creation of 3D models from high-resolution images usually results in an enormous number of triangles. This may prohibit interactive visualization or even in-memory storage. Therefore, a surface simplification step is required. In order to avoid an intermediate storage and retrieval step of the possibly huge intermediate triangulation, we chose to combine triangle generation and simplification in a single algorithm, using a streaming approach based on [ACSE05] and [WK03]. The triangle generation and the simplification steps are alternated. Generated triangles are fed into a triangle buffer, where the simplification takes place. The simplification is based on iterative edge contractions. Each edge eligible for contraction is assigned a cost and inserted into an edge heap, sorted by cost. The edge cost is computed using the well-known quadrics error metric (QEM)

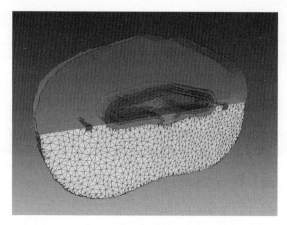

Fig. 9. Perspective view of a 3D model segment based on 10 automatically registered and segmented images of cross-sections of a barley grain (15-fold stretched in height). The segmentation results were achieved by a combination of results of two ANN classifiers (MLP, SRNG). The segmented structures were triangulated and the resulting polygon mesh was adaptively simplified from an initial 3.6M to 80K triangles.

[GH97]. During simplification, the edge with the lowest cost is removed from the heap and contracted, thereby merging the edge endpoints and removing the incident triangles. The position of the new vertex and the updated cost of the incident edges are computed as in [GH97]. Newly created vertices are not allowed to move, because this would cause a mismatch with the simplices which are incident to these vertices, but which are not yet created. Therefore, newly created vertices are marked as a boundary vertices. Edges incident to such vertices are not inserted in the edge heap, but in a waiting queue. After each triangulation step, the list of boundary vertices is updated. Edges not incident to one of those vertices are removed from the waiting queue and inserted into the edge heap.

The user can determine the stopping criterion for simplification. For an in-core result, e.g. to be directly visualized, the user specifies the desired number of triangles T_{final} of the final mesh. The size of the triangle buffer T_{buffer} should be at least $T_{final} + T_{boundary} + T_{new}$, where $T_{boundary}$ is the number of triangles having at least two boundary vertices, and T_{new} is the number of triangles added to the buffer in a single triangulation step. $T_{boundary}$ can be kept small by enforcing a coherent boundary, which can easily be accomplished by traversing the labeled volume in array order. By triangulating slice-wise for example, an upper bound is the number of cells per slice times the maximum number of boundary triangles per cell. An upper bound of T_{new} is the maximum number of triangles per cell times the number of cells processed in one triangulation step. As the average number of generated triangles per cell will be much smaller than the maximum and by processing less cells per step, T_{new} can also be kept small. In practice, the influence on the final surface of the value of T_{buffer} beyond the mentioned minimum seems negligible (although no measurements were made). After triangles have been added to the buffer, the buffer is simplified until there is enough space to hold another T_{new} triangles. The triangulation and simplification are repeated until the whole volume has been processed. Then, a final simplification step reduces the number of triangles to T_{final}.

Instead of the final triangle count, the user may also specify a maximum error E_{max}. In this case, the buffer is simplified until the cost of the next edge contraction would exceed E_{max}. Then as many triangles are written to disk as required to be able to hold another T_{new} triangles in the buffer. Triangles which have been in the buffer longest are written to disk first. To make sure that such triangles are not modified anymore, their in-core vertices are marked as boundary vertices, in analogy to the newly generated vertices above. Edges incident to boundary vertices are removed from the heap. Again, the triangulation and simplification steps are alternatingly repeated, until the whole volume has been processed. An optional remeshing step can be applied to create a more regular mesh for visualization or numerical purposes using for example the method of [ZLZ07] for non-manifold in-core meshes (see Fig. 8(c)). For out-of-core meshes [AGL06] can be used.

3 Results

We presented a coherent set of methods for the nearly automatic creation of 3D surface models from large stacks of 2D sectional images. Experience from the past, where such models have been created using almost solely manual tools [S. 07], motivated our efforts towards automation of the generally very laborious tasks. In practice, image artifacts turned out to be a complicating factor. Therefore, emphasis was put on the robustness of the automatic methods and an interactive environment is provided in most cases as a fall-back option. Another problem addressed is the management of large data sets, which do not fit into the main memory of common desktop computers. All presented methods are able to work on out-of-core data (stitching and segmentation operate on one image at a time). Our main achievements for each step in the geometry reconstruction pipeline are summarized below.

Automatic image stitching method, based on the normalized cross-correlation metric in combination with an efficient search algorithm. In an experiment, 600 cross-sections were assembled from 2-6 sub-images each, resulting in a total of 2720 investigated sub-image pairs. In 2680 cases, the automatic result was correct. In 40 cases, the algorithm was not able to automatically find the correct translations. In 30 cases this was due to bad image quality so an interactive pattern selection by the user was necessary, 10 times this was due to non-overlapping sub-images (photographic error). However, problems were in all cases reliably detected by the algorithm and reported, so the user could take appropriate action.

Effective and efficient slice alignment, based on a combination of automatic and interactive tools. The automatic rigid registration method is fast and leads to satisfying results when the objects do not exhibit extreme distortions. If the nature of the data prohibits an entirely automatic alignment, the user interface provides effective means to make corrections, inspect the current status and to interactively control the entire process. With the simple but effective *inverted view* visualization method the alignment quality of two slices can be assessed. The possibility to display orthogonal cross-sections is a valuable tool for judging and improving the alignment quality of the entire stack.

The *elastic image stack registration* compensates for the deformations, which occurred during the cutting and flattening process. We extended an existing method to make it more robust against imaging artifacts. The 3D coherence of tissues is restored to a large extent. This may help in the subsequent segmentation step and results, in the end, in a smoother and more realistic 3D model. As the elastic transformation is generally very smooth, small features are unlikely to disappear.

Development of an *automated segmentation tool*. While so far segmentation was mainly achieved using manual tools [S. 07], now nearly automatic segmentation becomes more feasible. Figure 7 illustrates the first promising segmentation results utilizing ANNs. We obtained a classification accuracy of constantly about 90%, which is a good starting point for the postprocessing procedure. It turned out, that a different feature vector as well as different classification systems have quite a high impact on the segmentation results. Consequently and in order to respect the different properties of the cross-sections throughout the entire grain, a modular solution containing pairwise optimized feature vectors/classification systems is desired.

Simultaneous surface reconstruction and simplification makes the creation and visualization of high-quality 3D polygonal models from large, high-resolution voxel data sets feasible. The algorithm enables the user to compute an in-core polygonal model from an out-of-core labelled volume without having to store the usually very large intermediate triangulation. By incrementally processing the labeled voxels, discontinuities at boundaries, occurring when the data is processed block-wise, can be avoided. The resulting surface model can then directly be visualized.

All methods (except for segmentation) are general in their nature, so they can be applied to data sets of different objects, obtained by different imaging modalities.

4 Outlook

The presented pipeline enables users to efficiently create detailed, high-quality three-dimensional surface models from stacks of histological sections. This ability will greatly facilitate the creation of spatio-temporal models and atlases, that are key tools in developmental biology, in particular because

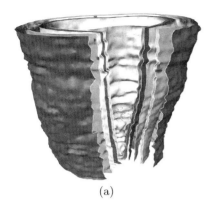 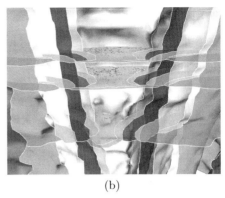

(a) (b)

Fig. 10. (a) Integration of experimental data into a (part of a) 3D model. The pattern results from a color reaction after binding of an antibody in order to localize a particular protein. Multiple cross-sections were treated this way, and subsequently photographed and segmented. The correct z-position of each section in the model was determined by hand. As the marker data and the 3D model stem from different individuals, the former must be elastically registered to fit into the model. Shown is the first promising result of the integrated marker (in red) after registration. (b) Close-up of the model in (a).

they potentially serve as reference models for the integration of experimental data [R. 03, M. 97, DRJ01].

The integration of such marker data is however a non-trivial problem. As such data usually stems from a different individual than the model, an elastic registration step is required to establish the spatial correspondence and to transform the marker data accordingly before it fits into the model. For 2D stained histological sections, this means that first the correct position in the 3D model needs to be found. Another problem may be that the model and the marker data may have been imaged using different modalities. Fig. 10 shows the result of a first attempt to integrate protein localization data in a 3D model. The integration of experimental data into reference models will be an important research topic for the near future. Other future research topics include the use of multimodal data for reconstruction, in particular 3D NMR data to support the alignment of the 2D histological sections.

Acknowledgements

We would like to thank Marc Strickert for the training and application of the SRNG classifier and Steffen Prohaska for his work on large data sets. The authors are also grateful to Renate Manteuffel and Ruslana Radchuk for providing the antibody for immuno-labelling. Parts of this work were supported by BMBF-grant (BIC-GH, FKZ 0312706A) and by BMBF-grant (GABI-Seed, FKZ 0312282).

References

[A. 04] A. MacKenzie-Graham et al. A Multimodal, Multidimensional Atlas of the C57BL/6J Mouse Brain. *J. Anatomy*, 204(2):93–102, 2004.

[ACSE05] D. Attali, D. Cohen-Steiner, and H. Edelsbrunner. Extraction and Simplification of Iso-surfaces in Tandem. In *Symposium on Geometry Processing*, pages 139–148, 2005.

[AGL06] M. Ahn, I. Guskov, and S. Lee. Out-of-core Remeshing of Large Polygonal Meshes. *IEEE Transactions on Visualization and Computer Graphics*, 12(5):1221–1228, 2006.

[BSS06] C. Brüß, M. Strickert, and U. Seiffert. Towards Automatic Segmentation of Serial High-Resolution Images. In *Proc. BVM Workshop*, pages 126–130. Springer, 2006.

[CS04] T. Czauderna and U. Seiffert. Implementation of MLP Networks Running Backpropagation on Various Parallel Computer Hardware Using MPI. In Ahmad Lotfi, editor, *Proc. 5th Int. Conf. Recent Advances in Soft Computing (RASC)*, pages 116–121, 2004.

[dCZ04] L. Nunes de Castro and F.J. Von Zuben. *Recent Developments In Biologically Inspired Computing*. Idea Group Publishing, Hershey, PA, USA, 2004.

[DPH05] V.J. Dercksen, S. Prohaska, and H.-C. Hege. Fast Cross-sectional Display of Large Data Sets. In *IAPR Conf. Machine Vision Applications 2005, Japan*, pages 336–339, 2005.

[DRJ01] M. Dhenain, S.W. Ruffins, and R.E. Jacobs. Three-Dimensional Digital Mouse Atlas Using High-Resolution MRI. *Developmental Biology*, 232(2):458–470, 2001.

[FM99] B. Fischer and J. Modersitzki. Fast Inversion of Matrices Arising in Image Processing. *Numerical Algorithms*, 22:1–11, 1999.

[For04] N. Forbes. *Imitation of Life - How Biology is Inspiring Computing*. The MIT Press, Cambridge, MA, USA, 2004.

[GH97] M. Garland and P.S. Heckbert. Surface Simplification Using Quadric Error Metrics. In *SIGGRAPH'97 Conf. Proc.*, pages 209–216, 1997.

[Gli06] S.M. Glidewell. NMR Imaging of Developing Barley Grains. *J. Cereal Science*, 43:70–78, 2006.

[GW02] R.C. Gonzalez and R.E. Woods. *Digital Image Processing*. Prentice-Hall, Upper Sadle River, New Jersey, 2nd ed. edition, 2002.

[H.-97] H.-C. Hege et al. A Generalized Marching Cubes Algorithm Based on Non-Binary Classifications. Technical report, ZIB Preprint SC-97-05, 1997.

[Has03] J. Haseloff. Old Botanical Techniques for New Microscopes. *Biotechniques*, 34:1174–1182, 2003.

[HSV05] B. Hammer, M. Strickert, and T. Villmann. Supervised Neural Gas with General Similarity Measure. *Neural Processing Letters*, 21(1): 21–44, 2005.

[J.C97] J.C. Mazziotta et al. Atlases of the Human Brain. In S.H. Koslow and M.F. Huerta, editors, *Neuroinformatics - An overview of the human brain project*, pages 255–308, 1997.

[J.F99] J.F. Brinkley et al. Design of an Anatomy Information System. *Computer Graphics and Applications*, 19(3):38–48, 1999.

[J.P05] J.P. Carson et al. A Digital Atlas to Characterize the Mouse Brain Transcriptome. *PLoS Computational Biology*, 1(4), 2005.

[J.T06] J.T. Johnson et al. Virtual Histology of Transgenic Mouse Embryos for High-Throughput Phenotyping. *PLos Genet*, 2(4):doi: 10.1371/journal.pgen.0020061, 2006.

[Ju05] T. Ju. *Building a 3D Atlas of the Mouse Brain*. PhD thesis, Rice University, April 2005.

[K. 92] K. H. Höhne et al. A 3D Anatomical Atlas Based on a Volume Model. *IEEE Computer Graphics Applications*, 12(4):72–78, 1992.

[K. 06] K. Lee et al. Visualizing Plant Development and Gene Expression in Three Dimensions Using Optical Projection Tomography. *The Plant Cell*, 18:2145–2156, 2006.

[KOS02] H. Kuensting, Y. Ogawa, and J. Sugiyama. Structural Details in Soybeans: A New Three-dimensional Visualization Method. *J. Food Science*, 67(2):721–72, 2002.

[LC87] W.E. Lorensen and H.E. Cline. Marching Cubes: A High Resolution 3D Surface Construction Algorithm. In *Proc. ACM SIGGRAPH'87*, volume 21(4), pages 1631–169, 1987.

[LHJ99] E.C. LaMar, B. Hamann, and K.I. Joy. Multiresolution techniques for interactive texture-based volume visualization. In David S. Ebert, Markus Gross, and Bernd Hamann, editors, *IEEE Visualization '99*, pages 355–361, Los Alamitos, California, 1999. IEEE, IEEE Computer Society Press.

[Lip87] R. P. Lippmann. An Introduction to Computing with Neural Nets. *IEEE ASSP Magazine*, 4(87):4–23, 1987.

[LW72] C. Levinthal and R. Ware. Three Dimensional Reconstruction from Serial Sections. *Nature*, 236:207–210, 1972.

[M. 97] M. H. Kaufman et al. Computer-Aided 3-D Reconstruction of Serially Sectioned Mouse Embryos: Its Use in Integrating Anatomical Organization. *Int'l J. Developmental Biology*, 41(2):223–33, 1997.

[Mod04] J. Modersitzki. *Numerical Methods for Image Registration*. Oxford University Press, New York, 2004.

[MP69] M. Minsky and S. Papert. *Perceptrons: An Introduction to Computational Geometry*. MIT Press, Cambridge, 1969.

[MR94] K. Montgomery and M.D. Ross. Improvements in semiautomated serial-section reconstruction and visualization of neural tissue from TEM images. In *SPIE Electronic Imaging, 3D Microscopy conf. proc.*, pages 264–267, 1994.

[MV98] J. Maintz and M. Viergever. A Survey of Medical Image Registration. *Medical Image Analysis*, 2(1):1–36, 1998.

[PH04] W. Pereanu and V. Hartenstein. Digital Three-Dimensional Models of Drosophila Development. *Current Opinion in Genetics & Development*, 14(4):382–391, 2004.

[R. 03] R. Baldock et al. EMAP and EMAGE: A Framework for Understanding Spatially Organized Data. *Neuroinformatics*, 1(4):309–325, 2003.

[R. 05] R. Brandt et al. A Three-Dimensional Average-Shape Atlas of the Honeybee Brain and its Applications. *J. Comparative Neurology*, 492(1):1–19, 2005.

[RHW86] D. E. Rumelhart, G. E. Hinton, and R. J. Williams. Learning Internal Representations by Error Propagation. In D. E. Rumelhart et al., editor, *Parallel Distributed Processing: Explorations in the Microstructure of Cognition*, pages 318–362, Cambridge, 1986. MIT Press.

[RZH99] K. Rein, M. Zöckler, and M. Heisenberg. A Quantitative Three-Dimensional Model of the Drosophila Optic Lobes. *Current Biology*, 9:93–96, 1999.

[S. 04] S. Wirtz et al. Super-Fast Elastic Registration of Histologic Images of a Whole Rat Brain for Three-Dimensional Reconstruction. In *Proc. SPIE 2004, Medical Imaging*, 2004.

[S. 07] S. Gubatz et al. Three-Dimensional Digital Models of Developing Barley Grains for the Visualisation of Expression Patterns. In prep., 2007.

[Sei04] U. Seiffert. Artificial Neural Networks on Massively Parallel Computer Hardware. *Neurocomputing*, 57:135–150, 2004.

[SHB99] M. Sonka, V. Hlavac, and R. Boyle. *Image Processing, Analysis, and Machine Vision*. Brooks//Cole Publishing Company, Upper Sadle River, New Jersey, 2nd ed. edition, 1999.

[SHH97] C. Studholme, D. L. G. Hill, and D. J. Hawkes. Automated Three-Dimensional Registration of Magnetic Resonance and Positron Emission Tomography Brain Images by Multiresolution Optimization of Voxel Similarity Measures. *Medical Physics*, 24(1):25–35, 1997.

[STA96] G. Subsol, J.-Ph. Thirion, and N. Ayache. Some Applications of an Automatically Built 3D Morphometric Skull Atlas. In *Computer Assisted Radiology*, pages 339–344, 1996.

[SVM96] J.A.K. Suykens, J.P.L. Vandewalle, and B.L.R. De Moor. *Artificial Neural Networks for Modelling and Control of Non-Linear Systems*. Kluwer Academic Publishers, Den Haag, The Netherlands, 1996.

[SWH05] D. Stalling, M. Westerhoff, and H. C. Hege. Amira - a Highly Interactive System for Visual Data Analysis. In C.D. Hansen and C.R. Johnson, editors, *Visualization Handbook*, pages 749–767. Elsevier, 2005.

[SWM97] J. Streicher, W.J. Weninger, and G.B. Müller. External Marker-Based Automatic Congruencing: A New Method of 3D Reconstruction from Serial Sections. *The Anatomical Record*, 248(4):583–602, 1997.

[Sze05] R. Szeliski. Image Alignment and Stitching. In N. Paragios, Y. Chen, and O. Faugeras, editors, *Handbook of Mathematical Models in Computer Vision*, pages 273–292. Springer, 2005.

[T. 06] T. Ju et al. 3D volume reconstruction of a mouse brain from histological sections using warp filtering. *J. of Neuroscience Methods*, 156:84–100, 2006.

[U. 05] U. D. Braumann et al. Three-Dimensional Reconstruction and Quantification of Cervical Carcinoma Invasion Fronts from Histological Serial Sections. *IEEE Transactions on Medical Imaging*, 24(10):1286–1307, 2005.

[Wes03] M. Westerhoff. *Efficient Visualization and Reconstruction of 3D Geometric Models from Neuro-Biological Confocal Microscope Scans*. Phd thesis, Fachbereich Mathematik und Informatik, Freie Universität Berlin, Jan. 2003.

[Whi00] R.T. Whitaker. Reducing Aliasing Artifacts in Iso-Surfaces of Binary Volumes. In *Proc. 2000 IEEE Symposium on Volume Visualization*, pages 23–32. ACM Press, 2000.

[WK03] J. Wu and L. Kobbelt. A Stream Algorithm for the Decimation of Massive Meshes. In *Graphics Interface'03 Conf. Proc.*, 2003.

[Y. 01] Y. Ogawa et al. Advanced Technique for Three-Dimensional Visualization of Compound Distributions in a Rice Kernel. *J. Agricultural and Food Chemistry*, 49(2):736–740, 2001.

[ZF03] B. Zitova and J. Flusser. Image registration methods: a survey. *Image and Vision Computing*, 21(11):977–1000, October 2003.

[ZLZ07] M. Zilske, H. Lamecker, and S. Zachow. Remeshing of non-manifold surfaces. ZIB-Report 07-01, In prep., 2007.

A Topological Approach to Quantitation of Rheumatoid Arthritis

Hamish Carr, John Ryan, Maria Joyce, Oliver Fitzgerald, Douglas Veale, Robin Gibney, and Patrick Brennan

University College Dublin

Summary. Clinical radiologists require not only visual representations of MRI and CT data but also quantitative measurements representing the progression of chronic conditions such as rheumatoid arthritis of the knee. Since inflammation is confined to a thin irregularly shaped region called the synovial capsule it is necessary to segment a suitable approximation of the capsule, then compute quantitative measurements within the segmented region. We report preliminary results on applying topological tools to identify the desired region visually and to extract quantitative information, along with a protocol for clinical validation of the method.

1 Introduction

Visualization aims to aid scientific and medical insight by generating meaningful images representing salient features of data. This commonly involves exploratory interaction by a skilled user who is interested in qualitative rather than quantitative results. In contrast, clinical radiology aims to answer specific questions in a repeatable and controllable fashion, requiring quantitative rather than qualitative results. As a result, qualitatitve tools designed for visualisation are often inapplicable to clinical tasks.

Traditionally, clinical radiology has relied on single X-rays rather than volumetric data, in part due to radiologists familiarity with these images, and in part due to difficulties with segmentation: the identification and extraction of relevant objects from volumetric datasets. Yet it is clear to those involved that volumetric data has significant advantages over single images, provided that issues of segmentation and processing time can be addressed.

In particular, clinical situations arise where quantitative and qualitative results are intermingled. One example of this is clinical tracking of rheumatoid arthritis of the knee. This involves the quantitative measurement of inflammation in the synovial capsule of the knee. Unfortunately, the inflamed areas do not show up readily on an MRI scan unless a contrast agent such as gadolinium is introduced via the bloodstream. While this increases the intensity of

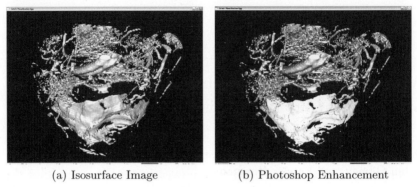

(a) Isosurface Image   (b) Photoshop Enhancement

Fig. 1. A Starting Point: Highlighting the Synovial Capsule with Adobe Photoshop. Note the large number of irrelevant contours which must be excluded from quantitative computation, and the thin irregular shape of the desired contour.

inflamed voxels, it also increases the intensity of voxels in blood vessels. Thus, in order to obtain a clinically reliable measurement of inflammation, a qualitative decision is required to restrict quantitation to the synovial capsule proper. Since this capsule is highly irregular in shape and quite thin, it is difficult to define a suitable neighbourhood inside which quantitation can occur.

While isosurface (i.e. threshold) segmentation was clinically identified as appropriate, it was quickly apparent that this resulted in a large number of objects extracted, most of which were irrelevant to the task. Moreover, radiologists had no difficulty identifying the synovium visually, initially by highlighting the relevant isosurface component with Adobe Photoshop, as shown in Figure 1.

What is required, therefore, is a tool whereby the radiologist can quickly and easily identify an approximate boundary of the synovium, combined with an efficient mechanism for computing quantitative results inside that boundary. Recent work in visualisation has resulted in the *flexible isosurface* in [CS03, CSvdP04, CSvdP07] which combines a simple user interface for choosing isosurface components with a wealth of geometric and topological information.

In this paper, we report preliminary results that confirm that this approach is applicable to the clinical problem, and details of the protocol necessary for full validation: full clinical results will necessarily be reported in the medical literature rather than visualisation venues.

We describe the clinical problem in more detail in Section 2, identify relevant previous work in Section 3, then describe our approach in more detail in Section 4. Our preliminary results confirming the validity of the approach are then presented in Section 5, followed by details of the clinical protocol necessary for full validation in Section 6, conclusions in Section 7 and comments on future directions in Section 8.

2 Rheumatoid Arthritis

Rheumatoid Arthritis (RA) is one of the most important chronic inflammatory disorders affecting approximately 0.5-1.5% of the worlds population [AVD05]. RA is an autoimmune disease characterised by chronic inflammation of the synovium (a lubricating and shock-absorbing structure in the knee) and of extra-articular sites within the joint cavity. The joints are usually affected in a symmetrical pattern and the majority of patients will develop radiographic evidence of damage in the joints of their hands and wrists within the first years of the disease [KvLvR+97]. It is a disabling condition frequently resulting in swelling, pain and loss of motility. The exact cause of RA is unknown but it may result as a combination of genetic, environmental and hormonal factors. The main site of irreversible tissue damage occurs at the junction of the synovium lining the joint capsule and the pannus, the area where the capsule joins the cartilage and bone [Mar06]. The cells of the pannus migrate over the underlying cartilage and into the bone, causing the subsequent erosion of these tissues [FBM96]. As the pathology progresses the inflammatory activity leads to erosion and destruction of the joint surface, which impairs the range of movement and leads to deformity.

RA is an incurable disease and treatment is aimed at reducing inflammation and preventing joint damage, and at relieving symptoms, especially pain. Prevention of bone erosion is one of the main objectives of treatment because it is generally irreversible. Treatment involves a combination of medication, rest, exercise, and methods of protecting the joints. There are three general groups of drugs commonly used in the treatment of rheumatoid arthritis: non-steroidal anti-inflammatory agents (NSAIDs), corticosteroids, and disease modifying anti-rheumatic drugs (DMARDs). Traditionally, rheumatoid arthritis was treated in stages starting with NSAIDs and progressing through more potent drugs such as glucocorticoids, DMARDs, and more recently biologic response modifiers such as tumour necrosis factor (TNF) antagonists. Currently however, it has been established that early use of DMARD therapy in patients with developed RA not only controls inflammation better than less potent drugs but also helps to control synovitis and prevent joint damage [RBSB06].

MR imaging permits differentiation of inflamed synovia when used with a contrast agent called gadolinium, and several studies have proposed this as a potentially valuable tool in evaluating synovitis [ØSLN+98]. While subjective assessment of synovial enhancement on this type of MR images is an acknowledged characteristic suggestive of inflammation, accurate grading of disease stage and severity and any response to treatment requires a more objective quantification. Currently, there are manual and semi-automated segmentation techniques available to quantify the degree of synovitis in the arthritic knee. Even with advances in image analysis software, these semi-quantitative techniques are time consuming and unsuitable for the regular evaluation of synovitis. A qualified clinician is required to highlight the inflamed areas using a computer mouse or graphics tablet on a slice by slice basis. This takes a significant amount of time and as such is rarely implemented in a clinical setting. These methods also suffer from inter and intra observer variability. Another implication of this method is that users perception of the inflamed synovial limits may vary. In reality, a semi-quantitative assessment is usually performed with the radiologist visually grading the degree of synovitis. In order to avail of the true diagnostic potential of musculoskeletal 3D MRI in the clinical setting, an automatic method is needed to both expedite this process as well as remove the possibility of user error.

3 Previous Work

As described above, two principal tasks are required for clinical use of MRI scans in tracking rheumatoid arthritis - easy interactive segmentation of the synovial capsule, and efficient computation of quantitative results inside the capsule.

Given the gadolinium marker, which is preferentially taken up by inflamed regions, isosurface extraction reliably extracts the synovium, as shown in Figure 1, although only one of the surfaces (contours) extracted is relevant. While isosurface extraction has been used for twenty years [LC87], it is only recently that a user interface has been described that allows a user direct control over individual contour surfaces [CS03]. This is because the original Marching Cubes paper [LC87] generated individual polygonal fractions of multiple contours, resulting in *polygon soup*, a collection of polygons whose sequence was dictated by their spatial position in the dataset rather than the objects they bounded. And, although post-processing can unite these fragments into the constituent components of the isosurface, acceleration methods focussed on maximising the speed with which polygons were computed and transferred to the graphics hardware rather than treating them as representations of coherent objects.

In contrast to this spatial incoherence, the *continuation* method [WMW86] sought to extract individual surfaces by propagation from an initial seed point. Since objects of clinical interest often show up as independent surfaces, this method has the advantage of aligning the extraction and rendering method with the underlying application task. However, the disadvantage is that making this work successfully depends on having systematic access to suitable seed points for every possible surface, and the ability to identify which surfaces are of interest. While seed points could be generated [IK95, vKvOB$^+$97] for individual surfaces, the ability to identify individual surfaces visually and systematically was introduced in the *flexible isosurface* interface [CS03], based on a topological abstraction of the data called the *contour tree*.

Figure 2 shows an example of the operation of the flexible isosurface interface. Initially, a single isosurface is displayed (a): here, the exterior surface of the head *occludes* or blocks the view of all interior surfaces. Using the conventional point-and-click metaphor of graphic user interfaces, this surface is

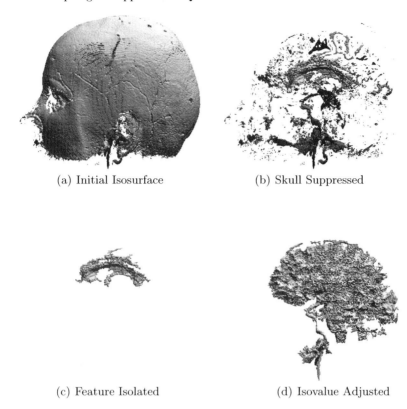

Fig. 2. Example of Flexible Isosurface Editing to Isolate the Brain. Initially, an isosurface is extracted (a). The occluding surface of the head is selected with the mouse, and subtracted from the view (b), then a small interior structure is selected and isolated by subtracting all other surfaces (c). Finally, the isovalue defining this structure is manipulated while continuing to suppress other surfaces, until the desired structure is extracted (d).

selected and suppressed to provide a view of all interior surfaces (b). Next, a single structure is selected and isolated, eliminating irrelevant and distracting surfaces to focus the attention (c). Finally, the isovalue of this surface is adjusted until a desired structure is extracted (d).

At its simplest, then, the flexible isosurface interface allows interactive identification and manipulation of single isovalued surfaces representing clinically significant objects, with simple operations (deletion, isolation, isovalue adjustment). This is based on the observation that the contour tree can be used as an index to all possible isovalued surfaces in the data set and as a representation of the relationships between them.

Details of the contour tree, its implementation, and of the flexible isosurface have been reported elsewhere [CSvdP07]. Here, it suffices to observe that the contour tree not only supports this interactive identification of individual surfaces, but also provides efficient computation of geometric and statistical properties of the objects defined by these boundaries. Most relevant to the task at hand is that these geometric and statistical properties allow us to answer quantitative questions, such as: "how many voxels inside this boundary have more than a given amount of inflammation?"

These quantitative measurements were first precomputed for isosurfaces in the contour spectrum interface [BPS97], which computed properties such as bounded area and contour length by sweeping a cutting (hyper-)plane through a 2-D data set. These properties were then displayed in the interface in graph form as a function of isovalue to assist in isoline/isosurface selection. Subsequent work [CSvdP04] extended these computations to geometric properties of individual contours such as enclosed volume and hypervolume computed by sweeping single contours through the volume using the contour tree as a guide. Moreover, this contour sweep through the contour tree passes through all voxels inside the boundary, allowing simple computation of relevant quantitative measurements.

4 Topological Quantitation

Clinically, two measurements were identified as being important: the number of voxels inside the synovium that exceed a given threshold, and the total intensity of these voxels. Moreover, it was identified that the threshold used to segment the synovium need not be the same as the threshold used to identify the voxels of interest. The exact values of these thresholds will depend on clinical protocol and such factors as variable intensity in the radiological procedures. It was therefore decided to compute histograms of all of the voxels inside the synovium. From these histograms, cumulative histograms from various sub-thresholds can be computed to obtain the number of voxels satisfying the desired criteria, and to compute the total intensity.

While it has been shown recently [CDD06] that histograms are imprecise quantifications of isovalue properties, they give a first approximation suitable for the preliminary nature of the results reported herein. Moreover, the total number of voxels of interest and the total intensity are the result of summing the histogram measurements, which reduces the impact of histogram quality on the results.

Since storage of histograms for each edge in the contour tree is prohibitively expensive, we compute them by propagating a contour upwards (i.e. inwards) from the selected contour, processing the voxels as they are encountered in the contour tree: details of this sweep procedure can be found in [CSvdP07].

This computation then results in a quantitative measurement bounded qualitatively by a user- selected contour surface.

5 Preliminary Results

Before planning a protocol for clinical tests, initial tests were performed on seven MRI datasets already available, consisting of between . These datasets were acquired with significant variation of radiological procedures, resulting in inconsistent intensity ranges. While this limits the validity of our results to date, they are sufficient to confirm that the approach outlined has significant merit warranting full clinical study.

In each of these 8-bit datasets, a contour was identified visually using the flexible isosurface interface shown in Figure 3 and numerical results computed

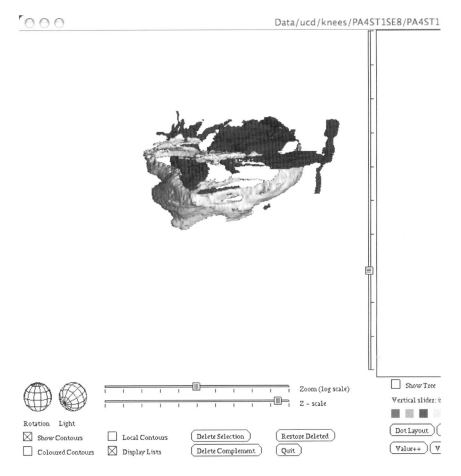

Fig. 3. Flexible Isosurface Interface used for Preliminary Results. The dataset shown is PA4, and the contour shown in light grey in the leftmost panel has been selected as representing the synovium. The contour tree has been neither simplified nor displayed, to reduce the cognitive load on non-expert users. In the clinical study, a simplified interface will be used.

Table 1. Initial Quantitative Results for 8-bit MRI Scans Tested. See text for discussion. Ranking gives the clinical determination of the relative ordering of severity of inflammation.

Dataset	Slices	Threshold	Voxels	Intensity	Ranking
L1	21	91	19,034	2,119,850	0
PA3	30	51	92,397	6,447,491	1
PA5	23	87	23,934	2,712,580	2
F1	28	102	48,568	6,542,193	3
PA4	33	86	26,424	2,697,229	4
PA4 (revised)	33	72	55,654	5,154,226	4
PA2	32	102	48,070	5,580,760	5
PA1	30	109	64,526	7,849,215	6

as shown in Table 1. Simultaneously and independently, these datasets were ranked according to the degree of clinical inflammation apparent. We then inspected images of the contours initially chosen, and identified the following problems:

1. Dataset PA4 had been extracted at too high an isovalue, in part because the intensity in this dataset was approximately 75% of the intensity in the other datasets. A lower isovalue was selected which generated a more appropriate contour, and the intensity values scaled up accordingly for comparison purposes.
2. Dataset PA3 had significant artifacts which resulted in the synovium not occurring as a suitable contour at any isovalue. Since the artifacts were radiological in nature and would be avoidable by designing a suitable protocol, this dataset was excluded from further comparisons.

In addition, it was noted that the datasets consisted of differing numbers of slices positioned in different places relative to the knee, and that the synovium commonly extended outside the dataset. We therefore corrected by dividing the voxel and intensity computations by the number of slices. As we see in Table 2, both voxel counts (volume) and intensity correspond roughly with the ranking we identified.

6 Clinical Protocol

As a result of our preliminary results, we have concluded that development of a targetted clinical tool based on the existing software is merited, and should be tested as one of the diagnostic tools in an upcoming clinical trial.

Two open, proof-of-concept studies have now been completed at the Rheumatology Department, St. Vincents University Hospital (SVUH)

Table 2. Adjusted Quantitative Results for 8-bit MRI Scans Tested. See text for discussion.

Dataset	Slices	Threshold	Voxels (scaled)	Intensity (scaled)	Ranking
L1	21	91	906	100,945	0
PA5	23	87	1,041	117,938	2
F1	28	102	1,735	233,650	3
PA4 (revised)	33	72	1,686	206,252	4
PA2	32	102	1,502	174,399	5
PA1	30	109	2,151	261,641	6

where patients with active Psoriatic arthritis (PsA) were treated with the Interleukin-1 Receptor Antagonist, Kineret (n=12) (Study A) or with the anti-TNF binding protein, Etanercept (n=15) (Study B). Following ethical approval and with patient consent, all patients underwent a synovial biopsy and a contrast-enhanced MRI scan of an affected knee joint prior to and again following 12 weeks of therapy. These MRI datasets provide well defined before and after results which can be compared with the results from a semi-quantitative analysis by a musculo-skeletal radiologist.

In study A, 10 patients completed the study. Two male patients withdrew due to non-compliance with the study protocol and lack of efficacy respectively. Baseline physical examination revealed psoriasis and active synovitis in all patients. While 6 of the 10 patients (60%) met the primary clinical response criteria (PsARC) at week 12, all patients had on-going active joint inflammation and none chose to remain on Kineret after the study was completed. Interestingly, there were no changes in synovial immunohistochemical cellular markers (CD3, CD68, Factor VIII) or in semi-quantitative MRI scores following treatment.

The last patient in Study B has just completed the treatment period and analysis of clinical, immunohistochemical and MRI responses are now beginning in a blinded fashion. As Etanercept has been licensed for treatment of PsA, we would anticipate a greater response than with Kineret. Thus, we would anticipate having available for study 27 paired MRI scans which can be subjected to our new quantitative MRI measurement of enhancing synovial volume. This data can then be compared with both clinical and immunohistochemical responses making for a very powerful analysis of treatment responses.

In order to verify observations in a larger cohort of patients with inflammatory arthritis, patients about to start new disease-modifying or biologic treatments will be asked to undergo a contrast-enhanced MRI scan of an affected knee joint both prior to and 12 weeks following treatment initiation (Study C). 30 patients will be recruited and standard clinical measures of response (joint counts, ESR, CRP, DAS assessments) will be documented at both time points. Changes in quantitative MRI assessments of synovial volume will also be calculated and will be compared with clinical measures.

A specific MRI acquisition protocol will be designed to optimize the clarity of data. Artefacts will be minimized by following this protocol. Moreover, a limitation of our preliminary results was that the region of interest was not always fully included, so the field size will be fixed to include the whole synovial capsule. All of these datasets will be semi-quantitatively scored by a musculoskeletal radiologist.

7 Conclusions

We conclude from the results reported in Section 5 that our approach, of topological identification of the synovium followed by quantitation of the region bounded by the contour selected, gives sufficiently reliable results to warrant full clinical trials, and have reported the appropriate clinical protocol designed as a result of this work. We expect to perform these clinical trials and report results in the appropriate medical venues.

8 Future Work

As noted in [CDD06], histograms are imperfect representations of isovalue distributions, and we intend to perform the full clinical trials with appropriate isosurface statistics. We are also investigating the substitution of level set segmentation for isosurface segmentation, which may improve consistency of results. Finally, we note that, for suitable clinical protocols, it may be possible to identify the synovium automatically in the contour tree, reducing or removing the necessity for user exploration of the data.

9 Acknowledgements

Thanks are due to Science Foundation Ireland for research grant funding and to University College Dublin for equipment grant funding.

References

[AVD05] Y Alamanos, PV Voulgari, and AA Drosos. Rheumatoid arthritis in southern europe: epidemiological, clinical, radiological and genetic considerations. *Current Rheumatology Reviews*, 1:33–36, 2005.

[BPS97] Chandrajit L. Bajaj, Valerio Pascucci, and Daniel R. Schikore. The Contour Spectrum. In *Proceedings of Visualization 1997*, pages 167–173, 1997.

[CDD06] Hamish Carr, Brian Duffy, and Barry Denby. On histograms and isosurface statistics. *IEEE Transactions on Visualization and Computer Graphics*, 12(5):1259–1265, 2006.

[CS03]   Hamish Carr and Jack Snoeyink. Path Seeds and Flexible Isosurfaces: Using Topology for Exploratory Visualization. In *Proceedings of Eurographics Visualization Symposium 2003*, pages 49–58, 285, 2003.

[CSvdP04]   Hamish Carr, Jack Snoeyink, and Michiel van de Panne. Simplifying Flexible Isosurfaces with Local Geometric Measures. In *Proceedings of Visualization 2004*, pages 497–504, 2004.

[CSvdP07]   Hamish Carr, Jack Snoeyink, and Michiel van de Panne. Flexible isosurfaces: Simplifying and displaying scalar topology using the contour tree. Accepted for publication in Computational Geometry: Theory and Application, 2007.

[FBM96]   M Feldmann, FM Brennan, and RN Maini. Role of cytokines in rheumatoid arthritis. *Annual Review of Immunology*, 14:397–440, 1996.

[IK95]   Takayuki Itoh and Koji Koyamada. Automatic Isosurface Propagation Using an Extrema Graph and Sorted Boundary Cell Lists. *IEEE Transactions on Visualization and Computer Graphics*, 1(4):319–327, 1995.

[KvLvR+97] HH Kuper, MA van Leeuwen, PL van Riel, ML Prevoo, PM Houtman, WF Lolkema, and MH van Rijswijk. Radiographic damage in large joints in early rheumatoid arthritis: relationship with radiographic damage in hands and feet, disease activity, and physical disability. *British Journal of Rheumatology*, 36:855–860, 1997.

[LC87]   William E. Lorenson and Harvey E. Cline. Marching Cubes: A High Resolution 3D Surface Construction Algorithm. *Computer Graphics*, 21(4):163–169, 1987.

[Mar06]   C Marra. Rheumatoid arthritis: a primer for pharmacists. *American Journal of Health-System Pharmacy*, 63: Supp 4:S4–S10, 2006.

[ØSLN+98]   M Østergaard, M Stoltenbeg, P Lvgreen-Nielsen, B Volck, S Sonne-Holm, and IB Lorenzen. Quantification of synovitis by mri: correlation between dynamic and static gadolinium-enhanced magnetic resonance and microscopic and macroscopic signs of the synovial inflammation. *Magnetic Resonance Imaging*, 16:743 754, 1998.

[RBSB06]   K Raza, CE Buckley, M Salmon, and CD Buckley. Treating very early rheumatoid arthritis. *Best Practice & Research Clinical Rheumatology*, 20(5):849–863, 2006.

[vKvOB+97] Marc van Kreveld, René van Oostrum, Chandrajit L. Bajaj, Valerio Pascucci, and Daniel R. Schikore. Contour Trees and Small Seed Sets for Isosurface Traversal. In *Proceedings of the 13th ACM Symposium on Computational Geometry*, pages 212–220, 1997.

[WMW86]   Brian Wyvill, Craig McPheeters, and Geoff Wyvill. Animating Soft Objects. *Visual Computer*, 2:235–242, 1986.

3D Visualization of Vasculature: An Overview

Bernhard Preim[1] and Steffen Oeltze[1]

Otto-von-Guericke University Magdeburg, Germany
preim@isg.cs.uni-magdeburg.de, stoeltze@isg.cs.uni-magdeburg.de

Summary. A large variety of techniques has been developed to visualize vascular structures. These techniques differ in the necessary preprocessing effort, in the computational effort to create the visualizations, in the accuracy with respect to the underlying image data and in the visual quality of the result. In this overview, we compare 3D visualization methods and discuss their applicability for diagnosis, therapy planning and educational purposes. We consider direct volume rendering as well as surface rendering.

In particular, we distinguish *model-based approaches*, which rely on model assumptions to create "idealized" easy-to-interpret visualizations and *model-free approaches*, which represent the data more faithfully. Furthermore, we discuss interaction techniques to explore vascular structures and illustrative techniques which map additional information on a vascular tree, such as the distance to a tumor. Finally, navigation within vascular trees (*virtual angioscopy*) is discussed. Despite the diversity and number of existing methods, there is still a demand for future research which is also discussed.

1 Introduction

Due to its relevance, the visualization of vascular structures is an important and established topic within the broader field of visualizing medical volume data. In tumor surgery, for example, surgeons are interested in the effect of surgical strategies on the blood supply of an organ. Vascular surgeons require 3D visualizations to precisely locate stenotic regions and to interactively plan a stent implantation or bypass surgery.

General visualization techniques, such as slice-based viewing, direct volume rendering and surface rendering, are applicable in principle to display vascular structures from contrast-enhanced CT or MRI data. However, to recognize the complex spatial relations of a vascular tree and its surrounding structures more clearly, dedicated techniques are required. The suitability of a visualization technique depends on some general criteria, such as performance, and primarily on the specific purpose. In particular, the required accuracy of the visualizations is different for diagnostic purposes, such as the search and analysis of vascular abnormalities, and therapy planning scenarios, where the vascular structures, themselves are not pathologic. Therefore, we distinguish *model-based approaches*, which rely on model assumptions to create "idealized" easy-to-interpret visualizations and *model-free approaches*, which represent the data more faithfully.

Besides the visualization of the vessel anatomy, we also discuss how additional nominal or scalar data may be mapped on the vessel's surface. There are many different situations where it is desirable to encode additional information: the distance from the vessel surface to another structure, e.g. a tumor should be conveyed, the different vascular territories, supplied by parts of a vascular tree should be indicated or results of a CFD (Computational Fluid Dynamics) simulation, such as the shear stress inside an aneurysm should be displayed. For the sake of brevity, this overview is focussed on 3D visualization techniques. Despite their clinical relevance, 2D visualization techniques such as slice-based viewing and anatomic reformations, e.g. multiplanar reformation and curved planar reformation [KFW+02] are not discussed.

2 Visualization of Vasculature with Direct Volume Rendering

2.1 Projection Methods

The most common volume rendering technique is the maximum intensity projection (MIP) which basically displays the voxel with the highest image intensity for every ray. Based on the enhancement with a contrast agent, vessels are usually the brightest structures and can thus be selectively visualized with this method. This basic strategy fails if contrast-enhanced vessels and skeletal structures, which also exhibit high intensity values, are close to each other. The removal of bony structures is a necessary preprocessing step in such cases (see [FSS+99] for a discussion of bone removal techniques). More general, the 3D visualization of vascular structures benefits from an enhancement of elongated vessel-like structures which may be accomplished by a shape analysis [FNVV98]. Figure 1 shows the effect of such preprocessing on MIP visualizations.

Since static MIP images do not provide any depth perception, they are either interactively explored or presented as animation sequences. This, however, does not account for another problem related to MIP: small vascular structures are often suppressed since they are primarily represented by border voxels—voxels which contain also portions of surrounding tissue. The averaging effect of the image acquisition, the so-called partial volume effect, leads to the situation that those voxels appear less intense than the inner voxels of large vascular structures. In order to display small vessels in front of larger ones, the MIP method was modified by a threshold which specifies that the

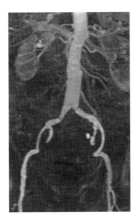
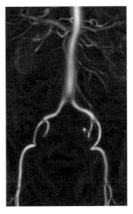

Fig. 1. A MIP image of an MRA dataset is generated (left). The "vesselness" filter is applied to suppress other structures. (From: [FNVV98])

first local maximum above the threshold should be depicted instead of the global maximum. This method is known as either local MIP or closest vessel projection (CVP) [SRP91, Zui95]. However, this refined technique requires to specify an appropriate threshold which may be tedious without support. A histogram analysis may be performed to "suggest" an appropriate threshold, for example a certain percentage of the global maximum of the image intensity.

2.2 Compositing

In contrast to projection methods, direct volume rendering (e.g., raycasting, splatting) provides realistic depth-cues by blending data considering the "in-front-of" relation. However, the direct application of general volume visualization algorithms to rather small vascular structures leads to severe aliasing artifacts. Pommert et al. [PBH92] point out that the gray-level gradients simply do not correspond to reality resulting in a jaggy appearance. They suggest to resample the volume data with a factor of four in each dimension for overview visualizations and to resample with factor eight for detail views. Pommert el al. also illustrate the potential of presenting 3D visualizations of vasculature in the context of other structures, such as brain tissue. Later, they constructed a model of cerebral vasculature and used it as a building block for their anatomy teaching system *VoxelMan* [UGS+97].

Volume rendering of vascular structures usually involves only a small fraction of the overall data size. Frequently, the transfer function is adjusted such that only 1 or 2% of the voxels become visible [HHFG05]. Therefore, volume rendering may be strongly accelerated by means of empty space skipping.

The ability of 1D transfer functions to discriminate vascular structures from its surrounding is particularly limited in the neighborhood of skeletal

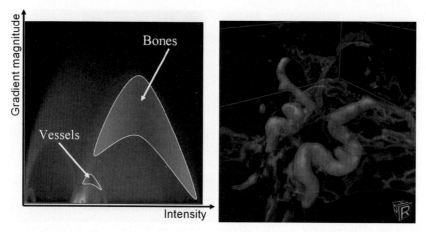

Fig. 2. In the 2D histogram (image-intensity and gradient magnitude) vessels and skeletal structures may be distinguished. With an appropriate 2D transfer function, skeletal structures might be effectively suppressed (right image). (Courtesy of Fernando-Vega and Peter Hastreiter, University of Erlangen-Nuernberg)

structures which exhibit the same range of image intensity values. Therefore, 2D transfer functions have been explored [VST+03], where gradient magnitude has been used as second dimension in addition to image intensity (Figure 2). This is in principle, a promising strategy. However, the adjustment of 2D transfer functions is difficult to accomplish and cannot be completely standardized since gradient values differ strongly for different acquisition devices and modalities. Vega et al. [HHFG05] present an advanced approach to standardize 2D transfer functions. They consider the transfer function specification of one dataset as a template which may be reused for other datasets by adapting it to the joint histogram of gradient magnitude and intensity values.

3 Model-free Surface Visualization of Vasculature

There are two classes of surface visualization methods: methods which strictly adhere to the underlying data and methods which are based on certain model assumptions and force the resulting visualization to adhere to these assumptions—at the expense of accuracy. In this section, we describe the first class of methods and refer to them as *model-free* visualization techniques. These techniques are primarily useful for diagnostic purposes where accuracy is crucial.

In general, surface visualization provides an alternative way of displaying vascular structures. The most common surface reconstruction technique is Marching Cubes with an appropriate threshold. Unfortunately, the quality of the resulting visualizations is relatively low which is due to inhomogeneities in the contrast agent distribution and to the underlying interpolation (linear interpolation along the edges compared to trilinear interpolation in most direct volume rendering realizations). Textbooks on radiology warn their readers of the strong sensitivity of the selected isovalue on the visualization illustrating that a small change of 1 or 2 Hounsfield units would lead to a different diagnosis (e.g. presence or absence of a severe stenosis—a region where the vessel diameter is strongly reduced).

3.1 Smoothing Surface Visualizations

In particular, if binary segmentation results are employed, severe aliasing artifacts arise in the surface reconstruction which gives rise to a subsequent smoothing step. It must be noted that appropriate smoothing of vascular structures is challenging and often does not lead to the desired results. Simple smoothing procedures such as Laplacian smoothing should not be used at all because they may easily change the topology of vascular trees. Better results are achieved with error-controlled smoothing approaches, such as LowPass-filtering [Tau95] and the improved Laplacian smoothing introduced in [VMM99]. However, even these advanced approaches may lead to severe errors such as topological inconsistency [CL01]. Figure 3 illustrates the results of LowPass-filtering.

3.2 Visualization of Segmented Vasculature

The quality may be enhanced if the relevant vessels are explicitly segmented prior to surface reconstruction. The segmentation step involves a certain amount of user interaction, such as selecting seedpoints and providing other parameters for region-growing variants. The segmentation process is also a possible source of uncertainty and error. Compared to direct volume rendering

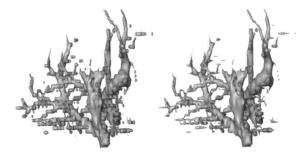

Fig. 3. General mesh smoothing techniques are not appropriate to post-process vascular structures created with Marching cubes (left). The LowPass-filter was carefully adjusted (10 iterations, weighting factor 0.5) but no parameter adjustment lead to satisfying results (right). Vascular structures look very unnatural and the accuracy of the resulting visualizations is very low. (Courtesy of Ragnar Bade, University of Magdeburg)

(recall Section 2), the segmentation and thus the visualization is in general not complete since segmentation methods tend to miss details in the periphery.

Despite these problems, better visualizations may be achieved. However, the Marching Cubes algorithm is not dedicated to visualize binary volume data which arise from the segmentation (0 represents background and 1 represents voxels of the vascular structure). Strong terracing artifacts are typical for surfaces generated with Marching Cubes from binary segmentation results. Only very few smoothing techniques are dedicated to these typical artifacts. Even advanced feature-preserving smoothing approaches completely fail since they would recognize the staircases as features which should be preserved.

The best general method to smooth visualizations of binary segmentation results are Constrained Elastic Surface Nets [Gib98]. This iterative process restricts the movement of vertices to the cell they belong to according to the segmentation result. As a result, this method provides a very good trade-off between accuracy and smoothness. However, for small vascular structures, the constraint to a particular cell is not sufficient, since these are often only one voxel wide. Such structures degenerate after smoothing to a line.

Visualization with MPU Implicits

Recently, a surface visualization based on the segmentation result was developed which provides a superior quality compared to constrained elastic surface nets [SOB+07]. This method is based on MPU Implicits [OBA+03], a technique originally created to visualize point clouds with implicit surfaces. The points are generated for the border voxels of the segmentation result. At thin and elongated structures, additional subvoxels are included and more points are generated (Figure 4). Thus, it can be ensured that enough points are available to fit a quadric surface to. The pointset is locally sampled and approximated with quadric surfaces which nicely blend together. The implicit surfaces, in general, are able to represent a given geometry smoothly without explicitly constructing the geometry (Figure 5). In addition to the number of boundary voxels, the number of voxels that belong to thin structures affects the computational complexity of the point cloud generation. The generation

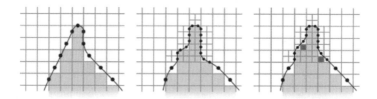

Fig. 4. Points are generated according to the segmentation result (left). To avoid artifacts at small details like thin branches, additional points are generated (middle). Since the additional points do not provide the desired results, additional subvoxels need to be included before the generation of points (right). (From: [SOB+07])

Fig. 5. A segmented liver tree is visualized with Marching Cubes (left) and with MPU Implicits (right). The appearance is strongly improved and the accuracy of the visualization is maintained. (From: [SOB+07])

and polygonization of the implicit surface mainly depends on the complexity of the surface in terms of curvature. To put it in a nutshell, the complexity of the MPU implicit surface generation depends on the size of the segmentation result and the complexity of the described object (fraction of details and thin structures) due to its adaptive nature.

As a drawback, MPU Implicits are about three times slower compared to Constrained Elastic Surface Nets, at least in the current implementation [SOB+07]. Note, that this is a rough estimate, since the performance of MPU Implicits strongly depends on the surface topology, whereas Constrained Elastic Surface Nets depend on the number of boundary voxels and the number of iterations.

4 Model-based Surface Visualization of Vasculacture

For therapy planning (in the case that not the vessels themselves are affected) and for educational purposes, model-based techniques are appropriate. Model-based techniques require another preprocessing step: beyond noise reduction and segmentation, the vessel centerline and the local vessel diameter have to be determined. For this purpose, skeletonization algorithms are used. There are different approaches to skeletonization: Some methods firstly determine the skeleton, guided by a few points specified by the user, and the local cross-sections and thus the segmentation result (these are referred to as *direct skeletonization*) [KQ04, SPSHOP02]. Other methods use the segmentation result as input and apply, for example a modified iterative erosion, to determine the vessel centerline. The drawback of this additional analysis step is that it introduces another source of error. As an example, whether a side-branch is suppressed or represented as part of the centerline critically depends on the selected algorithm and its parameters. Despite these problems, the basic shape is well-characterized with skeletonization approaches. The basic concept and a first solution towards model-based surface visualization based on radiological data of an individual patient was provided by Gerig et al. [GKS+93].

Ehricke et al. [EDKS94] provided an in-depth discussion of the pipeline for model-based vessel visualization.

4.1 Explicit Construction of Vascular Models

The central model assumption is usually that the cross-section is always circular.[1] This assumption gave rise to techniques which construct a geometry explicitly with graphics primitives which exhibit a circular cross-section. Cylinders [MMD96] and truncated cones [HPSP01] have been employed. The latter allows to faithfully represent how the diameter diminishes towards the periphery. The quality depends on the number of vertices used to approximate the circular cross-section as well as on the number of graphics primitives attached along the skeleton (see Figure 6 for a comparison of the visualization by means of truncated cones in different resolutions). Explicit methods cannot completely avoid artifacts and discontinuities at branchings where graphics primitives are fitted together. This problem becomes annoying if close-up views are generated, e.g. for fine-grained surgical planning. The major advantage of these methods is their speed.

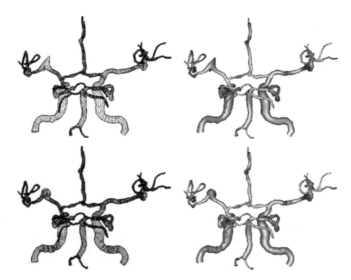

Fig. 6. Cerebral blood vessels are reconstructed from MRI data and sampled with different resolutions along the skeleton. The wire-frame visualizations (left) correspond to the Gouraud-shaded visualizations (right). In the top row, a moderate sampling rate is chosen whereas in the bottom row the maximum sampling rate results in a fine-grained mesh and a smooth surface visualization. (Courtesy of Horst Hahn, MeVis Research Bremen)

---

[1] Other possible assumptions are that the diameter shrinks from the root of a vessel tree to the periphery and that the vessels do not end abruptly [HPSP01].

4.2 Parametric and Implicit Surfaces of Vascular Models

To achieve smooth transitions at branchings, a variety of methods has been investigated. B-spline surfaces have been proposed in [HPea00] to approximate small vascular structures and nerves which could not be completely segmented (see Figure 7).

Felkel et al. [FWB04] describe a method based on the subdivision of an initial coarse base mesh. The advantage of their method is that they provide a trade-off between speed and quality which may be used to achieve high quality also at branchings (see Figure 8 for a sketch and Figure 9 for an application to clinical data). Simplex meshes can also be used for a high-quality visualization of vascular structures, as Bornik et al. [BRB05] recently showed. Simplex meshes may be adapted to the segmentation result and thus produce more accurate visualizations. As for model-free visualization of vascular structures, implicit surfaces may be used for model-based visualization. Convolution surfaces, developed by Bloomenthal [BS91], allow to construct a scalar field along skeletal structures. With an appropriate filter for the convolution, the scalar field can be polygonized [Blo94] in such a way that the vessel diameter is precisely represented. The filter selection must also avoid the so-called "unwanted effects", such as unwanted blending and bulging which are typical for implicit surfaces. This basic approach enhanced with sophisticated acceleration strategies has been presented in [OP04]. Figure 10 gives an example.

A general problem of the high-quality visualization methods is that they tend to be slower than explicitly constructed vascular trees (previous subsection). For larger datasets, the visualization with convolution surfaces takes 20 to 50 seconds (recall [OP04] for an in-depth discussion) compared to 3 to 5 seconds with truncated cones. Another problem is that the accuracy of most of these methods has not been carefully investigated. Comparisons of different methods with respect to the resulting surfaces and their distances are necessary to state how reliable the results actually are. A validation of

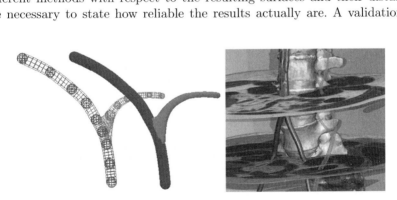

Fig. 7. Small anatomic tree structures are modelled by placing ball-shaped markers and represented as a special variant of splines. Due to partial volume effects, these structures could not be completely segmented from the Visible Human dataset. (From: [PHP+01])

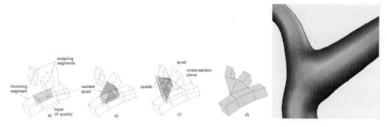

Fig. 8. Left: The subdivision method generates quad strips and modifies the process at branchings to achieve smooth transitions. The right image shows a typical result. (Courtesy of Petr Felkel)

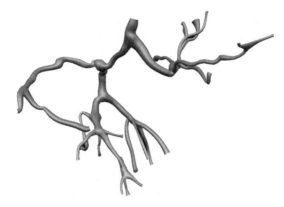

Fig. 9. Subdivision surfaces applied to a vessel tree derived from clinical data. (Courtesy of Petr Felkel)

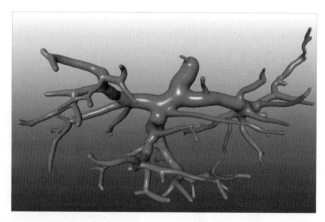

Fig. 10. Visualization of the portal vein derived from a clinical CT dataset with 136 edges of the branching graph and 125 000 polygons. (From: [SOB[+]07])

Fig. 11. A segmented cerebral aneurysm is visualized with Marching Cubes (left, 166 K polygons), with MPU Implicits (middle, 180 K polygons) and convolution surfaces (right, 201 K polygons). Despite the high visual quality, the convolution surface is less appropriate than MPU Implicits due to the strong deviation from the circular cross-section. (From: [SOB+07])

convolution surfaces was accomplished and indicates that the accuracy of this method is close to the accuracy of the truncated cones method [OP05]. However, it must be noted that the good results (mean deviation: 0.5 times the diagonal size of a voxel) relate to datasets without any pathologies. Figure 11 compares visualizations of an aneurysm with Marching Cubes, MPU Implicits and convolution surfaces.

4.3 Encoding Information on Vessel Surfaces

In the introduction, we briefly discussed scenarios where additional information should be mapped on the anatomy of vascular structures. In general, the mapping requires that the vessels are explicitly segmented. Moreover, their centerline and their branching structure are determined. Therefore, this issue is firstly discussed in the context of model-based visualization techniques which also require this information. The most obvious and wide-spread visualization technique is to color-code the data on the vessel's surface (see Figure 12 for two examples). Since variations of the brightness occur due to shading computations, hue and saturation are the major color components for this purpose. In general, the use of color is restricted to encode one attribute, such as the membership of a branch to a particular anatomic structure. If more attributes should be depicted, textures or other illustrative techniques should be applied (see Sect. 5).

5 Illustrative Visualization of Vasculature

Illustrative techniques, such as hatching and silhouette generation, serve two purposes in the context of vessel visualization: they enhance the shape perception and they may encode additional information on the vessel's surface.

Enhancing shape perception. With respect to shape perception, silhouettes and hatching lines have been explored [RHP+06]. Hatching lines require

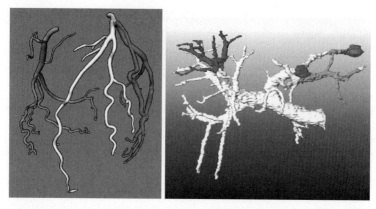

Fig. 12. Left: Colors encode to which major coronary artery the branches belong. The vascular model is visualized with convolution surfaces. Right: Colors encode the discretized distance to a tumor. Red indicates distances below 5 mm, yellow 5-10 mm and green 10-15 mm

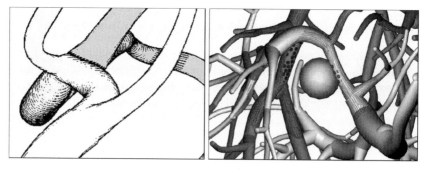

Fig. 13. Illustrative rendering techniques applied to visualizations with truncated cones. Left: Hatching lines support the shape perception. At a crossing, the depth perception is enhanced by interrupting hatching lines at the more distant vessel branch. Right: Two colors serve to distinguish two vascular trees. Different textures are mapped onto the surface to indicate distances to a simulated tumor (yellow sphere). (Courtesy Christian Hansen, MeVis Research Bremen)

an appropriate parametrization of the vessel surface. Usually, hatching lines are guided by principal curvature directions. So far, such a parametrization has only been performed for vessels represented by concatenated truncated cones ([Han06] and [RHP+06]). They presented a texture-based and hardware supported approach to hatching vessels composed of truncated cones (recall [HPSP01]). Figure 13 presents two examples of their work. Hatching was selectively employed to indicate distances between two branches at bifurcations. In a user study, Hansen and Ritter could demonstrate that the desired effect

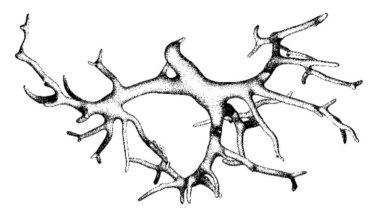

Fig. 14. With an efficient stippling approach it is possible to apply stippling textures with minimal distortions to surface models of vascular trees. In this example, the surface model was generated with convolution surfaces. (Courtesy of Alexandra Baer, University of Magdeburg)

on shape perception was actually achieved [RHP+06]. The generation of expressive hatching lines for model-free visualizations has not been investigated.

Encoding information on vessel surfaces. Illustrative techniques bear a great potential to encode additional information on vessel surfaces. The density and orientation of hatching lines, for example, are attributes which can be efficiently distinguished. Thus, hatching can be used to encode two or even three nominal values. Stippling can also be used to adapt the occurrence, size and density of stippling points to the value of different variables. Baer et al. introduced a texture-based stippling approach which may be used to efficiently stipple vascular structures (Figure 14 and [BTBP07]). Thus, the use of colors can be restricted to a few regions which are particularly relevant, such as the branches of a vascular tree close to a tumor.

6 Interactive Exploration and Modification of Vasculature

The exploration of vascular structures has not been investigated in the same level of detail as the visualization methods. In particular, therapy planning in the context of large vascular trees requires facilities for exploration. Powerful exploration techniques are based on vessel analysis results. In particular, the centerline with associated diameter information and the branching graph are crucial for the exploration. As an example, subtrees may be collapsed based on the branching graph. *Visual exploration* and *quantitative analysis techniques* can be distinguished.

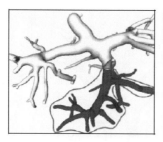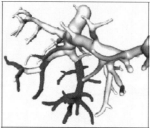

Fig. 15. Image-based selection (LassoSelection) often yields undesired results due to the complex depth relations. A rotation (right image) indicates that not only one subtree was selected. (From: [vH06])

6.1 Visual Exploration

Visual exploration relates to techniques which emphasize subtrees or subregions, e.g., regions close to a tumor. Usually, the image-based selection within the viewport is not sufficient since the branches strongly overlap and after a rotation it often turns out that too many branches were selected (see Figure 15). Subtrees may be selected e.g., using a hierarchy measure or a criterion which relates to the vessel diameter.

For therapy planning it is often essential to visually distinguish regions of a different diameter (less than 2 mm: irrelevant for planning, 2-5 mm: relevant, larger than 5 mm: vessels should not be damaged or - if this is not feasible - must be reconstructed). Detail-and-context visualizations which allow to show a relevant portion of a vascular tree enlarged and unoccluded, support diagnostic as well as therapy planning tasks. Synchronization facilities between the two views further enhance the visual support. Visual exploration techniques are briefly discussed in [HPSP01]. An interesting combination of direct volume rendering and curved planar reformation is the *VesselGlyph* described in [SCC+04]. This technique ensures the visibility of important vascular structures in the context of other anatomic structures.

6.2 Quantitative Analysis

Quantitative analysis techniques relate to distances along the vessel centerline or the vessel surface (*geodesic distance*), radius distribution or bifurcation angles. Distance measures are relevant, e.g., for endoscopic interventions where a catheter is introduced into vascular structures. Distance measures along the centerline require that the points selected by the user on the surface are projected to the skeleton. A dedicated effort on quantitatively analyzing vascular structures by means of 3D surface visualizations has been accomplished by Reitinger [RSBB06]. They provide measurement tools to evaluate angles at bifurcations and distances between selected branches in an augmented reality environment with stereo rendering and bi-manual input.

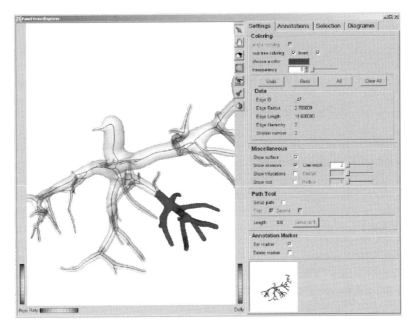

Fig. 16. The VesselExplorer integrates facilities to explore vascular trees. A subtree is selected and the vessel centerline is included. Based on the selection various analysis techniques may be enabled. For example, distances may be measured, histograms of cross-sections may be displayed and annotations may be added. (From: [vH06])

A thorough investigation has recently been accomplished [vH06] and also includes quantitative analysis techniques (Figure 16). These techniques are integrated into the *VesselExplorer*. With this system, detail-and-context visualizations are provided, annotation and labelling is possible and special branching types, such as trifurcations, are emphasized. The exploration of complex vascular structures might benefit from 3D input devices. Wischgoll et al. [WMM05] investigate the use of different 3D input devices, including those primarily developed for steering games, to the exploration of complex cardiovascular structures.

6.3 Interactive Modification of the Vessel Geometry

The modification of the vessel geometry is primarily motivated by intervention planning scenarios. Interventional procedures, such as coiling and stenting, as well as bypass operations, change the vessel geometry. It is desirable to study different variants of such interventions preoperatively. The modification may be supported with different facilities, e.g. with a facility to modify the vessel centerline and a facility to specify the local vessel diameter along the centerline. Morphological operators applied to the vessel segmentation result also allow to modify the vessel geometry. For certain vascular diseases, such as aneurysms, it is essential to investigate the blood flow resulting from a modification of the vessel geometry. The complex correlation between vessel geometry and blood flow may be effectively analyzed by means of CFD simulations [CL01, CCA+05].

In general, exploration facilities deserve more attention: the exploration of complex vascular structures includes challenging 3D interaction problems which have not been carefully investigated so far.

7 Virtual Angioscopy

For educational purposes, for some diagnostic tasks as well as for planning vascular interventions, virtual fly throughs are a useful option. This is a special instance of *virtual endoscopy*, called *virtual angioscopy*. Some of the described visualization methods, e.g., model-based approaches using cylinders or truncated cones are not appropriate for virtual angioscopy at all due to the construction of polygons inside the vessel lumen. Other model-based approaches facilitate a fly-through, however their accuracy is not sufficient for the planning of a vascular surgery. Instead, model-free approaches, such as MPU implicits or smoothed meshes based on a Marching Cubes visualization are better suited. It is essential to combine detailed views of the inner structures with an overview of the anatomic structures (see Figure 17). A dedicated application of virtual angioscopy, namely the diagnosis of neurovascular diseases has been described in [BSWS99] and [BSSW99]. The navigation inside vascular structures is usually accomplished by guided navigation concepts, used in other areas of virtual endoscopy, e.g. virtual colonoscopy. Bartz et al. [BSWS99] compute a distance field based on the segmentation result and

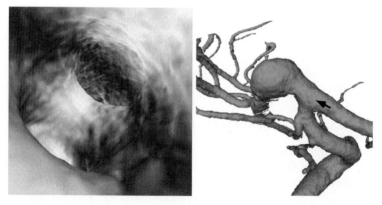

Fig. 17. Virtual angioscopy for the diagnosis of a cerebral aneurysm at the Arteria Cerebri Media. The arrow in the right overview image indicates the position and the orientation of the camera in the left image. (Courtesy of Dirk Bartz, University of Leipzig)

interpret this as a *potential field* to provide an effective collision avoidance and navigation.

Virtual angioscopy is also an important aspect of the Virtual Vasculature project [ACea98, PZGT97] which is dedicated to the exploration of cerebral vessels. Virtual angioscopy in this project is combined with vascular fluid dynamic simulations.

8 Concluding Remarks

The visualization of vascular structures for advanced diagnostic or therapeutic problems requires dedicated visualization techniques. The accuracy of the visualization as well as the removal of annoying discretization artifacts are the two major and conflicting goals. For diagnostic processes, accuracy is more important, as it gives rise to techniques without model-assumptions. For therapy planning, minor deviations are usually acceptable and therefore model-based techniques are appropriate. Among the model-based techniques, noticeable differences in the visual quality exist. Methods based on subdivision surfaces, simplex meshes and convolution surfaces produce better results at the expense of increased computational effort (e.g., the generation of convolution surfaces leads to a 2-3 times larger rendering times compared to truncated cones). Advanced model-based techniques avoid any discontinuities at branchings as well as inner polygons. However, the accuracy of some of these methods has not been analyzed so far. Volume visualization methods are attractive since no explicit segmentation is required. In particular, in emergency situations, this advantage is crucial. Any 3D visualization of vasculature benefits from high-quality stereo visualizations [ACea98, RSBB06]. The interactive exploration and navigation inside of vascular structures may be further enhanced by providing virtual environments and 3D input devices [WMM05, RSBB06]. These high-end solutions are primarily aimed at medical research or advanced education, not at routine clinical tasks. For a quantitative comparison between MPU Implicits, Constrained Elastic Surface Nets, and Convolution Surfaces with respect to accuracy, geometric complexity and reconstruction time, we refer to [SOB$^+$07]. A comparison between truncated cones, Marching Cubes and convolution surfaces may be found in [OP05].

A large portion of the visualization and exploration techniques presented here have been realized in the MeVisLab platform (www.mevislab.de).

Future Work. Despite the quantity and diversity of the existing methods for visualizing vascular structures, there is still a demand for further research. Volume visualization of segmented vascular structures has not been investigated well. The visualizations based on MPU Implicits and Convolution surfaces produce very good results for diagnostic and therapy planning tasks, but they are simply too slow for wide-spread clinical use. Both methods would probably benefit from an adaptive curvature-dependent polygonization and an implementation which better exploits graphics hardware. The illustrative visualization and interaction techniques are currently limited to a particular vessel visualization method. It should be attempted to provide a more general parametrization of vascular trees in order to combine illustrative and interaction techniques with more than one visualization method.

So far, only individual visualization methods have been applied and analyzed with respect to their properties. There is a need to come up with hybrid combinations between different visualization methods. Based on image analysis results, it can be inferred in which regions certain model assumptions are fulfilled and in which regions the deviation to the assumptions is too large for applying a model-based technique. The challenge of such hybrid combinations is to provide seamless transitions between the individual visualization methods. This overview was focussed on 3D visualizations. Synchronized 2D and 3D visualizations are a promising extension, in particular for advanced therapy planning tasks. Finally, it is obvious that vessel visualization methods should be analyzed in more depth with respect to specific clinical problems. Eventually, methods have to be parameterized in a specific way or have to be fine-tuned. In particular, the intraoperative use of vessel visualization techniques deserves more attention. A major challenge for future work is to provide effective integrated visualizations of vascular geometry and CFD simulation results.

Acknowledgements. We would like to thank Christian Schumann, Christian Hansen, and Verena von Hintzenstern for their engaged work—this overview is largely based on their master's theses. We thank Ragnar Bade for systematically evaluating mesh smoothing techniques for vascular structures and Alexandra Baer for investigating stippling on vascular surfaces. Finally, we thank MeVis for the chance to use their MeVisLab system (www.mevislab.de) and their method for visualizing vascular structures, realized by Horst Hahn and Olaf Konrad. Felix Ritter contributed to the illustrative visualization techniques realized in Christian Hansen's Master thesis. Dirk Bartz provided insight into the virtual angioscopy and provided appropriate images. Many thanks also to K. Bühler, P. Felkel, A. Frangi, P. Hastreiter, A. Pommert and F. Vega for providing their images and advice how to describe them.

References

[ACea98] Gassan Abdoulaev, Sandro Cadeddu, and Giovanni Delussu et al. ViVa: The virtual vascular project. *IEEE Transactions on Information Technology in Biomedicine*, 22(4):268–274, December 1998.

[Blo94] Jules Bloomenthal. An Implicit Surface Polygonizer. In *Graphics Gems IV*, pages 324–349. Academic Press, Boston, 1994.

[BRB05] Alexander Bornik, Bernhard Reitinger, and Reinhard Beichel. Reconstruction and Representation of Tubular Structures using Simplex Meshes. In *Proc. of Winter School of Computer Graphics (WSCG)*, pages 61–65, 2005.

[BS91] Jules Bloomenthal and K. Shoemake. Convolution Surfaces. *Computer Graphics (Proc. of ACM SIGGRAPH)*, 25:251–256, 1991.
[BSSW99] Dirk Bartz, Wolfgang Straßer, Martin Skalej, and Dorothea Welte. Interactive Exploration of Extra- and Intracranial Blood Vessels. In *Proc. of IEEE Visualization*, pages 389–392, 1999.
[BSWS99] Dirk Bartz, Martin Skalej, Dorothea Welte, and Wolfgang Straßer. 3D Interactive Virtual Angiography. In *Proc. of Computer-Assisted Radiology and Surgery*, pages 389–392, 1999.
[BTBP07] Alexandra Baer, Christian Tietjen, Ragnar Bade, and Bernhard Preim. Hardware-Accelerated Stippling of Surfaces Derived from Medical Volume Data. In *IEEE/Eurographics Symposium on Visualization, Eurographics*, 2007.
[CCA+05] Juan R. Cebral, Marcelo Adrián Castro, Sunil Appanaboyina, Christopher M. Putman, Daniel Millan, and Alejandro F. Frangi. Efficient pipeline for image-based patient-specific analysis of cerebral aneurysm hemodynamics: technique and sensitivity. *IEEE Transactions on Medical Imaging*, 24(4):457–467, 2005.
[CL01] Juan R. Cebral and Rainald Lohner. From Medical Images to Anatomically Accurate Finite Element Grids. *International Journal of Numerical Methdos in Engineering*, 51:985–1008, 2001.
[EDKS94] Hans-Heino Ehricke, Klaus Donner, Walter Killer, and Wolfgang Straßer. Visualization of vasculature from volume data. *Computers and Graphics*, 18(3):395–406, 1994.
[FNVV98] Alejandro F. Frangi, Wiro J. Niessen, Koen L. Vincken, and Max A. Viergever. Multiscale vessel enhancement filtering. In *Proc. of Medical Image Computing and Computer-Assisted Intervention*, volume 1496 of *Lecture Notes in Computer Science*, pages 130–137. Springer, 1998.
[FSS+99] Martin Fiebich, Christopher M. Straus, Vivek Sehgal, Bernhard C. Renger, Kunio Doi, and Kenneth R. Hoffmann. Automatic bone segmentation technique for CT angiographic studies. *Journal of Computer Assisted Tomography*, 23(1):155–161, 1999.
[FWB04] Petr Felkl, Rainer Wegenkittl, and Katja Bühler. Surface Models of Tube Trees. In *Proc. of Computer Graphics International*, pages 70–77, 2004.
[Gib98] Sarah F. Frisken Gibson. Constrained elastic surface nets: Generating smooth surfaces from binary segmented data. In *Proc. of MICCAI*, volume 1496 of *Lecture Notes in Computer Science*, pages 888–898. Springer, 1998.
[GKS+93] Guido Gerig, Thomas Koller, Gábor Székely, Christian Brechbühler, and Olaf Kübler. Symbolic description of 3-d structures applied to cerebral vessel tree obtained from mr angiography volume data. In *Proc. of Information Processing in Medical Imaging*, volume 687 of *Lecture Notes in Computer Science*, pages 94–111. Springer, 1993.
[Han06] Christian Hansen. Verwendung von Textur in der Gefävisualisierung. Master's thesis, Dept. of Computer Science, Magdeburg, 2006.
[HHFG05] Fernando Vega Higuera, Peter Hastreiter, Rudolf Fahlbusch, and Gunter Greiner. High performance volume splatting for visualization of neurovascular data. In *Proc. of IEEE Visualization*, 2005.
[HPea00] Karl-Heinz Höhne, Bernhard Pflesser, and Andreas Pommert et al. A Realistic Model of the Inner Organs from the Visible Human Data.

	In *Proc. of Medical Image Computing and Computer-Assisted Intervention*, volume 1935 of *Lecture Notes in Computer Science*, pages 776–785. Springer, 2000.
[HPSP01]	Horst K. Hahn, Bernhard Preim, Dirk Selle, and Heinz-Otto Peitgen. Visualization and Interaction Techniques for the Exploration of Vascular Structures. In *IEEE Visualization*, pages 395–402, 2001.
[KFW+02]	Armin Kanitsar, Dominik Fleischmann, Rainer Wegenkittl, Petr Felkel, and Eduard Gröller. CPR: curved planar reformation. In *Proc. of IEEE Visualization*, pages 37–44. IEEE Computer Society, 2002.
[KQ04]	Cemil Kirbas and Francis Quek. A Review of Vessel Extraction Techniques and Algorithms. *ACM Computing Surveys*, 36(2):81–121, June 2004.
[MMD96]	Yoshitaka Masutani, Ken Masamune, and Takeyoshi Dohi. Region-Growing-Based Feature Extraction Algorithm for Tree-Like Objects. In *Proc. of Visualization in Biomedical Computing*, volume 1131 of *LNCS*, pages 161–171. Springer, 1996.
[OBA+03]	Yutaka Ohtake, A. Belyaev, Marc Alexa, Greg Turk, and Hans-Peter Seidel. Multilevel Partition of Unity Implicits. *ACM Transactions on Graphics*, (22):463–470, 2003.
[OP04]	Steffen Oeltze and Bernhard Preim. Visualization of Vascular Structures with Convolution Surfaces. In *Proc. of IEEE/Eurographics Symposium on Visualization (VisSym)*, pages 311–320, 2004.
[OP05]	Steffen Oeltze and Bernhard Preim. Visualization of Vascular Structures with Convolution Surfaces: Method, Validation and Evaluation. *IEEE Transactions on Medical Imaging*, 25(4):540–549, 2005.
[PBH92]	Andreas Pommert, Michael Bomans, and Karl Heinz Höhne. Volume visualization in magnetic resonance angiography. *IEEE Computer Graphics and Application*, 12(5):12–13, 1992.
[PHP+01]	Andreas Pommert, Karl Heinz Höhne, Bernhard Pflesser, Ernst Richter, Martin Riemer, Thomas Schiemann, Rainer Schubert, Udo Schumacher, and Ulf Tiede. Creating a high-resolution spatial/symbolic model of the inner organs based on the visible human. *Medical Image Analysis*, 5(3):221–228, 2001.
[PZGT97]	Piero Pili, Antonio Zorcolo, Enrico Gobbetti, and Massimiliano Tuveri. Interactive 3D visualization of carotid arteries. *International Angiology*, 16(3):153, September 1997.
[RHP+06]	Felix Ritter, Christian Hansen, Bernhard Preim, Volker Dicken, Olaf Konrad-Verse, and Heinz-Otto Peitgen. Real-Time Illustration of Vascular Structures for Surgery. *IEEE Transactions on Visualization and Graphics*, 12(5):877–884, 2006.
[RSBB06]	Bernhard Reitinger, Dieter Schmalstieg, Alexander Bornik, and Reinhard Beichel. Spatial Analysis Tools for Medical Virtual Reality. In *Proc. of IEEE Symposium on 3D User Interface 2006 (3DUI 2006)*, 2006.
[SCC+04]	Mats Straka, Michal Cervenansk, Alexandra La Cruz, Arnold Köchl, Milos Sramek, Eduard Gröller, and Dominik Fleischmann. The VesselGlyph: Focus & Context Visualization in CT-Angiography. In *Proc. of IEEE Visualization*, pages 385–392, 2004.
[SOB+07]	Christian Schumann, Steffen Oeltze, Ragnar Bade, Bernhard Preim, and Heinz-Otto Peitgen. Model-free Surface Visualization of Vascular

	Trees. In *IEEE/Eurographics Symposium on Visualization*, Eurographics, 2007.
[SPSHOP02]	Dirk Selle, Bernhard Preim, Andrea Schenk, and Heinz-Otto-Peitgen. Analysis of Vasculature for Liver Surgery Planning. *IEEE Transactions on Medical Imaging*, 21(11):1344–1357, November 2002.
[SRP91]	James Siebert, T. Rosenbaum, and Joseph Pernicone. Automated Segmentation and Presentation Algorithms for 3D MR Angiography (Poster Abstract). In *Proc. of 10th Annual Meeting of the Society of Magnetic Resonance in Medicine*, page Poster 758, 1991.
[Tau95]	Gabriel Taubin. A signal processing approach to fair surface design. In *Proc. of ACM SIGGRAPH*, pages 351–358, 1995.
[UGS+97]	Markus Urban, Christoph Groden, Thomas Schiemann, Rainer Schubert, and Karl Heinz Höhne. A 3d model of the cranial vessels for anatomy teaching and rehearsal of interventions. In *Proc. of Computer Assisted Radiology and Surgery (CAR '97)*, volume 1134 of *Excerpta Medica International Congress Series*, pages 1014–1015. Elsevier, Amsterdam, 1997.
[vH06]	Verena von Hintzenstern. Interaktionstechniken zur Exploration von Gefäßbäumen. Master's thesis, Dept. of Computer Science, Magdeburg, 2006.
[VMM99]	Jörg Vollmer, Robert Mencel, and Heinrich Mueller. Improved Laplacian smoothing of noisy surface meshes. In *Proc. of EuroGraphics*, pages 131–138, 1999.
[VST+03]	Fernando Vega, Natascha Sauber, Bernd Tomandl, Christopher Nimsky, Gunter Greiner, and Peter Hastreiter. Enhanced 3D-Visualization of Intracranial Aneurysms Involving the Skull Base. In *Proc. of Medical Image Computing and Computer-Assisted Intervention*, volume 2879 of *LNCS*, pages 256–263. Springer, 2003.
[WMM05]	Thomas Wischgoll, Elke Moritz, and Jörg Meyer. Navigational Aspects of an Interactive 3D Exploration System for Cardiovascular Structures. In *Proc. of IASTED Int. Conference on Visualization, Imaging, and Image Processing*, pages 721–726, 2005.
[Zui95]	Karel Zuiderveld. *Visualization of Multimodality Medical Volume Data using Object-Oriented Methods*. PhD thesis, Universiteit Utrecht, 1995.

3D Surface Reconstruction from Endoscopic Videos

Arie Kaufman and Jianning Wang

Center for Visual Computing and Computer Science Department
Stony Brook University, Stony Brook, NY 11794, USA
{ari,jianning}@cs.sunysb.edu

Endoscopy is a popular procedure which helps surgeons investigate the interior of a patient's organ and find abnormalities (*e.g.*, polyps). However, it requires a great expertise using only a stream of 2D images of the interior, and there is a possibility that the physician will miss some polyps. Instead, a 3D reconstruction of the interior surface of the organ will be very helpful. It turns the stream of 2D images into a meaningful 3D model. The physicians could then spend more time scrutinizing the interior surface. In addition, the 3D reconstruction result will provides more details about the patient's organ (*e.g.*, concavity/convexity, a coordinate system and 3D measurements), and could be saved for later uses. In a related work, Helferty et al. [HZMH01, HH02] have used a CT-based virtual endoscopic registration technique to guide the bronchoscopic needle biopsy. Winter et al. [WSRW05] also proposed a reconstruction scheme enhanced by a data-driven filtering and a knowledge driven extension.

Shape from X (X={shading, motion, texture, etc.}) techniques, which have been studied for decades in the computer vision community, are good candidates for the 3D surface reconstruction. However, the specific problems associated with reconstruction from endoscopic video make traditional shape from X techniques inappropriate:

1. Local and moving light sources. The endoscope is often equipped with one or two spot light sources. The light sources are attached to the endoscope and moves together with it. In contrast, most shape from X techniques requires distant and static light sources.
2. Fluid inside the organ. Fluid causes light refraction and reflection.
3. Non-Lambertian surface. The specularity will lead to some highlighted regions.
4. Inhomogeneous materials. The surface is composed of several different materials (*e.g.*, blood vessels on the colon surface).
5. Nonrigid organs. The organ is moving non-rigidly during the endoscopic procedure (*e.g.*, spasm).

In this paper, in order to simplify these problems we make the following assumptions: (1) the object undergoes only rigid movements; (2) except for the highlighted (saturated) spots, the remaining regions are Lambertian; (3) except for some feature points, most regions are composed of homogeneous materials . We also assume that the intrinsic parameters of the camera are known. If this is not the case, these parameters could be estimated before the experiments with relative ease [Zha00].

1 Reconstruction Pipeline

Our method is a combination of a shape from shading (SfS) algorithm and a shape from motion (SfM) algorithm. We use a SfS algorithm to reconstruct the 3D geometry of the partial interior surface region for each frame. We then use a SfM algorithm to find the motion parameters of the camera as well as the 3D location of some feature points for the sake of integrating the partial surfaces reconstructed by the SfS algorithm. The SfS algorithm we have used [Pra04] handles the moving local light and light attenuation. This mimics the real situation during endoscopy inside the human organ. One additional advantage of this algorithm as compared with other SfS algorithms is that we could obtain an unambiguous reconstructed surface for each frame. We simply delete the non-Lambertian regions using a threshold on intensity to make the SfS algorithm work for the other regions.

We then integrate partial surfaces obtained from different frames using the motion information obtained by the SfM algorithm [CH96]. Inhomogeneous regions are identified as feature points. These features are used by the SfM algorithm to estimate the extrinsic parameters of the camera for each frame. This information would provide enhanced accuracy for the integration of the partial surface of various frames as compared with ICP (Iterative Closest Points) algorithms [RHL02], especially in the case of few geometric features on the partial surfaces.

We simulate the real optical endoscopic video using a virtual colon dataset, obtained from a CT scan of a real human. We fly through the centerline of the colon to mimic the movement of the endoscope. We use OpenGL to attach a single spot light to the virtual camera and to obtain each frame of the video. The material and shading are simplified as well at this stage. With the virtual colon dataset, we are able to evaluate the performance of our framework by comparing it to the surface reconstruction result, obtained from the segmented and cleansed CT dataset, using for example Marching Cubes, as done in virtual colonoscopy [HMKBH97, KLKB05].

Basically, we use the SfS algorithm [PF05] to obtain the detailed geometry for each frame. Then, the location of the cameras when the frames were taken are computed using the SfM algorithm. In addition, several 3D feature points are also recovered. With the motion parameters of the cameras, the results of the partial surfaces from the SfS algorithm are registered.

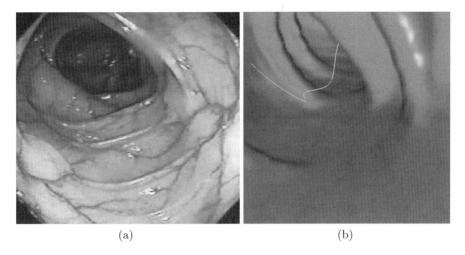

Fig. 1. (a) A frame from a real endoscopic video sequence of the colon. The image has some specular regions due to the existence of fluid. Blood vessels are also visible. (b) A frame from the synthesized endoscopic video of the colon. We use OpenGL to simulate the lighting and feature points. The green curve represents the centerline of the colon.

It is difficult to conduct the evaluation of our framework on the real videos due to the lack of ground truth. We only evaluate the performance of our proposed framework using a virtual colonoscopy dataset with synthetic lighting conditions. However, the fact that the real video has more detailed features (see Figure 1(a)) might help us design more robust and accurate reconstruction SfS and SfM algorithms. The major contribution of this paper is a novel framework to combine the SfS and the SfM algorithm, which offers the foundation for a complete solution for 3D reconstruction from endoscopic videos. In addition, we propose a feature matching algorithm for the SfM algorithm, which is the simplified version of the feature matching algorithm of Snavely et al. [SSS06]. At this stage, our algorithm is only applied to the virtual endoscopic video. Only the SfS algorithm has been applied to the optical endoscopic data.

To simulate the endoscopic environment, we attach to the virtual endoscope a single spot light source using OpenGL. This light source moves together with the virtual camera during the fly-through. We set the OpenGL variable `GL_CONSTANT_ATTENUATION` to 0 and set `GL_QUADRATIC_ATTENUATION` to 0.002, to mimic the quadratic attenuation of light with respect to the different distances between the center of projection and the surface points. Feature points are simplified to be blue points on the colon surface, which are implemented by texture mapping with the `GL_BLEND` mode. We let the virtual camera travel along the centerline of the colon [BKS01]. The viewing direction of the virtual camera is arranged to be perpendicular to its moving

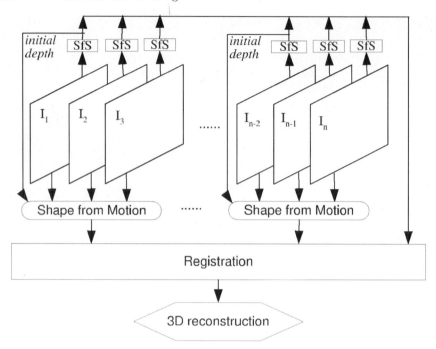

Fig. 2. The pipeline of our proposed framework which reconstructs a surface from an endoscopic video.

direction. With a user-specified step size, we could generate frames representing different regions of the colon surface. One frame of the synthesized video is shown in Figure 1(b) while Figure 1(a) shows a frame excerpt from a real optical endoscopic video.

We first feed each frame to the SfS algorithm and obtain the partial surfaces. After tracking the feature points on the frames, we use the SfM algorithm to compute the extrinsic parameters for each frame. Then, the 3D location of feature points and the motion parameters are fed into nonlinear optimization procedure. Note that the initial 3D location of the feature points are obtained from the partial surface for one frame. However, due to changing viewpoints and occlusions, we may lose the tracked feature points after several frames. Therefore, we use a small number (4 to 6) of contiguous frames, called chunks, for the SfM algorithm. After recovering the motion information for all the chunks, they are registered via a global optimization procedure. This reconstruction pipeline is illustrated in Figure 2.

2 Shape from Shading Process

We use the algorithm of Prados and Faugeras [PF05] to recover the shape from a single frame using the shading information. Traditional SfS algorithms

[ZTCS99] suffer from inherent ambiguity of the results. However, this algorithm features unambiguous reconstruction by taking $1/r^2$ light attenuation into account. Furthermore, it does not require any information about the image boundary, which makes it very practical.

With the spot light source attached at the center of projection of the camera, the image brightness $E = \alpha I \frac{\cos\theta}{r^2}$, where α is the albedo, r is the distance between the light source and the surface point, and θ is the angle between the surface normal and the incident light. The problem of recovering shape from shading information is formulated by partial differential equations (PDEs). The surface for a single view is then defined as: $S(x) = \frac{fu(x)}{|x|^2+f^2}(x,-f)$ where $u(x)$ is the depth value of the 3D point corresponding to pixel x and f is the focal length. $S(x)$ also represents the light direction because the spot light source is at the center of projection. Prados and Faugeras [PF05] further assume that the surface is Lambertian. They obtain the following PDE equation:

$$I(x)f^2 \frac{[f^2|\nabla u|^2 + (\nabla u \cdot x)^2] + u^2}{u} - u^{-2} = 0 \qquad (1)$$

where $Q(x) = \sqrt{f^2/(|x|^2 + f^2)}$. By replacing $\ln(u)$ with v, we have:

$$-e^{-2v(x)} + J(x)\sqrt{f^2|\nabla v|^2 + (\nabla v \cdot x)^2} + Q(x)^2 = 0 \qquad (2)$$

with the associated Hamiltonian:

$$H_F(x,u,p) = -e^{-2u} + J(x)\sqrt{f^2|p|^2 + (p \cdot x)^2} + Q(x)^2 \qquad (3)$$

where $J(x) = \frac{I(x)f^2}{Q(x)}$.

The authors [PF05] prove the well-posedness of this SfS problem and proposed a provably convergent numerical method, which requires no initial values for the image boundary. They showed that $H_F(x,u,p) = -e^{-2u} + \sup_{a \in A}\{-f_c(x,a) \cdot p - l_c(x,a)\}$, where A is the closed unit ball of $\mathbf{R}^2$. Please refer to [Pra04] for the detailed description of $f_c(x,a)$ and $l_c(x,a)$. A finite difference approximation scheme [BS91] is used to solve for u, such that $S(\rho, x, u(x), u) = 0$, where ρ is the underlying pixel grid.

Using the semi-implicit approximation scheme, we could iteratively solve for the new depth value:

$$S(\rho, x, t, u) = t - \triangle\tau e^{-2t} + \sup_{a \in A}\{-(1 - \triangle\tau \sum_{i=1}^{2} \frac{|f_i(x,a)|}{h_i})u(x)$$

$$- \triangle\tau \sum_{i=1}^{2} \frac{|f_i(x,a)|}{h_i} u(x + s_i(x,a)h_i \vec{e_i}) - \triangle\tau l_c(x,a)\}$$

where $\triangle\tau = (\sum_{i=1}^{2}|f_i(x,a_0)/h_i|)^{-1}$ and a_0 is the optimal control of the Hamiltonian: $H_C(x, \nabla x) \approx \sup_a \{\sum_{i=1}^{2} -f_i(x,a)\frac{u(x)-u(x+s_i(x,a)h_i}{-s_i(x,a)h_i} - l_c(x,a)\}$

In the following iterative algorithm, U is the depth of a pixel. Its superscript is the index of iterations and its subscript is the index of the pixel. Please refer to [Pra04] for more details about the implementation.

1. Initialize all $U_k^0 = -\frac{1}{2}\ln(I(x)f^2)$
2. Choose a pixel x_k and modify U_k^n, such that $S(\rho, x_k, U_k^{n+1}, U_k^n) = 0$
3. Use the alternating raster scan order to find the next pixel and go back to step 2

3 Shape from Motion Process

A SfM algorithm is often arranged to have three steps: (1) tracking feature points; (2) computing initial values; and (3) non-linear optimization.

Features are simply represented in the experiments as blue points. Therefore, pixels representing features could be identified easily in the RGB color space. These pixels are then clustered based on pixel adjacency, and the center of each cluster becomes the projection of a feature point. We assume the camera is moving slowly (*i.e.*, sufficient frame rate) during the fly-through and features are distributed sparsely in the image. The corresponding feature does not move too far away, and therefore matching could be simplified to a local neighborhood search. We remove the matching outliers using the approach of Snavely et al. [SSS06], where RANSAC (Random Sample Consensus) iterations are used to iteratively estimate the fundamental matrix.

We tried to obtain good initial values using the factorization algorithms [CH96, PK97]. However, they all failed because of the planar object shape and perspective effects [PK97]. Instead, we use the 3D location of the feature points on one frame (partial surface) as the initial guess for the 3D location of the feature points and set the initial guess for the Euler angles (for rotation) and the translation to 0, which are quite reasonable due to the small motion.

We then use a non-linear least squares optimization scheme to minimize the following error:

$$E = \sum_{f=1}^{F}\sum_{i=1}^{P} \|u_{fi} - KH_f(p_i)\|^2 \quad (4)$$

where u is the pixel location of the feature point, K is the intrinsic matrix, and H_f is the extrinsic matrix for frame f. Here, the parameters for the optimization are three Euler angles $(\alpha_f, \beta_f, \gamma_f)$, the translation vectors $\mathbf{T}_f$ (*i.e.*, H_f), and 3D points p_i. Note that the optimization process could be performed independently for each frame (6 motion parameters) and for each point (3 parameters).

A feature point could not always be tracked because it may be occluded for some frames. In order to obtain as many feature points for the SfM algorithm, we break the stream of frames into chunks. In our implementation, each chunk has 4-6 consecutive frames, and neighboring chunks have overlapping frames. For each chunk, we use Equation 4 to solve for the motion parameters and for each chunk it is an Euclidean reconstruction. However, it is expressed in the coordinate system of the specific chunk. Suppose a frame $\mathcal{F}$ is shared by one chunk (C_1) and the next chunk (C_2). We have two extrinsic matrices (H_1 and H_2) associated with $\mathcal{F}$, which are computed from C_1 and C_2, respectively. The coordinates (p_1 and p_2) of the same point are then related as $p_1 = H_1^{-1} H_2 p_2$, and the extrinsic matrix for each frame in C_2 becomes $H H_2^{-1} H_1$, where H is the original extrinsic matrix.

Now we could register all the chunks together under one framework. In case that a feature point is viewed by several chunks and the 3D locations, computed from different chunks, do not agree, we take their average as the result. In the end, we feed all the points and motion parameters for all the frames into Equation 4 for a global optimization. Even if the results from the chunks do not match well, the average of the parameters could still be a good estimate, and it is usually sufficient for use as the initial guess for the optimization. Using the updated motion parameters, we could finally integrate the partial surfaces into a complete model.

4 Results and Discussion

We apply the proposed algorithm to two types of data: virtual colonoscopy video and optical endoscopy video. We compute the colon surface out of the virtual colonoscopy dataset using the Marching Cubes algorithm [LC87]. We let the virtual camera travel along the centerline of the colon to acquire frames for the synthetic video. The optical endoscopy video is recorded from an actual colonoscopy procedure. We first apply the SfS algorithm to recover the partial surface for a frame excerpted from a colonoscopy video (see Figure 3). This image is degraded by motion blur. The colon surface seems wet, and thus this frame contains several specular regions. The reconstructed partial surface is compatible with the view. However, we have no means to verify the reconstructed surface and compute the reconstruction error due to the lack of the "ground truth" of the 3D information of the colon surface. The results are evaluated using reprojection errors and other means. Note the all the experiments are conducted on a Pentium 3.6G machine.

We further apply the SfS algorithm to recover the partial surface for a view of a simple synthetic surface, whose size is 200×200. It takes 60 iterations (81 seconds) for the algorithm to converge. We project the reconstructed surface under the same lighting condition and compare the pixel intensity to the input image. The maximal difference is 2 pixels. In Figure 4, we show the reconstruction result.

Then, we use the virtual colon dataset to evaluate the performance of the proposed framework. Each frame of the synthesized endoscopic video has a size of 500×500. It takes about 100 iterations (about 6 minutes) on average

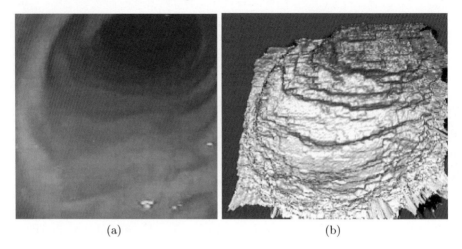

Fig. 3. The reconstruction result of the SfS algorithm using a frame from a colonoscopy video: (a) the input frame. The blurriness of this endoscopic image results from the motion of the endoscope; (b) the reconstructed surface.

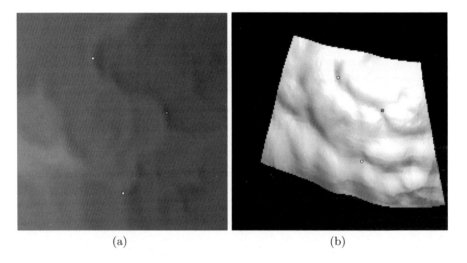

Fig. 4. The reconstruction result of the SfS algorithm using a simple synthetic model: (a) the input frame of size 200 × 200; (b) the reconstructed surface. Some corresponding feature points are also depicted.

for each frame to be reconstructed using the SfS algorithm. In Figure 5, we show an input frame and the corresponding reconstructed partial surface. The reconstruction error for the SfS algorithm is evaluated as the depth ratio error. Since we have the real surface of the synthetic colon, computed from the virtual colonoscopy dataset using the Marching Cubes algorithm [LC87],

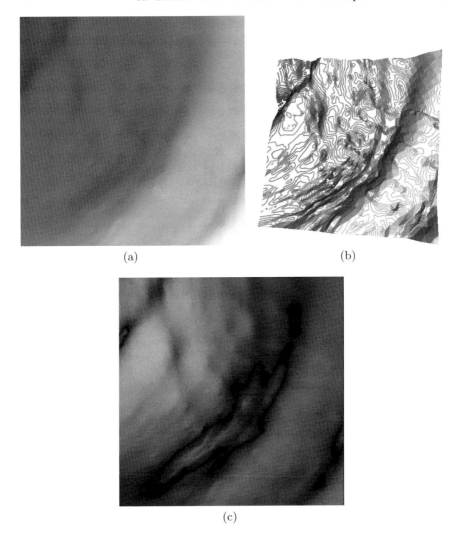

Fig. 5. The reconstruction result of the SfS algorithm: (a) the input frame; (b) the reconstructed partial surface; (c) the depth ratio error map, where brighter pixels represent larger errors. We multiply the errors by 10 for legibility.

we could obtain the depth value for each pixel during the stage of video generation. We compute the depth ratio S as the depth value at the center of the partial surface divided by the depth value at the center of the real surface (d_r^*). Suppose d_p and d_r are the depth values for the same pixel location, for the partial surface and the real surface, respectively. The error is then computed as $\frac{\|Sd_p - d_r\|}{d_r^*}$, which measures the deviation of the reconstructed surface from the actual one in the viewing direction. We found that on average the depth

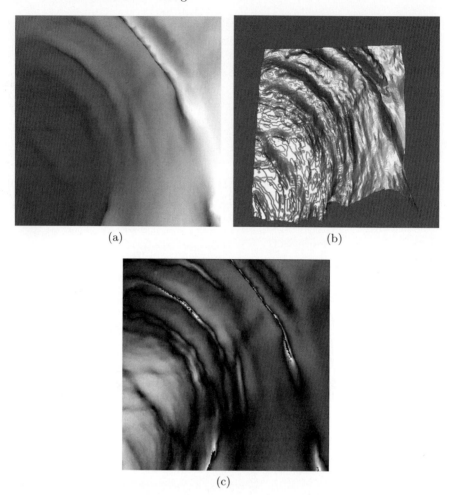

Fig. 6. The reconstruction result of the SfS algorithm for a frame with occlusion and creases: (a) an input frame; (b) the partial surface; (c) the depth ratio error map. We multiply the errors by 10 for legibility.

ratio error is 5%, and the maximal depth ratio error for frames with creases and occlusion is about 16%. Figures 5(a) and 5(b) show an input frame for the SfS algorithm and its corresponding partial surface. Figure 5(c) shows the corresponding depth ratio error for every pixel.

For frames with occlusion and creases, the SfS algorithm has a worse performance. For the example in Figure 6, the average depth ratio error becomes 8% and the maximal ratio error is 40%, and from Figure 6(c), we found that the largest errors are located around the creases.

We then track the feature points and use them to invoke the SfM algorithm. In our experiments, we broke 7 frames into 2 chunks. With the fourth frame shared by both chunks. Using the 3D location of feature points from the computed partial surface as the initial guess, the nonlinear optimization procedure minimizes the re-projection error. Chunk 1 has 32 feature points and the optimization takes 1.23 seconds. The average re-projection error for each feature point is 0.066 pixel^2. Chunk 2 has 19 feature points and the optimization takes 0.90 seconds. The average re-projection error is 0.091 pixel^2. A global optimization is then invoked to register the two chunks. The global optimization takes 5.22 seconds and the average re-projection error becomes

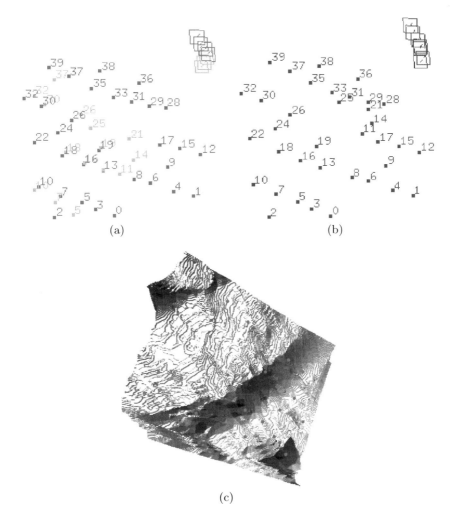

Fig. 7. Registration: (a) feature points and cameras before registration. The two chunks are shown in red and in green, respectively; (b) registered cameras and feature points; (c) registered 7 partial surfaces, where the feature points are shown as well.

0.61 pixel^2 for each feature point. Figure 7(a) shows the situation when the second chunk (green) is moved according to the computed rigid transformation between the two frameworks while the coordinate system of the first chunk (red) is used as the world framework. Figure 7(b) shows the feature points and camera locations after the global optimization. Figure 7(c) shows the registered partial surfaces.

5 Conclusions

In this paper, we proposed a novel framework to combine the SfM algorithm and the SfS algorithm, as an attempt to reconstruct inner surfaces of organs using endoscopic video. We first reconstruct partial surfaces from individual frames. Then, the motion of the camera is estimated using a SfM algorithm based on several feature points. Using this motion information, the partial surfaces are registered and are integrated into a complete model.

We have tested the proposed framework only under simplified situations. It is difficult to conduct the evaluations on the real optical videos with no further information about the organ surface. However, the real video has more details, which could be used to improve the SfS and SfM algorithms. This information has not been exploited though. The proposed algorithm mainly works on virtual endoscopy at this stage. There are many other challenges to be handled in our future work. The most significant one is the inaccuracy of the SfS algorithm, and that it is relatively slow. We plan to test other numerical approaches to solve the PDE in the SfS problem. We will mark the regions representing creases and occlusion in the frames and revise the pixel traversal order to enhance the accuracy of the SfS algorithm under those circumstances. In real situations, it often occurs that some regions in a frame receive saturated illuminations. The SfS algorithm will fail on these regions. We will need to mark them as well. Another major future direction is the handling of non-rigid motions, which poses a great challenge for SfM and registration. We plan to extend here the idea of Vicente et al. [BHB00].

Acknowledgements

This work has been partially supported by NSF grant CCR-0306438 and NIH grants CA082402 and CA110186. We disclose that one of the authors owns a fraction of Viatronix Inc. shares. The CT colon datasets are courtesy of Stony Brook University Hospital.

References

[BHB00] Bregler, C., Hertzmann, A., Biermann, H.: Recovering non-rigid 3D shape from image streams. IEEE Conference on Computer Vision and Pattern Recognition, 690–696 (2000)

[BKS01] Bitter, I., Kaufman, A.E., Sato, M.: Penalized-distance volumetric skeleton algorithm. IEEE Transactions on Visualization and Computer Graphics, **7**, 195–206 (2001)

[BS91] Barles, B., Souganidis, P.: Convergence of approximation schemes for fully nonlinear second order equations. Journal of Asymptotic Analysis, **4**, 271–283 (1991)

[CH96] Christy, S., Horaud, R.: Euclidean shape and motion from multiple perspective views by affine iterations. IEEE Transactions on Pattern Analysis and Machine Intelligence, **18**, 1098–1104 (1996)

[HH02] Helferty, J.P., Higgins, W.E.: Combined endoscopic video tracking and virtual 3d ct registration for surgical guidance. IEEE International Conference on Image Processing, II: 961–964 (2002)

[HMKBH97] Hong, L., Muraki, S., Kaufman, A., Bartz, D., He, T.: Virtual Voyage: interactive navigation in the human colon. SIGGRAPH, 27–34 (1997)

[HZMH01] Helferty, J.P., Zhang, C., McLennan, G., Higgins, W.E.: Videoendoscopic distortion correction and its application to virtual guidance of endoscopy. IEEE Transactions on Medical Imaging, **20** 605–617 (2001)

[KLKB05] Kaufman, A., Lakare, S., Kreeger, K., Bitter, I.: Virtual colonoscopy. Communications of the ACM, **48**, 37–41 (2005)

[LC87] Lorensen, W.E., Cline, H.E.: Marching cubes: a high resolution 3D surface construction algorithm. SIGGRAPH, 163–170 (1987)

[MHOP01] Mahamud, S., Hebert, M., Omori, Y., Ponce, J.: Provably-convergent iterative methods for projective structure from motion. IEEE Conference on Computer Vision and Pattern Recognition, 1018–1025 (2001)

[PK97] Poelman, C.J., Kanade, T.: A paraperspective facttorization method for shape and motion recovery. IEEE Transactions on Pattern Analysis and Machine Intelligence, **19**, 206–218 (1997)

[Pra04] Prados, E.: Application of the theory of the viscosity solutions to the shape from shading problem. Ph.D thesis, University of Nice Sophia-Antipolis, France (1996)

[PF05] Prados, E., Faugeras, O.: Shape from shading: a well-posed problem?. IEEE Conference on Computer Vision and Pattern Recognition, 870–877 (2005)

[RHL02] Rusinkiewicz, S., Hall-Holt, O., Levoy, M.: Real-time 3D model acquisition. SIGGRAPH, 438–446 (2002)

[SSS06] Snavely, N., Seitz, S.M., Szeliski, R.: Photo tourism: exploring photo collections in 3D. SIGGRAPH, 835–846 (2006)

[TK92] Tomasi, C., Kanade, T.: Shape and motion from image streams under orthography: a factorization method. International Journal of Computer Vision, **9**, 137–154 (1992)

[WSRW05] Winter, C., Scholz, I., Rupp, S., Wittenberg, T.: Reconstruction of tubes from monocular fiberscopic images - application and first results. Vision, Modeling, and Visualization, **20** 57–64 (2005)

[ZTCS99]   Zhang, R., Tsai, P., Cryer, J.E., Shah, M.: Shape from shading: a survey. IEEE Transactions on Pattern Analysis and Machine Intelligence, **21**, 690–706 (1999)

[Zha00]   Zhang, R.: A flexible new technique for camera calibration. IEEE Transactions on Pattern Analysis and Machine Intelligence, **22**, 1330–1334 (2000)

Part II

Geometry Processing in Medical Applications

A Framework for the Visualization of Cross Sectional Data in Biomedical Research

Enrico Kienel[1], Marek Vančo[1], Guido Brunnett[1], Thomas Kowalski[1], Roland Clauß[1], and Wolfgang Knabe[2]

[1] Chemnitz University of Technology, Straße der Nationen 62, D-09107 Chemnitz, Germany {enrico.kienel, marek.vanco, guido.brunnett, roland.clauss}@informatik.tu-chemnitz.de
[2] AG Neuroembryology, Dept. of Anatomy and Embryology, Georg August University, Kreuzbergring 36, D-37075 Göttingen, Germany wknabe@gwdg.de

Summary. In this paper we present the framework of our reconstruction and visualization system for planar cross sectional data. Three-dimensional reconstructions are used to analyze the patterns and functions of dying (apoptotic) and dividing (mitotic) cells in the early developing nervous system. Reconstructions are built-up from high resolution scanned, routinely stained histological serial sections (section thickness = 1 μm), which provide optimal conditions to identify individual cellular events in complete embryos. We propose new sophisticated filter algorithms to preprocess images for subsequent contour detection. Fast active contour methods with enhanced interaction functionality and a new memory saving approach can be applied on the pre-filtered images in order to semiautomatically extract inner contours of the embryonic brain and outer contours of the surface ectoderm. We present a novel heuristic reconstruction algorithm, which is based on contour and chain matching, and which was designed to provide good results very fast in the majority of cases. Special cases are solved by additional interaction. After optional postprocessing steps, surfaces of the embryo as well as cellular events are simultaneously visualized.

1 Introduction

Imaging techniques have achieved great importance in biomedical research and medical diagnostics. New hardware devices permit image formation with constantly growing resolution and accuracy. For an example, whole series of digital photos or images from computed tomography (CT) or magnetic resonance imaging (MRI) are consulted to analyze human bodies in three dimensions.

Three-dimensional reconstructions also can be favorably used to investigate complex sequences of cellular events which regulate morphogenesis in the early developing nervous system. Having established the high resolution scanning system *Huge image* [SWKK02] as well as new methods for the alignment of histological serial sections [KWBK02] and for the triangulation of large embryonic surfaces [BVH+03, Kow05], we have applied three-dimensional reconstruction techniques to identify new patterns and functions of programmed cell death (apoptosis) in the developing midbrain [KKW+04], hindbrain [KWBK04], and in the epibranchial placodes [WOB+05].

Our three-dimensional reconstructions are based on up to several hundred routinely stained histological serial sections, which are scanned at maximum light microscopic resolution ($\times 100$ objective). For each serial section, all single views of the camera are automatically reassembled to one *huge image*. The size of the resulting huge image may amount to 1 gigapixel.

Due to the size of these images we do not consider volume rendering techniques, which have extremely high memory requirements [SWKK02, vZBW+06]. Instead, we construct polygonal surfaces from contours that are extracted from the sections. Further references are given in the respective sections.

The present work first aims to briefly describe the whole pipeline. For this purpose, Section 2 provides an overview of our complete reconstruction and visualization system. Here it will be shown how individual components of the system which, in some cases, are variants of existing methods, interact to produce adequate reconstructions of the patterns of cellular events in the developing embryo. Section 3 demonstrates new methods introduced to improve specific parts of the framework in more detail, e.g. new image processing filters which eliminate artifacts, and thus facilitate automatic contour extraction. Furthermore, it is shown how the active contour model can be equipped with an elaborate interaction technique and a new memory saving approach to efficiently handle huge images. Finally, we present our new reconstruction algorithm that is based on heuristic decisions in order to provide appropriate results very fast in the majority of cases. We demonstrate the successful application of our system by presenting representative results in Section 4. Section 5 concludes this paper with a summary and with an outlook on future research activities.

2 The Framework

In this section we describe how the 3D models of embryonic surfaces are built-up and visualized.

2.1 Image Acquisition

Resin-embedded embryos are completely serially sectioned (section thickness = 1 μm). Consecutive histological sections are alternately placed on two sets of slides. For cytological diagnosis, sections from the *working series* are routinely stained with hematoxylin (Heidenhain). During the staining procedure, contours of the resin block are lost. Consequently, corresponding unstained sections of the *reference series* are used to semiautomatically extract these external fiducials in AutoCAD [KWBK02]. Thereafter, manually vectorized contours and cellular events from sections of the working series as well as vectorized contours of the embedding block from corresponding sections of the reference series are fused to *hybrid sections* in AutoCAD. Reconstructions are built-up from hybrid sections that were realigned with the methods described in [KWBK02].

Histological sections are scanned with the high resolution scanning system *Huge image* [SWKK02]. Stained sections of the working series are scanned with maximum light microscopic resolution ($\times 100$ objective). Each single field of view of the camera provides one single small image with a resolution of 1300×1030 pixels. To facilitate reassembly of all small images to one *huge image* and to speed-up further processing of the resulting huge images, all small images are scaled down to a resolution of 650×515 pixels [SWKK02]. Total resolution of one huge image may amount to approximately 1 gigapixel. Acquisition of block contours from unstained sections of the reference series does not require high resolution scanning. Hence, these sections are scanned with the $\times 5$ objective to reduce scanning times [KWBK02]. The following filtering steps are exclusively applied to images that were derived from high resolution scanned sections of the working series.

Figure 1 demonstrates how apoptotic bodies can be identified in images from high resolution scanned sections of the working series.

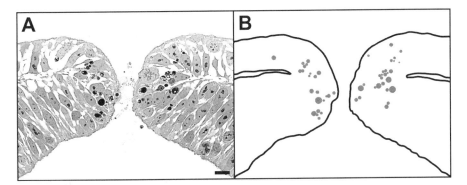

Fig. 1. 13-day-old embryo of *Tupaia belangeri*. High resolution scanned stained sections of the working series (**A**) are required for the unambiguous identification of apoptotic bodies ((**B**), grey) in the embryonic brain and surface ectoderm; scale bar = 10 μm

2.2 Image Preprocessing

Reconstruction of a given embryonic surface requires extraction of the corresponding contours from images of stained sections of the working series at appropriate intervals. Adequate filtering of reassembled images improves the subsequent semiautomatic contour detection. Furthermore, artifacts that may have been introduced during the scanning process are removed. To these ends, three major filters are applied on the images.

The *antigrid filter* is applied to remove a grid pattern which, in some few huge images, may have resulted from blebs in the immersion oil used during high resolution scanning. Hence, time-consuming repetition of such scans is no longer needed (Fig. 2a).

Often, embryonic surfaces of interest delimit *cavities*, which may contain *particles*, e.g. blood cells which are transported in blood vessels. If an active contour is applied inside the vessel to detect its inner contour, the deforming snake would be obstructed by the particle boundaries, which represent unwanted image edges. To avoid these problems, a *particle filter* was designed, which eliminates such particles and provides a uniformly filled cavity background (Fig. 2b).

These two novel filters are described in more detail in Section 3.1. They are applied to high resolution scanned huge images in order to get as exact results as possible. Compared with cytological diagnosis, which requires

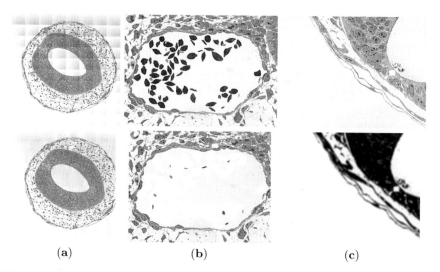

(a) (b) (c)

Fig. 2. Images from high resolution scanned sections of the working series (overview or detail) prior to (top row) or after the application of filters (bottom row); (**a**) The *antigrid filter* removes grid patterns, which sometimes may result from the scanning process; (**b**) The *particle filter* removes immature blood cells inside an embryonic vessel, which would obstruct moving active contours; (**c**) Several common image filters are used to enhance the contrast and to blur the image

maximum light microscopic resolution, neither the extracted contours nor the reconstructed surfaces of an embryo need to have this level of detail for a visualization of sufficient quality. Therefore, huge images are resampled to lower scales allowing an easier handling with common desktop hardware in terms of the available main memory. Obviously, a trade-off between a sufficient quality of contour extraction and the highest possible speed or at least available memory has to be made. Presently, huge images are resampled to sizes between 20 and 200 megapixels.

After resampling we apply a combination of several common image filters. We use *thresholding* and *histogram stretching* to enhance contrast of the images. Furthermore, a *Gaussian* is used to blur the image (Fig. 2c). Application of these filters results in the intensification of image edges and enlarges their attraction range for active contours.

2.3 Contour Extraction

Presently, the huge images are imported in AutoCAD to manually extract contours of large embryonic surfaces as well as to manually mark individual cellular events using graphic tablets [KWBK02]. Prior to surface reconstruction, vectorized embryonic contours are realigned mainly by external fiducials. For this purpose, *hybrid sections* are created which combine anatomical information from the embryos and contours of the embedding block [KWBK02].

Manual extraction of the contours of large embryonic surfaces delivers the most exact results but takes a considerable amount of time and, for technical reasons, may introduce errors in the composition of polylines. Approaches providing fully automatic segmentation algorithms, like region growing or watershed segmentation, often suffer from over- and undersegmentation, of course, depending on the image quality. Since it is impossible to extract contours of interest fully automatically, the present work demonstrates how time-consuming manual extraction of the contours of selected embryonic surfaces can be assisted by semiautomatic contour detection (*active contours* or *snakes*), and how huge images should be preprocessed to facilitate this procedure.

Two major active contour classes can be distinguished, the classical parametric snakes [KWT87] and geometric snakes [CCCD93, MSV95], which are based on front propagation and level sets [OS88]. The major advantage of the latter approach is a built-in handling of topological changes during the curve development. However, we prefer the parametric model for reasons of computational efficiency and the possibility to incorporate user interaction [MT95]. In the following, we consider snakes to be parametric active contours.

Snakes can be applied to selected embryonic contours of interest, e.g. the inner surface of the brain or the outer surface of the ectoderm. In these cases, our semiautomatic approach works faster than the manual contour extraction. Furthermore, the integration of user control allows overcoming problems resulting in incorrectly detected contours, i.e. snakes can be guided to detect the contours in complicated parts of the embryonic surfaces.

A snake can be defined as a unit speed curve $\mathbf{x} : [0,1]_\mathbb{R} \to \mathbb{R}^2$ with $\mathbf{x}(s) = (x(s), y(s))^T$ that deforms on the spatial domain of an image by iteratively minimizing the following energy functional:

$$E = \int_0^1 \frac{1}{2}\left(\alpha \left|\frac{\partial \mathbf{x}}{\partial s}\right|^2 + \beta \left|\frac{\partial^2 \mathbf{x}}{\partial s^2}\right|^2\right) + E_{ext}(\mathbf{x})ds \qquad (1)$$

where α and β are weighting parameters. The first two terms belong to the internal energy E_{int} controlling the smoothness of the snake while the external energy $E_{ext} = E_{img} + E_{con}$ consists of a potential energy E_{img} providing information about the strength of image edges and possibly a user-defined constraint energy E_{con}.

Despite their many advantages, active contours also suffer from some drawbacks [Ker03]. Due to the limited capture range of image edges, the snakes need to be initialized rather close to the embryonic contour of interest. They tend to shrink as they minimize their length and curvature during the deformation process. In the original formulation, active contours are not able to detect concave boundary regions. To overcome these problems, different extensions and improvements of the method have been proposed. In our implementation, we use the following well-known approaches. The *Gradient Vector Flow* (GVF) field [XP97] represents alternative image forces providing a larger attraction range of image edges. Pressure forces [Coh91] are additional forces that act along the normals and enable the active contours to be inflated like balloons. Both approaches ease the task of contour initialization and permit the detection of concave boundaries. Further enhancements of the active contour model are described in Section 3.2.

The snakes can be successfully applied on the images of high resolution scanned sections of the working series to detect selected embryonic contours, e.g. the inner surface of the brain and the outer surface of the ectoderm. Other embryonic contours that face or even contact cells from neighboring tissues (e.g. outer surface of the brain, inner surface of the ectoderm) are presently undetectable by snakes. Further limitations include the fact that production of hundreds of hematoxylin-stained semithin serial sections (section thickness = 1 μm) requires refined technical skills and, therefore, is not routinely performed in biomedical laboratories. Compared with semithin sections, routinely-cut and routinely-stained sections from paraffin-embedded embryos (section thickness = 7 to 10 μm) provide much lower contrast of cells which delimit embryonic surfaces. Consequently, further improvements of active contours are required for semiautomatic contour extraction from paraffin sections.

2.4 Reconstruction

Realigned contours of a given embryonic surface are rapidly reconstructed by using a new heuristic algorithm, which triangulates adjacent sections.

Provided that serial sections are chosen at intervals that do not exceed 8 μm, surface reconstruction with the heuristic approach works well in the majority of cases due to the presence of many simple 1:1 contour correspondences. More complicated cases are handled separately with an extended algorithm. Section 3.3 describes this algorithm in detail.

2.5 Postprocessing

The reconstructed meshes often suffer from sharp creases and a wrinkled surface that partly result (a) from unevitable deformations of embryonic and/or block contours during the cutting procedure, (b) from the mechanical properties of the scanning table, and (c) from the fact that serial sections are chosen at rather small intervals (8 μm) to build-up three-dimensional reconstructions. Furthermore, non-manifold meshes can arise due to very complicated datasets.

For these reasons, we have implemented several postprocessing and repairing approaches. Meshes are analyzed and different errors can be highlighted. Bad triangles (acute-angled, coplanar, intersections) as well as singular vertices and edges can be identified and removed from the mesh. Resulting holes can be closed with the aid of simulated annealing [WLG03]. Furthermore, the triangles can be oriented consistently, which is desirable for the computation of triangle strips that provide an efficient rendering. The implemented Laplacian, Taubin and Belyaev smoothing methods [BO03] can be applied to reduce the noisiness of the surface. Subdivision (Sqrt3, Loop) and mesh reduction algorithms allow a surface refinement or simplification, respectively.

2.6 Visualization

Identification of functionally related patterns of cellular events in different parts of the embryonic body is facilitated by simultaneous visualization of freely combined embryonic surfaces and cellular events [KKW+04, KWBK02, KWBK04, WOB+05]. Furthermore, material, reflectivity and transparency properties can be adjusted separately for each single surface. The scene is illuminated by up to three directional light sources. The visualization module supports vertex buffer objects and triangle strips [vKdSP04]. In the former case, the vertex data is kept in the graphics card's memory. Consequently, the data bus transfer from main memory to the graphics hardware is saved in each rendered frame. Strips are assemblies of subsequently neighbored triangles that are expressed by only one vertex per triangle except for the first one. Thus, it is not necessary to send three vertices for each triangle to the rendering pipeline. Programmable hardware is used to compute nice per-pixellighting, where the normal vectors instead of the intensities of the vertices are interpolated and determined for each pixel. Stereoscopic visualizations amplify the three-dimensional impression on appropriate displays. Finally, the render-to-image mode allows the generation of highly detailed images of an arbitrary resolution in principle.

3 New Methods Integrated into the Framework

In this section we present some novel image filters, further improvements of the active contour model and our new heuristic reconstruction algorithm.

3.1 New Image Filters

Antigrid Filter

The antigrid filter analyzes the brightness differences along the boundary of two adjacent single small images of a given huge image to detect scanning artifacts as shown in Fig. 2a. This comparison is based on areas along the boundaries (instead of just lines) to achieve robustness against noise. Four discrete functions are defined for every single small image – one for each edge – which describe how the brightness has to be changed on the borders in order to produce a smooth transition. These high-frequency functions are approximated by a cubic polynomial using the least squares method. The shading within a single image can be equalized by brightening the single pixels with factors obtained by the linear interpolation of the border polynomials.

The presented algorithm produces sufficient quality images in about 95% of the affected cases. Due to the non-linear characteristics of light, the linear interpolation is not suitable for the remaining cases. Currently, the algorithm assumes a user-defined uniform image size. However, it may be possible to automatically detect the image size in the future.

Particle Filter

The elimination of particles, as shown in Fig. 2b, can be generally divided into two work steps, firstly, the detection of the particles (masking) – the actual particle filter – and secondly, their replacement by a proper background color or image (filling).

The masking of the particles is not trivial because it must be ensured not to mask any objects of interest in the image, particularly parts of any embryonic surface that might be extracted in the further process. Thus, at first the filter detects all pixels that belong to the foreground, and then it uses heuristics to classify them into either objects of interest or particles. The detection of foreground objects is done in a scanline manner by a slightly modified breadth first search (BFS) assuming a N_8 neighborhood, which subsequently uses every pixel as seed pixel. A user-defined tolerance indicates, whether two adjacent pixels belong to the same object or not. Marking already visited pixels ensures this decision to be made only once for every pixel. Thus, the detection globally has a linear complexity. The background, assumed to be white with another user-defined tolerance, is initially masked out in order to speed up the particle detection.

While common algorithms mostly take advantage of depth first search (DFS), we use BFS in order to keep the memory requirement as low as possible. In the worst-case (all pixels belong to one foreground object), all pixels must be stored in the DFS stack, while the BFS only needs to keep the actual frontier pixels in a queue. The speed of the filter considerably benefits from the fact that typically the whole queue can be kept in the CPU's cache.

A foreground object O with a size $A = |O|$ is heuristically analyzed immediately after its complete detection. Too small or too big objects are rejected due to a user-specified particle size range. For the remaining ones, the center of gravity M and diameter $D = 2R$ are determined:

$$M = \frac{1}{|O|} \sum_{P \in O} P \qquad R = \max_{P \in O} ||P - M|| \qquad (2)$$

where P denotes the position of any pixel belonging to the object O. With this information we are able to create a particle mask, which finally contains only those objects that have the shape of elliptical disks and that are not too longish. The ratio $r = \frac{d}{D}$ can be used to define the roundness of an ellipse, where d and D are the lengths of the semi-minor and semi-major axes, respectively. If $r = 1$ the ellipse is a circle. The smaller r the more longish is the ellipse. With the equation for the area of an ellipse, $4\pi A = Dd$, we can estimate the length of the semi-minor axis by $d = \frac{4\pi A}{D}$. Objects with $r < \hat{r}$, where $\hat{r}$ is a user-defined threshold, are discarded, i.e. they stay in the image unchanged. The remaining particles are easily replaced by the background color. Note if objects are not elliptical at all, e.g. if they have a complex and concave shape, they must have a small area A with respect to a big diameter D. In this case, they are discarded as they are interpreted as being too longish ellipses internally. Due to this procedure it is possible to prevent lines and other long but small objects from being eliminated.

3.2 Enhancement of the Active Contour Model

Due to the large size of the images used, the model had to be adapted in terms of speed and memory requirements. Some of these modifications have been presented in [KVB06]. In the following paragraphs we propose a new memory saving method and a sophisticated interaction technique.

The precomputation of the image energy, especially the GVF, requires lots of memory for the employment of active contours on huge images. Since it is defined for every pixel, it can easily exceed the available main memory of a contemporary desktop PC. We refer to [Kie04] for more detailed information concerning the memory requirements of involved data structures. In order to save memory, we only evaluate the image energy locally in regions the snake currently captures. To this end, we use a uniform grid that is updated as the snake deforms (Fig. 3). For every grid cell the snake enters, its deformation is paused until the image energy computation for that cell has been completed.

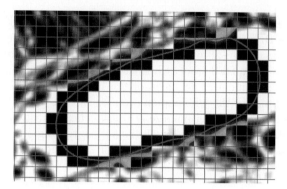

Fig. 3. The image energy is exclusively present in grid cells (*black*) the snake currently explores

On the other hand, if a snake leaves a grid cell the corresponding image energy information is discarded and the memory gets deallocated. Of course, the on-the-fly computation of the image energy considerably slows down the whole deformation, but this is accepted since larger images can be handled and hence, a higher level of contour extraction detail can be achieved. Furthermore, we note that pixels of adjacent grid cells surrounding the current one must be taken into account for the correct grid-based GVF computation. I.e. if the GVF of a $n{\times}n$ sized grid cell is required, we have to take $(n+m){\times}(n+m)$ pixels overall, where m is the number of iterations for the GVF computation, because the discrete partial derivatives have to be approximated at the boundary pixels. However, we found that grid cell sizes of 256×256 or 512×512 offer a good trade-off between quality and speed.

The approach of fixing segments during the deformation process, as described in [KVB06], provides the possibility to perform a refinement in a postprocess. If the semiautomatic extraction has finished with an insufficient result, the operator is able to pick a vertex of the snake in a region where a better result has been expected. A segment is formed around the vertex by the addition of vertices belonging to its local neighborhood while the rest of the snake is fixed. The snake deformation algorithm is subsequently evaluated for the picked segment allowing the user to drag the center vertex to the desired position in the image while the segment's remaining vertices follow their path according to the actual snake parameters. Care has been taken, if the user drags the segment too far. In this case, the segment borders might have to be updated, i.e. some vertices of the deforming segment must be fixed while others have to be unfixed, e.g. if the user drags the center vertex near the fixed part of the snake. However, during this refining procedure the length of the deforming segment is kept constant and the center vertex is updated with the cursor position.

3.3 Heuristic Reconstruction Approach

In several publications the problem of how to reconstruct a polygonal mesh from parallel cross-sectional data has been studied. For a detailed overview of existing work see [BCL96, MSS92]. In the next paragraphs we will specify those recent methods that are most closely related to our work.

When two adjacent cross sections are very different, it may be difficult to obtain a topologically correct and natural tiling. Especially, in branched objects additional polygonal chains have to be inserted in order to obtain a proper and manifold tiling. Oliva et al. [OPC96] solved this problem by using *angular bisector network* of *areas of differences* of two corresponding cross sections. Barequet et al. [BGLSS04] improved the approach by Oliva et al. for some special cases, and Barequet and Sharir [BS94] based their approach on tiling of matched chains and triangulation of remaining parts (*clefts*) using dynamic programming.

Our new fast triangulation algorithm [Kow05] which is also based on contour and chain matching belongs to the category of *heuristic methods* [BJZ04, CP94] that have been developed for fast tiling. Instead of an extensive searching of the "optimal" surface, which yields good results, but is computationally very expensive, algorithms of this class use sophisticated heuristics in order to speed up the generation of the triangulation.

In the following, a section pair $(\mathcal{S}_L, \mathcal{S}_U)$ consists of a lower section $\mathcal{S}_L$ and an upper section $\mathcal{S}_U$. Both sections may contain several contours. In these sections we compute the bounding boxes, the convex hulls and the kD-trees of every contour. Furthermore, we determine the nesting level for the contours, i.e. the outer contour has level zero. Our present implementation can handle only one nesting level and therefore all contours with a higher nesting level are deleted.

The next step deals with the correspondence problem of the different contours, i.e. which contour of $\mathcal{S}_L$ should be connected to which contour of $\mathcal{S}_U$. This is done based on orthogonal projection of the contours in $\mathcal{S}_L$ onto $\mathcal{S}_U$. The correspondence probability of contours with a big overlapping area is assumed to be high. Thus, the bounding boxes and the convex hulls are consulted to approximate the amount of overlap of two contours $\mathcal{C}_L$ and $\mathcal{C}_U$ from different sections. If the bounding box test fails, i.e. $\mathcal{BB}(\mathcal{C}_L) \cap \mathcal{BB}(\mathcal{C}_U) = \emptyset$, it is assumed that the two contours do not correspond to each other. Otherwise, the convex hulls are checked and the relative overlapping coefficient μ is computed:

$$\mu(\mathcal{C}_1, \mathcal{C}_2) = \frac{\mathcal{A}(\mathcal{H}(\mathcal{C}_1) \cap \mathcal{H}(\mathcal{C}_2))}{\mathcal{A}(\mathcal{H}(\mathcal{C}_1))} \qquad (3)$$

where $\mathcal{C}_1$ and $\mathcal{C}_2$ are two contours, $\mathcal{H}$ denotes the convex hull and $\mathcal{A}$ the area. For convenience, we write $\mu_L = \mu(\mathcal{C}_L, \mathcal{C}_U)$ and $\mu_U = \mu(\mathcal{C}_U, \mathcal{C}_L)$, respectively. Note that in general $\mu_L \neq \mu_U$. The correspondence between $\mathcal{C}_L$ and $\mathcal{C}_U$ is initially classified as follows:

Fig. 4. Convex hulls (right) are used to approximate the overlap in order to determine contour correspondences

$$\textit{reliable} \text{ matching} \Leftrightarrow \mu_L > \hat{\mu} \land \mu_U > \hat{\mu}$$
$$\textit{potential} \text{ matching} \Leftrightarrow \mu_L > 0 \land \mu_U > 0$$
$$\textit{no} \text{ matching} \Leftrightarrow \mu_L = \mu_U = 0$$

where $\hat{\mu}$ is a user-defined threshold. The matching type is calculated for all possible contour combinations. Reliable matchings can produce 1:1 connections as well as branchings. In contrast, potential matchings can only generate simple 1:1 correspondences. These are accepted only, if no reliable matchings have been found. To this end, all correspondences are sorted and processed in descending order. That way, incorrect correspondences or branchings can be sorted out, which are due to the computation of overlap based on convex hulls instead of contours (Fig. 4). Nested contours are preferred to be matched with other nested contours in the same way as the outer contours. Only in exceptional cases they are matched with an outer contour. These cases are handled separately. When the contour correspondences are finally determined, the section pair can be triangulated.

The surface or at least one part of the surface begins or ends in a particular section, if a contained contour has matchings in only one direction, i.e. with either the lower or the upper section. Such contours are termed *end contours*, which are processed at first. We use Shewchuk's library [She96] to compute the *Constraint Delaunay Triangulation (CDT)* of these end contours in order to close the surface in the current section.

The triangulation of unambiguous 1:1 contour correspondences is based on chain matching. Randomly, a starting point is chosen on one of the contours. The nearest neighbor point on the other contour can be found quickly with the aid of the precomputed kD-tree. We refer to these point correspondences as *edges*, since they are potential edges in the triangulation. Starting with the determined edge, we form triangles by creating further edges between the contours. For these edges we use the succeeding contour points according to the given orientations. In the case of displaced contours (Fig. 5), the length of these edges will grow very fast. Therefore, we restrict the length of the edges by a user-defined maximum. If this threshold is exceeded within a few steps, displaced contours are assumed, the triangulation is discarded, and a new starting edge must be determined. If this threshold is exceeded after a higher number of steps, the created triangulation is saved. Again, a

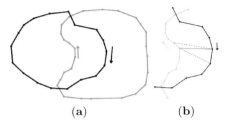

Fig. 5. (a) Displaced contours. The arrows denote the orientation. (b) Unfavorable starting edge. According to the orientation, the length of the edges increases very fast

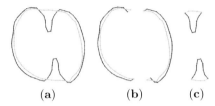

Fig. 6. (a) Two corresponding contours. (b) Matched chains. (c) Unmatched chains

new starting edge must be chosen. If, finally, no more edges can be found at all, the maximum tolerable length can be iteratively increased. In this way, the existing matchings are extended before we search for new starting edges. This procedure terminates, when all points of both contours are completely matched, or when a predefined relative total length of the contours could be matched successfully. Thus, matched chains grow step by step yielding more stable triangulations, especially for very different contours.

The remaining unmatched chains (Fig. 6) are triangulated in the same way, but without the maximum length constraint. Beginning at the boundary vertices of the chains, they are easily stitched together. In each iteration, a set of possible new edges on both triangulation frontiers is analyzed. The edge with the smallest length is chosen for the insertion of a new triangle, and the algorithm advances on the corresponding contour. An additional incidence constraint prevents the degeneracy of the mesh due to vertices with a high valence producing fans of acute-angled triangles. This can happen, if the edge direction of one contour rapidly changes: an example is illustrated in Fig. 7. By counting the incidences of each vertex on-the-fly, a fan can be detected immediately, as shown at the vertex $\mathbf{P}_i$ in Fig. 7a. In this case, the corresponding triangles are deleted and we force advancing on the contour with the incidence defect, from $\mathbf{P}_i$ to $\mathbf{P}_{i+1}$, see Fig. 7b. For the correction, we temporarily extend the search distance for the next best fitting edge: the edge $\mathbf{P}_{i+1}, \mathbf{Q}_j$ in Fig. 7b. After the determination of the next edge, the resulting hole is getting triangulated without the incidence constraint. In Fig. 7c, the vertex $\mathbf{P}_{i+1}$ violates the incidence criterion again. Therefore, the advancing

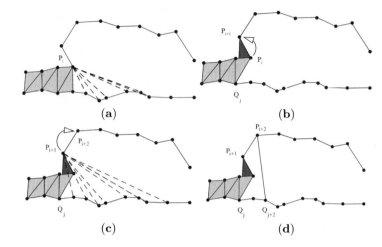

Fig. 7. Incidence constraint.

Fig. 8. Triangulation of shifted contour parts: (a) simply based on the Euclidean distance and (b) with tangent direction weighted distance

is forced to vertex $\mathbf{P}_{i+2}$ and the corresponding nearest vertex on the opposite contours is found - $\mathbf{Q}_{j+2}$. The hole between $\mathbf{P}_i$, $\mathbf{P}_{i+2}$ and $\mathbf{Q}_i$, $\mathbf{Q}_{i+2}$, Fig. 7d, is triangulated using only the minimal length criterion. This heuristic is controllable by several user-defined parameters, e.g. the incidence maximum and the search distance extension.

Consulting only the Euclidean distance to evaluate the admissibility of possible edges could lead to unexpected triangulations, especially for shifted contours (Fig. 8). Therefore, we propose to replace the length by a more sophisticated weight that considers the Euclidean distance as well as the tangent direction:

$$w(\mathbf{p}_1, \mathbf{p}_2) = \frac{\|\mathbf{p}_1 - \mathbf{p}_2\|}{(\langle \mathbf{t}_1, \mathbf{t}_2 \rangle + 2)^\lambda} \qquad (4)$$

where $\mathbf{p}_1$ and $\mathbf{p}_2$ are vertices on different contours, $\mathbf{t}_1$ and $\mathbf{t}_2$ the corresponding normalized tangent vectors that are approximated by simple central differencing, and λ a user-defined parameter controlling the tangents' influence.

The complex n:m correspondences are exceptional cases, where the surface branches between two slices or included contours merge with their outer contours. To this end, we insert additional structures that divide the involved contours and reduce the branching problem to simple 1:1 correspondences. The Voronoi Diagram (VD) is computed for these contours applying again

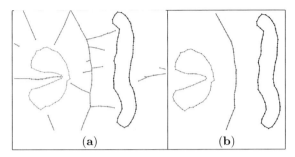

Fig. 9. (a) The EVS extracted from the Constrained Delaunay Triangulation, (b) Pruned EVS

Shewchuk's library [She96]. The External Voronoi Skeleton [Boi88] (EVS) can be extracted from the VD by deleting all vertices and edges within any contour. Thus, the EVS is a subgraph of the VD, which is used to separate contours in a section. An example for an EVS is shown in Fig. 9a. We use the EVS of a cross section for subdividing contours in the adjacent sections if n:m correspondences were detected for the respective section pair. In order to obtain a simple dividing structure all the redundant edges of the EVS are truncated, as shown in Fig. 9b. Furthermore, the EVS is smoothed by a simple subdivision algorithm, and additional points are inserted to adapt to the contours point sampling. Afterwards, the final EVS is projected into the adjacent section and intersected with all involved contours, which is the most time-consuming step. The relative overlapping coefficient of the resulting sub-contours is computed by (3), which is used to create unambiguous 1:1 correspondences that can be triangulated as described above. In very unfavorable cases, the automatically computed EVS is not suited to generate intuitively expected sub-contours leading to undesirable triangulations. Therefore, our software allows to manually draw a better dividing line. Finally, all triangulated contours are removed and the whole process is repeated for the remaining nested contours.

Adjusting some of the parameters in our software requires a visual inspection of the results. A key feature of the implementation is the possibility to retriangulate only those sections, which are involved in a detected artifact. Applying the algorithm to a section pair with adjusted parameters or manually drawn dividing structures generates new primitives that are inserted in the original triangulation seamlessly.

4 Results

The proposed framework of Section 2 is successfully applied in practice. In this section, we give a brief summary of the results that could be achieved with our software. All tests were performed on a standard PC with a Dualcore Intel Pentium D 3.0 GHz processor and 2 GB DDR2-RAM.

Reassembling and down-scaling of 1740 single small images, each having a final size of 650 × 515 pixels, can be done in 6 minutes, approximately. The resulting huge image has a total size of approximately 0.6 gigapixels. Applying the particle filter on an image with a resolution of 37700 × 15450 pixels took less than 13 minutes, while 47 minutes were necessary for the antigrid filter due to permanent paging operations. Nevertheless, the batch mode provides the possibility to filter several images subsequently over night.

Most probably, semiautomatic snake-based contour detection in semithin histological sections will replace manual extraction at least of selected embryonic contours in the near future. Thanks to the local grid-based image energy evaluation, we are able to handle images with resolutions of several hundred megapixels, which was not possible before due to the limited available main memory[3]. Segments that can be dragged around turned out to be a very comfortable and fast tool for contour refinement in a postprocess. Figure 11 shows a successfully detected final contour of the surface ectoderm and endoderm of a 14-day-old embryo of *Tupaia belangeri* containing about 5.500 points.

Reconstruction with our new algorithm could be considerably sped up compared to our previous implementation based on Boissonnat's approach [Boi88, BVH+03]. The triangulation of a dataset with 90.000 points took about five and a half minutes with our old implementation, while it takes only one second now, and is thus several hundred times faster. Our biggest dataset – 770.000 points, 422 sections with 587 manually extracted contours, thereof 13 inclusions and 120 branchings – could be reconstructed with more than 1.5 million triangles in 131 seconds (Fig. 10). In a few difficult cases, the results are not suitable or consistent. After visual inspection, the affected sections can be identified and retriangulated with modified parameters where appropriate. Currently, we have reconstructed about 500 consistently triangulated datasets[4].

Several embryonic surfaces of interest (manually extracted in our example) as well as cellular events can be simultaneously visualized (Fig. 12). The scene is highly customizable in terms of different rendering parameters. The optimizations mentioned in the previous section provide appealing visualizations at interactive frame rates for scenes with moderately sized meshes. Thus, real-time inspection and exploration (rotation, translation, zoom) next to the stereoscopic support provide adequate visual feedback concerning the studied patterns of cellular events.

---

[3] Contour extraction was limited to images with a maximum of approximately 20 megapixels having 1 GB of main memory.
[4] Embryos of *Tupaia belangeri* in different developmental stages

Fig. 10. Reconstruction of the surface ectoderm of a 17-day-old embryo of *Tupaia belangeri*

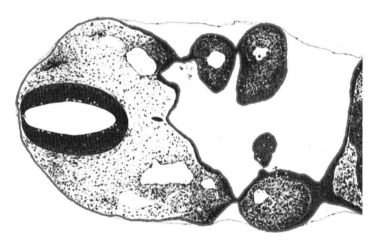

Fig. 11. 14-day-old embryo of *Tupaia belangeri*: Contour extraction result (surface ectoderm and endoderm shown in *blue*) of a quite complex embryonic surface

5 Conclusions and Future Work

We here present our reconstruction and visualization system that has been developed to study patterns and functions of cellular events during embryonic development. Unambiguous identification of cellular events by structural

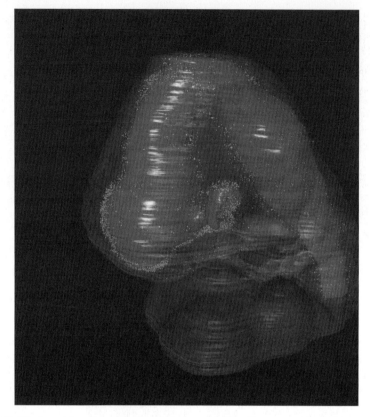

Fig. 12. 15-day-old embryo of *Tupaia belangeri*: Semitransparent surface ectoderm (*grey*, manually extracted), inner surface of the brain (*blue*, manually extracted), and cellular events (*red* and *yellow*; modeled as spheres) can be visualized in combination

criteria, e.g. apoptosis, requires images with a very high resolution, which are captured in a complex scanning process. We have shown how the resulting huge images have to be pre-filtered for the successful application of active contours in order to detect selected embryonic contours semiautomatically. The active contour approach has been optimized in different ways for the work with huge images, and a speedup could be achieved compared to the time-consuming manual extraction. We also introduce a new reconstruction algorithm that works fine for almost all datasets tested so far. Thanks to the heuristic nature of the algorithm, we are able to reconstruct large datasets within seconds. Defects in the triangulation can be corrected locally by a feature that allows the retriangulation of single section pairs. After mesh postprocessing, the visualization of whole compilations consisting of relevant surfaces with individual rendering properties and relevant cellular events allows a proper exploration to obtain useful information about the patterns and functions of cellular events during embryonic development.

In the future, we want to investigate possibilities for parallelization due to the upcoming multicore architectures, particularly with respect to the mentioned image filters. The memory management for these filters has to be further optimized with regard to the high resolutions and the limit of available main memory. Furthermore, we want to find ways to advance to a higher level of automation for the contour extraction or to ease the task of determining adequate parameters for the snake development. It would be desirable to support the manual alignment procedure by automatic decisions. Finally, the reconstruction algorithm should be able to handle even very complex contour constellations correctly and automatically.

References

[BCL96] Chandrajit Bajaj, Edward J. Coyle, and Kwun-Nan Lin. Arbitrary topology shape reconstruction from planar cross sections. *Graphical Models Image Processing*, 58(6):524–543, 1996.

[BGLSS04] Gill Barequet, Michael T. Goodrich, Aya Levi-Steiner, and Dvir Steiner. Contour interpolation by straight skeletons. *Graphical Models*, 66(4):245–260, 2004.

[BJZ04] Sergey Bereg, Minghui Jiang, and Binhai Zhu. Contour interpolation with bounded dihedral angles. In *9th ACM Symposium on Solid Modeling and Applications*, pages 303–308, 2004.

[BO03] Alexander Belyaev and Yutaka Ohtake. A comparison of mesh smoothing methods. In *The 4th Israel-Korea Bi-National Conference on Geometric Modeling and Computer Graphics*, pages 83–87, Tel-Aviv, 2003.

[Boi88] Jean-Daniel Boissonnat. Shape reconstruction from planar cross sections. *Computer Vision Graphics Image Processing*, 44(1):1–29, 1988.

[BS94] Gill Barequet and Micha Sharir. Piecewise-linear interpolation between polygonal slices. In *10th Annual Symposium on Computational Geometry*, pages 93–102. ACM Press, 1994.

[BVH+03] Guido Brunnett, Marek Vančo, Christian Haller, Stefan Washausen, Hans-Jürg Kuhn, and Wolfgang Knabe. Visualization of cross sectional data for morphogenetic studies. In *Informatik 2003, Lecture Notes in Informatics, GI 2003*, pages 354–359. Köllen, 2003.

[CCCD93] V. Caselles, F. Catte, T. Coll, and F. Dibos. A geometric model for active contours. *Numerische Mathematik*, 66:1–31, 1993.

[Coh91] Laurent D. Cohen. On active contour models and balloons. *CVGIP: Image Understanding*, 53(2):211–218, 1991.

[CP94] Y. K. Choi and Kyo Ho Park. A heuristic triangulation algorithm for multiple planar contours using extended double-branching procedure. *Visual Computer*, 2:372–387, 1994.

[Ker03] Martin Kerschner. *Snakes für Aufgaben der digitalen Photogrammetrie und Topographie*. Dissertation, Technische Universität Wien, 2003.

[Kie04] Enrico Kienel. *Implementation eines Snake-Algorithmus*. Studienarbeit, Technische Universität Chemnitz, 2004.

[KKW+04] Wolfgang Knabe, Friederike Knerlich, Stefan Washausen, Thomas Kietzmann, Anna-Leena Sirén, Guido Brunnett, Hans-Jürg Kuhn, and Hannelore Ehrenreich. Expression patterns of erythropoietin and its receptor in the developing midbrain. *Anatomy and Embryology*, 207:503–512, 2004.

[Kow05] Thomas Kowalski. *Implementierung eines schnellen Algorithmus zur 3D-Triangulierung komplexer Oberflächen aus planaren, parallelen Schnitten*. Diplomarbeit, Technische Universität Chemnitz, 2005.

[KVB06] Enrico Kienel, Marek Vančo, and Guido Brunnett. Speeding up snakes. In *First International Conference on Computer Vision Theory and Applications*, pages 323–330. INSTICC Press, 2006.

[KWBK02] Wolfgang Knabe, Stefan Washausen, Guido Brunnett, and Hans-Jürg Kuhn. Use of "reference series" to realign histological serial sections for three-dimensional reconstructions of the position of cellular events in the developing brain. *Journal of Neuroscience Methods*, 121:169–180, 2002.

[KWBK04] Wolfgang Knabe, Stefan Washausen, Guido Brunnett, and Hans-Jürg Kuhn. Rhombomere-specific patterns of apoptosis in the tree shrew *Tupaia belangeri*. *Cell and Tissue Research*, 316:1–13, 2004.

[KWT87] Michael Kass, Andrew Witkin, and Demetri Terzopoulos. Snakes: Active contour models. *International Journal of Computer Vision*, 1(4):321–331, 1987.

[MSS92] David Meyers, Shelley Skinner, and Kenneth Sloan. Surfaces from contours. *ACM Transaction on Graphics*, 11(3):228–258, 1992.

[MSV95] Ravikanth Malladi, James A. Sethian, and Baba C. Vemuri. Shape modeling with front propagation: A level set approach. *IEEE Transactions on Pattern Analysis and Machine Intelligence*, 17(2):158–175, 1995.

[MT95] Tim McInerney and Demetri Terzopoulos. Topologically adaptable snakes. In *Fifth International Conference on Computer Vision*, pages 840–845. IEEE Computer Society, 1995.

[OPC96] Jean-Michel Oliva, M. Perrin, and Sabine Coquillart. 3d reconstruction of complex polyhedral shapes from contours using a simplified generalized voronoi diagram. *Computer Graphics Forum*, 15(3):397–408, 1996.

[OS88] Stanley Osher and James A. Sethian. Fronts propagating with curvature-dependent speed: algorithms based on hamilton-jacobi formulations. *Journal of Computational Physics*, 79(1):12–49, 1988.

[She96] Jonathan Richard Shewchuk. Triangle: Engineering a 2d quality mesh generator and delaunay triangulator. In *FCRC '96/WACG '96: Selected papers from the Workshop on Applied Computational Geometry, Towards Geometric Engineering*, pages 203–222, London, UK, 1996. Springer-Verlag.

[SWKK02] Malte Süss, Stefan Washausen, Hans-Jürg Kuhn, and Wolfgang Knabe. High resolution scanning and three-dimensional reconstruction of cellular events in large objects during brain development. *Journal of Neuroscience Methods*, 113:147–158, 2002.

[vKdSP04] O.M. van Kaick, M.V.G. da Silva, and H. Pedrini. Efficient generation of triangle strips from triangulated meshes. *Journal of WSCG*, 12(1–3), 2004.

[vZBW+06] Joris E. van Zwieten, Charl P. Botha, Ben Willekens, Sander Schutte, Frits H. Post, and Huib J. Simonsz. Digitisation and 3d reconstruction of 30 year old microscopic sections of human embryo, foetus and orbit. In A. Campilho and M. Kamel, editors, *Image Analysis and Recognition, Proc. 3rd Intl. Conf. on Image Analysis and Recognition (ICIAR 2006)*, volume LNCS 4142 of *Lecture Notes on Computer Science*, pages 636–647. Springer, 2006.

[WLG03] Marc Wagner, Ulf Labsik, and Günther Greiner. Repairing non-manifold triangle meshes using simulated annealing. *International Journal on Shape Modeling*, 9(2):137–153, 2003.

[WOB+05] Stefan Washausen, Bastian Obermayer, Guido Brunnett, Hans-Jürg Kuhn, and Wolfgang Knabe. Apoptosis and proliferation in developing, mature, and regressing epibranchial placodes. *Developmental Biology*, 278:86–102, 2005.

[XP97] Chenyang Xu and Jerry L. Prince. Gradient vector flow: A new external force for snakes. In *Proceedings of the 1997 Conference on Computer Vision and Pattern Recognition (CVPR '97)*, pages 66–71. IEEE Computer Society, 1997.

Towards a Virtual Echocardiographic Tutoring System

Gerd Reis[1], Bernd Lappé[2], Sascha Köhn[1], Christopher Weber[2], Martin Bertram[3], and Hans Hagen[1,2]

[1] DFKI Kaiserslautern `reis@dfki.uni-kl.de`
[2] Technische Universität Kaiserslautern
[3] Fraunhofer ITWM Kaiserslautern

Summary. Three integral components to build a tutoring system for echocardiography are presented. A mathematical time-varying model for vessel-representations of the human heart, based on cubic B-Splines and wavelets facilitating the extraction of arbitrarily detailed anatomical boundaries. A dedicated ontology framework the model is embedded into enabling efficient (meta-)data management as well as the automatic generation of (e.g. pathologic) heart instances based on standardized cardiac findings. A simulator generating virtual ultrasound images from instances of the heart transformed into isotropic tissue representations.

1 Introduction

Ultrasound is a very important visual examination tool in medicine since it is a safe and non-invasive imaging modality and does not put patients to great inconvenience during an examination. However, ultrasound has some limitations (i.e. a comparably low resolution, high image anisotropy and significant imaging artifacts) that call for a very intensive education. In many countries like Germany the education of young medical doctors and physicists respectively is done during the clinical practice. This is of advantage in that students can learn under qualified supervision (e.g. real pathologies are examined, a discussion of special cases can be done during the examination, etc). On the other hand there is a bunch of drawbacks, e.g. there is always need for a supervisor, the education cannot be performed at any time and of course there is need for real (often seriously ill) patients with respective pathologies. Indeed, studies have shown that during a standard education of approximately one year students can learn about 80% of the important pathologies, only. To remedy the situation we previously outlined a Virtual Echocardiography System [KvLR+04]. The present work contributes the development of three essential components towards the automatic generation of computer-animated 3D models and virtual ultrasound images from textual finding ontologies. Based on data from a standardized finding system [HBvLS01] [SM00] and several ontologies [vLKB+04], geometric models of a pathologic heart are created and transformed into tissue descriptions. Now students can examine the data set [RBvLH04] [SRvL05] which is blended into a mannequin using a pose tracked virtual transducer. Their findings are recorded using a second instance of the finding module such that measurements and other examination details (like the handling of the ultrasound transducer) as well as the diagnosis as such can be compared automatically (c.f. color plate (a)). Current research focuses on the automatic generation and instantiation of pathologic heart models [BRvLH05] and the real time generation of artificial ultrasound images [SRvL05].

The remainder of the paper is organized as follows: in Sect. 2 an overview over previous work is given. Sect. 3 describes a new method to generate time-varying heart geometries. In Sect. 4 the ontology framework to manage the model instantiation is presented. In Sect. 5 the enhanced ultrasound image simulator is presented. Finally, results are given in Sect. 6.

2 Previous Work

In the last decade a number of tutoring systems for ultrasound have emerged. The very first systems aimed at the question which of the various medical imaging modalities is best suited for the examination of a particular pathology [MA97], i.e. when and under which scope to use ultrasound. Since then tutoring systems especially tuned for particular clinical ultrasound applications such as gynaecology [Gmb05], obstetrics [Ehr98], mammography [NPA04] or cardiography [WWP+00] have been developed. A comparative review of such systems can be found in [MSB+04]. These training systems usually consist of a probe dummy which is tracked by a positioning device, a live-sized mannequin and a computer managing a case database of real ultrasound images. The positioning of the probe relative to the mannequin is continuously transmitted to the computer, which searches the database for an ultrasound image that corresponds best with the received data. The major drawback of these systems is the limited amount of data that can be managed compared to the unlimited amount of potential images. In fact it is infeasible to provide a suitable image for every possible probe positioning especially in the regard of their highly anisotropic nature. For this reason the movement of the transducer is heavily restricted in practice.

As a solution to the problem mentioned above the reconstruction of ultrasound images into a volume naturally comes to mind [BAJ+] [RN00] [TGP+02] [RBvLH04]. Therefore ultrasound images are recorded by a movable probe. Early systems relied on either external mechanical devices to guide the ultrasound scanner [PNJ92] or on specialized probes for tilt [FTS+95] and rotational scanning [TDCF96] [EDT+96]. Newer systems support the more general freehand scanning which is far better suited to cardiology. These systems operate either in display synchronization mode (up to 25Hz) [PGB99], or at the full ultrasound device frequency (up to 120Hz) [Rei05]. The so acquired data is embedded into (time varying) volumetric data sets.

However, these data sets should not be used with a tutoring system in a general setting, since ultrasound images are highly anisotropic as already mentioned, such that using arbitrary cut planes will result in significant representation errors which is not tolerable in an educational environment.

Concerning tutoring systems it is very important to display images that reflect the real transducer-mannequin-relation for arbitrary pathologies. Consequently an isotropic, geometric representation of the anatomy is needed that can first be tuned to arbitrary pathologies and subsequently be transformed into ultrasound images in real-time.

There is a vast amount of literature on modeling the human heart, however these models suffer from several important facts where the most important one is their static nature [BFGQ96] [Joh] [SWS+00]. Partially these models are quite accurate and represent the anatomy of a heart very well. Especially when textures were added these models are perfectly suited for anatomy studies which are concerned with the - technically speaking - layout or structure of the heart. Unfortunately the heart is a dynamically behaving organ and most pathologies are primarily reflected in a deviant behavior of heart dynamics. These pathologies cannot be handled with static heart models. In [BRvLH05] a dynamic heart model was generated based on hand segmented MR-images. From the segmented images the geometric model was deferred. This model can be efficiently handled and displayed picturing most of the relevant anatomical features. On the other hand it is composed of several triangle meshes, which are not mathematically related. Hence, it is extremely tedious to model even simple pathologies spread over several triangles such as an aneurism or an infarcted region.

Regarding the semantic modeling of anatomical and medical structures much related work has been presented. At first there are different approaches representing structured medical vocabularies like MeSH *(Medical Subject Headings)*[1] and UMLS *(Unified Medical Language System)*[2]. These systems, however, are no proper ontologies but only taxonomies. Although they provide most of the medical vocabulary they do not represent the necessary semantic information between the different terms. Only few types of relations are realized. A more promising approach is the *Foundational Model* of the University of Washington [MMR]. The Foundational Model is based on an ontology modeling anatomical structures. In contrast to our system it covers not only the cardiovascular system but the anatomy of the whole human body. Nonetheless it is - even in a reduced form - not suitable for our purposes, since it does not concentrate on spatial information and its granularity is not high enough

---

[1] http://www.nlm.nih.gov/mesh/
[2] http://www.nlm.nih.gov/research/umls/umlsmain.html

to e.g. reflect a cardiac finding. The models of *OpenGALEN*[3] and *Ontolingua*[4] turned out to have only a partial ability to represent anatomical heart structures and to be ill-suited to model a finding structure.

Assuming an appropriate model from which an isotropic tissue representation can be deferred, the transformation into ultrasound images can be done in various ways. For example warped MR-images could be modulated with specific noise [Bro03]. Another way would be to simulate the acoustic behavior of ultrasound as can be found in [Jen96]. However, both methods are not well suited for tutoring systems. The former method is very fast, but the resulting images do not look like real ultrasound images. Artifacts common with ultrasound are not reproduced, hence students will not learn the intrinsic difficulties of ultrasound examinations. The latter method does respect some types of artifacts, but the time needed to calculate such a simulation takes even in the latest, heavily optimized version [JN02] several minutes and thus cannot be used in a tutoring system. In [LL88] [YLLL91] finite element models serve as a tool for examining the generation, propagation and interaction of elastic waves in solid materials for non-destructive evaluation and quality assurance, however these methods are not suitable for modeling the propagation of ultrasound in soft biological tissue and further cannot generate artificial ultrasound images in real-time.

In order to fuse the above mentioned components into a single framework enabling the standardized acquisition of echocardiographic findings, the knowledge based management of pathologies as well as the automatic derivation of a geometric heart model thereof, and the generation as well as visualization of artificial ultrasound images in a virtual tutoring environment the *Virtual Echocardiography System (VES)* outlined in [KvLR+04] is - at least to our best knowledge - the first and only system currently available.

3 Cardiovascular Structure Generation

As already mentioned in Sect. 1 the major drawback of most of the existing ultrasound tutoring systems is the fact that they rely on a fixed number of real ultrasound images. Hence, only a limited number of training cases is supported. Assuming a stenosis as an example the location and extend of the deformity is fixed for a particular image data set, whereas in reality the stenosis can occur at arbitrary locations and at various degrees. One of the central issues of VES is to provide a data set for every possible pathology.

In [BRvLH05] time-varying volumetric MRI data sets and the marching cubes algorithm were used to extract the various feature surfaces. Finally, 22 static meshes were constructed representing discrete time steps of a single heart beat. Unfortunately, these meshes have three main disadvantages,

---

[3] http://www.opengalen.org/
[4] http://www.ksl.stanford.edu/software/ontolingua/

namely the meshes are unstructured, they do not have a mathematical correspondences of points and triangles of successive time steps, and there is no mathematical description of the meshes that would permit a parametric modification. In fact it is extremely difficult to modify these meshes in such a way that e.g. a stenosis exposes the same degree and location at every time step, neglecting pulsatile deformations due to blood pressure changes.

To achieve the goal of a general heart model the heart has to be described mathematically by means of model parameters. In this section we present an algorithm to generate such a mathematical model from the given unstructured meshes. From this model meshes with arbitrary resolution can be deferred. The extraction process to generate the mathematical model is not related to e.g. sweeping methods used to extract iso-surfaces from volumetric data, although there are some conceptual similarities. The model generation can be assumed as a preprocessing step, whereas the mesh generation will be performed based on the model whenever a new heart instance is needed, i.e. whenever a student wants to use VES to examine a particular pathology. An important advantage is that the construction of a pathologic heart instance will not be done by altering meshes, but solely by a parameter modification of the mathematical model of the healthy standard heart.

The efficient mathematical structure of the model can also be used to reduce the data capacity, e.g. by applying a multi resolution analysis. Even Haar-wavelets yield very good results, since they are applied to the model parameters and not to geometric objects. By means of an uncomplicated generalization the algorithm though primarily developed for blood vessels with tube like topology can be applied to other parts of the heart like vessel bifurcations and ventricles.

Before the model extraction from raw meshes is described some details on the general static vessel model are given. A vessel consists of a three dimensional orientation curve defining the vessels course and an arbitrary but fixed number of cross-sections (CS) which have a unique position w.r.t. the orientation line.

As orientation curve a cubic B-spline curve $f : \mathbb{R} \supset [a, b] \to \mathbb{R}^3$ $(a, b \in \mathbb{R})$ is chosen. (c.f. Figs. 1 and 2). For every $x \in [a, b]$ a local left-handed coordinate-system is given by the derivative $e_1(x) := f'(x)$, a normal-vector $e_2(x)$ and the corresponding bi-normal $e_3(x) := e_1(x) \times e_2(x)$. Since Frenéts standard-normal is not defined if the second derivative vanishes, a *natural normal* is constructed for every $x \in [a, b]$ as follows. For $x = a$ the global Cartesian coordinate system $\{e^1, e^2, e^3\}$ is rotated such that e^1 coincides with $f'(a)$. Note that $\{e^1, e^2, e^3\}$ denotes the fixed global system while $\{e_1(x), e_2(x), e_3(x)\}$ represent a local coordinate system for a point x on the orientation line. Let $\alpha(a)$ be the angle between e^1 and $f'(a)$. Let $v(a)$ be a rotation axis defined as

$$v(a) := \begin{cases} e^2 & \text{if } f'(a) \parallel e^1 \\ e^1 \times f'(a) & \text{otherwise} \end{cases} \quad (1)$$

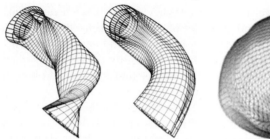
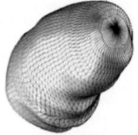

(a) Vessel modeled with Frenéts normal (left) and the natural normal (right)

(b) The left ventricle mesh derived from the vessel model

Fig. 1. Example meshes generated with the vessel model.

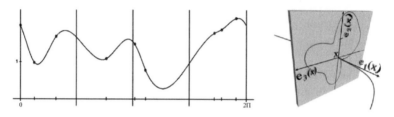

Fig. 2. A 2π-curve over $[0, 2\pi]$ and the same curve transferred into a CS-plane.

A local orthonormal system in $x = a$ with the natural normal $\eta(a) := R(\alpha(a), v(a)) \cdot e^2$ is then found by a rotation $R(\alpha(a), v(a))$. If $x > a$ an iteration is performed. Let $\eta(a)$ denote the start vector. Further let $0 < \varepsilon < b - a$. Then for every $i \in \mathbb{N}$ with $a + i \cdot \varepsilon < b$ equation (2) is evaluated:

$$\eta(a + i \cdot \varepsilon) = R\Big(\alpha(a + (i-1) \cdot \varepsilon), v(a + (i-1) \cdot \varepsilon)\Big) \cdot \eta(a + (i-1) \cdot \varepsilon) \quad (2)$$

Here $\alpha(a+(i-1)\cdot\varepsilon)$ denotes the angle between $f'(a+(i-1)\cdot\varepsilon)$ and $f'(a+i\cdot\varepsilon)$, and $v(a + (i-1) \cdot \varepsilon)$ is given by

$$v(a + (i-1) \cdot \varepsilon) := \begin{cases} v(a + i \cdot \varepsilon) & \text{if } f'(a + (i-1) \cdot \varepsilon) \parallel f'(a + i \cdot \varepsilon) \\ f'(a + (i-1) \cdot \varepsilon) \times f'(a + i \cdot \varepsilon) & \text{otherwise} \end{cases} \quad (3)$$

The natural normal always exists, since the curve f is continuously differentiable. The significant advantage of this construction over other methods is that the normal is continuous w.r.t. x and its rotation around the curve f is very small (c.f. Fig. 1(a)).

Now, two-dimensional cross-sections (CS) can be constructed. Every CS consists of two closed CS-curves, called the inner and outer layer. A CS-curve is defined as the boundary of an area with star-shaped topology (also known as Jordan curve). With regard to the multi-resolution analysis detailed below it is not useful to represent the cross sections by planar NURBS-curves. Instead we use polar-coordinates with respect to the center spline and hence one-dimensional interpolating 2π-periodic cubic B-spline curves (2π-curves). A 2π-curve representing a circle is a constant function described by a single value namely its radius. Every 2π-curve can be embedded into a two-dimensional plane (c.f. Fig. 2). Note that adding additional interpolation-points does not change the course of a 2π-curve if the points are located on the curve. This is important in the regard of defining well behaving spline curves by means of re-sampling, regarding the multi resolution analysis detailed below.

A CS can be positioned at any point $f(x)$ on the center spline by identifying the plane normal with the local tangent vector, the x-axis of the CS-plane with the natural normal and its y-axis using the corresponding bi-normal in $f(x)$ (c.f. Fig. 2). In case that the CS embedding plane cannot be positioned orthogonal to the local tangent vector $f'(x)$ two angle-parameters are defined allowing a general alignment. The alignment parameters become necessary to avoid intersections of the CSs during the model extraction procedure.

To model a bifurcation the participating CSs are bisected such that their CS-planes can be folded. Then a bifurcation is realized using three vessel-models with orientation curves ending at a common point. It is also possible to model arbitrary multi-furcations and non-tubular objects like the left ventricle (c.f. Fig. 1(b) and color plate (b)).

Following the model extraction procedure is described. Assume an unstructured mesh of a vessel. A cutting plane is defined for every CS of the model. The cutting planes need not to be parallel, but must not intersect each other inside the vessel. The edges of the mesh yield a finite number of intersection points with the cutting plane. Their center of gravity (CoG) is used as a control point of the orientation line and as origin of the respective CS-curves. After computing the orientation line, a local coordinate-system is given for every CoG. If the plane is non-orthogonal to the orientation line the two angle-parameters for a general alignment are computed. The intersection points are then transformed into a local polar coordinate system, where the angle coordinate is measured with respect to the natural normal. The data extraction for bifurcations is similar, since bisected planes can be treated as two simple planes. To construct a second layer a pre-specified constant scalar value is added.

To obtain a level-of-detail representation for real-time rendering and to reduce the number of control points for simple rule-based manipulation, a wavelet transform (Haar wavelet) is applied to the CS-control points. Note that the analysis is not performed on a geometric mesh, since this would collapse the mesh to a single line. Since the arguments of the interpolation points of the CS-curves were not uniformly distributed, the 2π-curves are uniformly re-sampled for 2^m (e.g. $m = 8$) arguments in the interval $[0, 2\pi)$. The corresponding radius coordinates are the interpolation points for a new 2π-curve s_m. The resulting 2^m-tuple can now be transformed via Haar-wavelets. (See [EJS], applied to $[0, 2\pi)$ instead of $[0, 1]$). For any $j < m$ a new 2π-curve

s_j is constructed by filtering the 2^{j+1}-tuple. Then s_j is a simplification of s_m, and s_0 is always a constant function yielding a CS with concentric circles (c.f. color plate (c)). Note that the orientation line is not simplified, since the simplification would yield a straight line in the limit. In contrast to the vessel shape this simplification would not represent the anatomy any more.

The above given definition and extraction method allows an efficient mathematical description of vessel structures and ventricles based on a given raw mesh and with regard to [BRvLH05] based on a time-varying volumetric MRI data set. Note that up to now no mesh was constructed. To generate a particular geometric heart instance of arbitrary resolution a mesh is generated based on the given mathematical description. This process is described in the remainder of this section.

For every CS a point with a fixed angle coordinate is chosen on a particular layer. The resulting points embedded in $\mathbb{R}$ can be used as interpolation-points for a new cubic B-spline curve, called *fibre*. For $N_X, N_Y \in \mathbb{N}$ with $N_X > 2$ and $N_Y > 3$ N_Y fibre points are extracted and interpolated by a spline curve, for both layers respectively. Then on every fibre N_X mesh points are extracted. This yields a regular mesh with the same resolution for every layer. Note that the level of detail (LoD) of the model is independent from the resolution of the resulting mesh, which depends on N_X and N_Y only.

To construct the time-varying model a discrete number of static models is generated using N_X and N_Y for every time step, hence the individual meshes can be mapped onto each other by means of a one-to-one correspondence of the mesh points.

4 Ontology Based Visualization

The generation of a *parametric heart paradigm* describing the human heart including the relationships between the various anatomical parts as well as their (inter-)functionality is essential for a virtual examination environment. Here, the expression paradigm is used to explicitly distinguish between a geometric model of the heart, consisting of e.g. triangular meshes or spline surfaces fused to generate a specific instance of a heart under the scope of e.g. visualization or simulation (c.f. Sect. 3) and the *knowledge based scheme* or plan of the human heart, which is the central point of this section.

In contrast to conventional modeling approaches which are usually image based, the underlying idea of this system is the generation of a geometric model instance of a particular heart using a *formal knowledge-based description*. More precisely the paradigm consists of a formal representation of the healthy human heart as well as a declarative representation of a standardized echocardiographic finding. Combining a selected finding from the finding database with the heart paradigm leads to a geometric model instance of a particular - not necessarily pathologic - human heart.

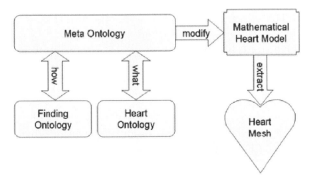

Fig. 3. The ontology framework. The Meta-ontology links the *how to modify* from a particular finding to the anatomical entities of the heart ontology, combines the information and alters the respective parameters of the mathematical heart model. From the altered model the particular mesh instance is generated.

Following this approach, the framework is realized using three different ontologies. The first one (called the *heart ontology*) represents the anatomical structure and the time-dynamic structure of the healthy human heart while a second one (named *finding ontology*) serves as a model for echocardiographic findings. A third ontology (*meta ontology*) working on a meta-level and as such combining heart and finding ontology, represents the relations between a particular finding and the heart anatomy by a set of rules. These rules map the individual parameters of a particular finding to their corresponding structures in the heart model (c.f. Fig. 3). Hence, the meta ontology describes the influence of a specific finding on the healthy standard anatomy.

To proof these ideas a system was developed that allows the input and administration of different findings as well as the semiautomatic generation of geometric models based on the stored findings and geometric modeling primitives. The system has access to the different ontologies using them to analyze the given structures and to generate a specific geometric heart model instance.

As development tool for the different ontologies the ontology editor Protégé-2000 was used. Protégé-2000 is developed by the Medical Informatics department at the Stanford University School of Medicine and serves for the modeling of domain knowledge. Protégé-2000 permits the definition of classes and derivation of respective class hierarchies therefrom, the definition of attributes and constraints on single attributes as well as the description of relations between classes and of properties of such relations. For a more detailed description of the underlying knowledge model and the usage of Protégé-2000 see [NFM00] and [NM01].

The following paragraphs give a short overview over the three ontologies their role in the framework and how they were realized. The chapter is closed by an example with the aim to clarify the process of generating a geometric model instance starting with a particular finding. As example we have chosen a stenosis of the *aorta descendens* which is a common deformity of vessels surrounding the heart. This example is easy to appreciate and used throughout the following sections to explain the basic ideas.

4.1 Finding Ontology

Based on the report "Quality Guidelines in Echocardiography" [Deu97], the German Cardiac Society proposed a documentation structure that was published in the Cardiology magazine as *consensus recommendation* for finding documentations in echocardiography [Deu00]. These recommendations facilitate the design of a finding ontology describing all important parameters in a formal and structured way. The basic structure of the finding ontology contains the anatomical structure of the heart in order to simplify the mapping of finding parameters to anatomical structures in further steps. For example a general finding is divided into sub-findings representing examination results for the atria, the ventricles, the aorta and so on (c.f. Fig. 4). The consensus recommendations as well as the ontology cover a wide range of measurements, constraints, descriptions and other criteria. These parameters can be numerical values, flags, textual descriptors or a collection of predefined values. Most of the parameters are equipped with default values which play an important role for the mapping of a finding to the heart anatomy. To give a more detailed example that will play an important role in the next steps we refer again to Fig. 4. On the bottom line of the diagram a sub-finding for the descending aorta can be found which is equipped with a slot named *stenosis* represented by a boolean value as well as a slot named *min diameter*. The latter allows a more detailed declaration using a multi-constrained numerical value.

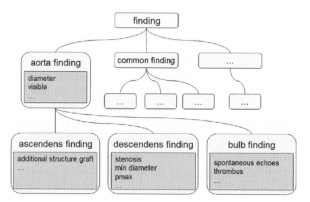

Fig. 4. The finding ontology describes how a general finding is structured. Additionally default values, constraints, flags and textual descriptors are provided. Individual findings are stored in a derived relational database.

From the ontology a relational database scheme was derived. Since the knowledge model of Protégé-2000 is not designed to hold large amounts of data this database is used to store the individual echocardiographic findings thus accounting for performance issues. The generation of the database scheme from the ontology is done by a Protégé-2000 plug-in that was developed during the term of the project. A great benefit of this approach is that a part of the system (namely the echocardiographic finding module) can be used as a stand-alone application for acquisition and management of findings in the clinical all day work. Since the database scheme is derived from the ontology it is possible to directly import findings collected during the hospital routine as instances of the finding ontology.

4.2 Heart Ontology

The heart ontology defines a paradigm of the healthy human heart and as such it serves two purposes. First, it represents a hierarchical description of the anatomical structures of the heart, e.g. a heart consists of a left ventricle, a right ventricle, two atria etc., while the wall of the left ventricle is subdivided into different wall segments (e.g. septal medial wall) and so on. Figure 5 gives a glimpse on the structure. Following the example, observe that the aorta is also part of the heart and is itself subdivided into three different parts namely the *ascending aorta*, the *descending aorta* and the *bulb of aorta*.

Secondly, the ontology represents the spatial relationships between the anatomical structures. Since the ontology provides a basis for the generation of a realistic time-varying heart model it contains references to geometric primitive models for each discreet time step of the heart cycle which are linked to the respective anatomical entities. The classification into anatomic

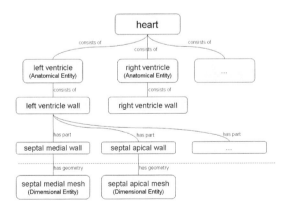

Fig. 5. The heart ontology describes the anatomy of the various heart components as well as their spatial relations. Only a very small part of the ontology is shown as an example.

and geometric entities is depicted in Fig. 5. In the initial version of the ontology the geometric entities refer to polygonal meshes, but the framework also allows the usage of arbitrary objects, e.g. point sets, spline surfaces or volumetric primitives.

4.3 Meta Ontology

A central point of the framework is the generation of an individual geometric heart model by combining an echocardiographic finding with the model of a healthy heart. The rules allowing the combination of these two models are defined by the meta ontology. To perform a mapping two kinds of questions are relevant: which anatomical structures are involved to represent a particular finding in a proper way and how shall the respective geometric objects be modified to comply with the finding. Accessing the content of the finding and heart ontology the rules of the meta ontology describe for every entity of a finding to which anatomical structure it relates and how this structure needs to be modified. Basically a rules fires if a parameter of a selected finding differs from the respective default value. In this case the corresponding anatomical structure is marked as relevant for the geometric model and is modified according to the matching rule. The principle procedure of handling rules in the meta ontology is shown in Fig. 6. In terms of the given example this means that there is a rule linking the *stenosis* slot mentioned in Sect. 4.1 to the *descending aorta* from Sect. 4.2 (c.f. Fig. 7).

An important fact is that the framework does not re-compute the whole 4D-model, but only those parts of the geometry which are affected by the pathology, hence the computational expense is kept at a minimum.

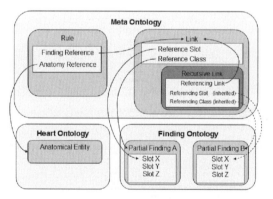

Fig. 6. The meta ontology combines the finding and the heart ontology and is used to perform the rule based generation of the final geometric model of a pathologic heart. Only a glimpse on scheme is given as example.

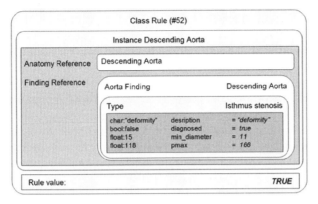

Fig. 7. Example Rule

4.4 Example

In this section the way from a particular finding to the geometric model is described. First a finding is selected from the finding database to be used as a base to generate the geometric heart model. In order to use the principles of the ontology framework the chosen finding is imported as an instance of the finding ontology while a copy of the healthy heart is created as an instance of the heart ontology. Now the system iterates over all rules of the meta ontology and checks for each of them whether it has to be applied to manipulate the heart instance. For the given example of a stenosis one of the rules signals the system to check if the value of the slot *stenosis* in the finding ontology differs from its default value *false*. In our example the rule carries the information that an object instance from the heart ontology has to be manipulated, namely the *descending aorta* (c.f. Fig. 7). Additionally the new vessel diameter for the stenosis is given, such that the mathematical model can be adapted accordingly. To construct more individual models we implemented a 4D-blood vessel editor that can be used to specify the stenosis in a more accurate way (c.f. color plate (d) right). This editor loads the blood vessel in scope together with the parameters of the finding. The user may then add more specific features that are not readily described by the finding (e.g. excentricity of the stenosis). The model is edited by manipulating the cross-sections and the control points of the mathematical model (c.f. Sect. 3). An example for defining a stenosis of a blood vessel is shown in color plate (d). Assuming that all other parameters of the finding are in a proper range the remaining parts of the heart model are taken from the standard heart model. Finally, the geometric mesh is generated.

5 Virtual Ultrasound

As already mentioned in Sect. 2 it is not a good idea for an ultrasound tutoring system to simply present ultrasound images acquired from real sessions due to the anisotropic character of those images. Hence, to support arbitrary viewpoints it is necessary to start from an isotropic representation and to transform this data into virtual ultrasound images presented to the user. This section provides information on how to generate virtual ultrasound images in real time. As such it is an extension of the simulator presented in [SRvL05].

In contrast to other methods developed to simulate ultrasound images like [Jen96] [JN02] [YLLL91] [LL88] this method aims on the generation of realistic looking artificial ultrasound images in real-time. The main difference to previous work is that the focus is not on modeling the actual physical sound theory, but only on generating images that look like ultrasound images. To achieve this goal the propagation and interaction of ultrasound in biological tissue is described following an optical model, i.e. we are defining *phonons* (assumed being sound particles) as acting entities.

The simulator currently consists of three core components capturing the most important effects found with medical ultrasound imaging. One component deals with the attenuation and scattering of ultrasound in biological tissue. A second component handles the effects due to reflection of ultrasound at tissue boundaries and the third one is responsible for creating the characteristic speckle pattern of ultrasound images.

To generate the artificial ultrasound scan a virtual ultrasound pulse is propagated over a tissue representation. The ultrasound pulse can be assumed as a set of phonons interacting with the tissue. For each of the core components mentioned above the tissue description contains factors describing the magnitude of respective sound interaction properties (attenuation, reflection and speckle development labeled T_a, T_r and T_s). Sound that does not undergo these interaction is assumed to be transmitted. During sound propagation each of the core components is evaluated. The brightness of a particular pixel in the simulation domain is determined by the strength of the backscattered sound energy.

Sound traveling through a medium is continuously attenuated. The attenuation is based on two separate effects namely absorption and reflection. Absorption is due to the transformation of sound energy (pressure) into heat and is a frequency dependant effect, i.e. the higher the sound frequency the stronger the attenuation. Let $\mathbf{u} := (x, y)$ be a position within the simulation domain, where x is the main sound propagation direction. The absorption is then defined on a local absorption coefficient which is represented by a transfer function to account for frequency dependency.

$$\mu_a(\mathbf{u}, f) := c_a(f) \cdot T_a(\mathbf{u}) \tag{4}$$

Similar measures are introduced for reflection and speckle development:

$$\mu_r(\mathbf{u}) := c_r \cdot F_r(\mathbf{u}, T_r)$$
$$\mu_s(\mathbf{u}) := c_s \cdot F_s(\mathbf{u}, T_s) \qquad (5)$$

The functions F_r and F_s describe the tissue formation dependency of reflection and speckle development. Reflection thereby depends on the local acoustic density gradient with respect to the propagation direction (c.f. Fig. 8(a)), whereas speckle formation is determined by means of local statistical properties.

Concerning speckle there are different types of scattering depending on the size of scatterers in relation to the frequency used, namely Mie scattering and Rayleigh scattering. Rayleigh scattering accounts for sound interaction at extremely small particles compared to the wave length, whereas Mie scattering accounts for the interaction of sound with small and medium sized particles. The total amount of scatter effects is determined by the factor T_s introduced earlier. Additionally every simulation cell contains three factors describing the distribution of very small, small, and medium scatter objects (p_r, p_{ms} and p_{mm}).

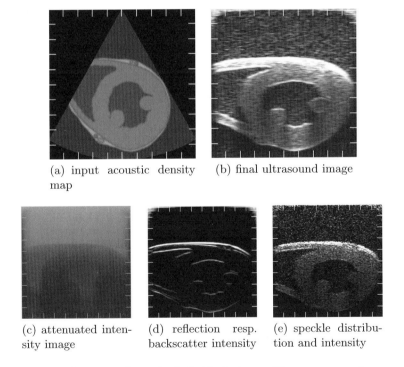

(a) input acoustic density map
(b) final ultrasound image
(c) attenuated intensity image
(d) reflection resp. backscatter intensity
(e) speckle distribution and intensity

Fig. 8. Example output from the individual stages. The upper row shows a simplified input image and the final output image prior to scan conversion, whereas the lower row from left to right shows the intensity, reflection and speckle images.

All the above mentioned effects contribute to the attenuation of sound energy traveling through the tissue. The overall attenuation of some initial sound energy I_0 for a given position $\mathbf{u}$ is then calculated as

$$I(\mathbf{u}, f) = I_o \exp\left(-\int_0^u \mu(u, f, t) dt\right), \qquad (6)$$

where μ is defined as the sum of μ_a, μ_r and μ_s. Let Δy be the sample separation in propagation direction, then the approximation can be rewritten recursively as

$$I(x, y, f) = I(x, y - \Delta y) \cdot \exp(-\mu(x, i\Delta y)\Delta y), \qquad (7)$$

where $I(x, 0) = I_0(x)$ is the pulse energy transmitted into the tissue at a particular location x.

Up to now the model is capable of computing the local, relative amounts of energy that is lost, transmitted (Fig. 8(c)) and reflected (Fig. 8(d)). Using this information it is possible to produce images that reflect a local mean brightness of ultrasound images, but these images do not look like ultrasound images since the grainy appearance (also known as *the speckle*) is not modeled yet. Speckle develops due to the interference of sound waves scattered at Mie or Rayleigh scatterers. Since speckle is based on numerous contributions within a simulation cell and from neighboring cells it is appropriate to model this effect statistically [WSSL83]. It was found empirically that the distribution of speckle splats based on a Gaussian distributed random variable results into quite realistic looking speckle patterns (c.f. [SRvL05] for further details).

We use tunable Gaussian splats to model the speckle. The splats can be tuned according to size (reflecting the frequency dependance of the speckle size) and according to their central height (reflecting the maximum amplitude of the sound wave and as such the brightness of the splat). Each simulation cell can hold a single splat, only (Fig. 8(e)). The probability of a splat at a particular cell depends on the factor T_s introduced earlier. The brightness is adjusted according to the amount of energy transformed into speckle. The size of the splat is determined by the tissue factors p_r, p_{ms} and p_{mm}. Finally, the speckle pattern develops by superimposing (blending) all the individual splats, which is performed by summing the contribution of all splats within a window (c.f. Fig. 8(b)). The blend window size can be adjusted and as default it is set to the double of the maximum splat size.

To produce the final intensity distributions the intermediate results (the speckle signal $S(\mathbf{u})$ (c.f. Fig. 8(c)) and the reflection signal $R(\mathbf{u})$ (c.f. Fig. 8(c))) generated by the core components have to be combined. Therefore we introduce tunable weights w_r and w_s representing the fractions for reflection and speckle in the final image and define

$$C(\mathbf{u}) = w_r \cdot R(\mathbf{u}) + w_s \cdot S(\mathbf{u}). \qquad (8)$$

At this point the image reflects the interaction of a portion of ultrasound energy with a tissue region, hence the backscattered energy is known. However, ultrasound undergoes further interaction with the tissue while traveling back to the transducer. We assume for simplicity that both travel directions (forth and back) use the same path, such that the *backward attenuation* can be define according to

$$B(\mathbf{u}) = C(\mathbf{u}) \cdot \prod_{i=1}^{n} \exp(-\mu(x, i\Delta y)\Delta y). \qquad (9)$$

The energy loss of ultrasound during its travel and hence the resulting brightness loss of an ultrasound image with depth is significant, such that one has to compensate for. In real ultrasound systems this procedure is called time gain compensation (TGC). This compensation is physically correct described using exponential (depth/time dependant) amplifiers, described by

$$A(\mathbf{u}) := exp(w_{tgc}(y) \cdot y), \qquad (10)$$

where $w_{tgc}(y)$ is a user adjustable parameter, which is heavily influenced by the attenuation properties of the tissue under examination and user preferences. As a very last step the images are slightly smoothed, scan converted and displayed. Figure 8 gives examples for the individual stages and a comparison between the original input image (acoustic density only) and the simulated output image prior to scan conversion.

6 Results

Concerning the model extraction we found the following computation complexity: the model extraction including the mesh generation takes only several seconds for small and medium sized raw data. Based on 20k triangles, up to 32 cross sections and a goal resolution of more than 20k triangles the complete extraction procedure takes approximately 4 minutes. The mesh generation itself takes less than a minute. The calculation times were found for a computer with a 1.2MHz CPU and 512MB RAM.

The automatic generation of a (pathologic) case representation takes less than a minute. More specific this is the time needed to construct a specific heart instance based on existing geometries. The extraction time needed to generate the heart models is not included. Some additional time is needed to fine tune the case parameters if these parameters do differ from a finding stored in the finding database.

When it comes to ultrasound image simulation our system can easily reach frame rates of 30fps on a nVidia 6800 GT graphics board, which is sufficient for immediate feedback during a virtual examination. These frame rates are reached independent from the type of data the simulation is based on, i.e. 2D, 3D or even dynamic volume data. Although the simulated ultrasound images are not perfect, they are far superior to existing methods and can be calculated in significantly less time. In contrast to real world ultrasound examinations

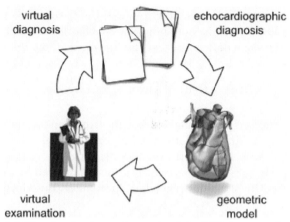

(a) The standard tutoring cycle of the VES framework. Based upon a cardiac finding an instance of the pathologic heart is created and presented to the student. The diagnosis of the virtual examination are stored in a second instance of the finding module and automatically compared to the original finding

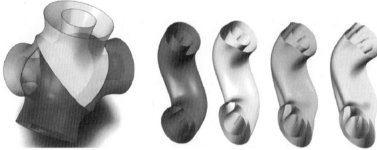

(b) A trifurcation modeled with the vessel model. All vessels are composed of an inner an outer layer

(c) Multi resolution analysis of the vessel model. All vessels are composed of the same number of mesh control points, but the detail increases from left to right. The leftmost vessel can be described by a single value per layer and cross-section, respectively

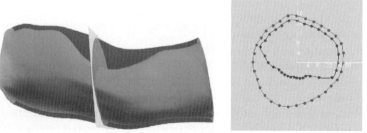

(d) Stenosis of a blood vessel and the respective view of the cross-section in the 4D-blood vessel editor

our system can provide insight into the geometric setting by visualizing the geometric representations and the ultrasound images side by side.

Concluding three integral components needed to build a general tutoring environment for echocardiography were presented. A method to generate a mathematical time-dynamic heart model based raw data given as unstructured triangle meshes and MR images respectively has been described. The integration of this model into an ontology framework to establish a general heart paradigm from which geometric representations of a pathologic heart can be derived automatically was outlined. Finally a real-time method to transform isotropic tissue descriptions into realistic looking virtual ultrasound images was detailed. Future work will concentrate on parameterizing further pathologies, as well as the automatic generation of tissue descriptions with respect to tissue altering pathologies like a cardiac infarction.

Acknowledgements

We want to acknowledge the valuable work of Rolf H. van Lengen and K. Schwenk for the VES project. The VES project was funded in part by the German Research Center For Artificial Intelligence (DFKI), the German Federal Ministry of Education and Research (BMBF) under contract number NR 01 IW A02, and by the German Research Foundation (DFG) through the project 4DUS.

References

[BAJ+]   C. Barry, C. Allot, N. John, P. Mellor, P. Arundel, D. Thomson, and J.Waterton. Three dimensional freehand ultrasound: Image reconstruction and image analysis. Ultrasound in Med. & Biol.

[BFGQ96] Th. Berlage, Th. Fox, G. Grunst, and K.-J. Quast. Supporting Ultrasound Diagnosis Using An Animated 3D Model of the Heart. In *(ICMCS'96)*, page 34, 1996.

[Bro03]   R. Brooks. Simulation and Matching of Ultrasound with MRI. Technical Report 110247534, McGill Centre for Intelligent Mashines, April 2003.

[BRvLH05] M. Bertram, G. Reis, R.H. van Lengen, and H. Hagen. Non-manifol Mesh Extraction from Time-varying Segmented Volumes used for Modeling a Human Heart. *Eurographics/IEEE TVCG Symposium on Visualizatzion*, pages 1–10, 2005.

[Deu97]   Deutsche Gesellschaft für Kardiologie-, Herz- und Kreislaufforschung. Qualitätsleitlinien in der Echokardiographie. *Z Kardiol*, 1997.

[Deu00]   Deutsche Gesellschaft für Kardiologie-, Herz- und Kreislaufforschung. Eine standardisierte Dokumentationsstruktur zur Befunddokumentation in der Echokardiographie. *Z Kardiol*, 2000.

[EDT+96] T. Elliot, D. Downey, S. Tong, C. McLean, and A. Fenster. Accuracy of prostate volume measurements in vitro using three-dimensional ultrasound. *Acad. Radiology*, 3:401406, 1996.

[Ehr98]   H.H. Ehricke. SONOSim3D: A Multimedia System for Sonography Simulation and Education with an Extensible Database. *European Journal of Ultrasound*, 7:225–300, 1998.

[EJS]   D. H. Salesin E. J. Stollnitz, T. D. DeRose. *Wavelets for Computer Graphics Theory and Applications*. Morgan Kaufmann Publishers Inc.

[FTS+95]   A. Fenster, S. Tong, S. Sherebrin, D. Downey, and R. Rankin. Three-dimensional ultrasound imaging. *Proc. SPIE*, 2432:176184, 1995.

[Gmb05]   SONOFIT GmbH. SONOFit SG3 ultrasound training system, 2005. http://www.sonofit.com.

[HBvLS01]   H. Hagen, T. Bähr, R.H. van Lengen, and O. Schweikart. Interlinguar in medizinischen Informationssystemen. In *Telemedizinführer Deutschland, Hrsg. A. Jäckel*, 2001. Deutsches Medizinforum.

[Jen96]   J.A. Jensen. Field: A Program for Simulating Ultrasound Systems. In *10th Nordic-Baltic Conference on Biomedical Imaging*, 1996.

[JN02]   J.A. Jensen and S.I. Nikolov. Fast Simuation of Ultrasound Images. In *IEEE International Ultrasonics Symposium*, 2002.

[Joh]   K. Johnes. Modelling the Human Heart. www.uniservices.co.nz/Modelling the Human Heart.pdf.

[KvLR+04]   S. Köhn, R.H. van Lengen, G. Reis, M. Bertram, and H. Hagen. VES: Virtual Echocardiographic System. In *IASTED VIIP'04*, pages 465–471, 2004.

[LL88]   R. Ludwig and W. Lord. A Finite Element Formulation for the study of Ultrasonic NDT Systems. *IEEE Trans. on Ultrasonics, Ferroelectrics and Frequency Control*, 35:809–820, 1988.

[MA97]   S. McRoy and S. Ali. Uniform Knowledge Representation for Language Processing in the B2 System. *Natural Language Engeneering*, 3(2):123–145, 1997. Cambridge University Press.

[MMR]   J. Michael, J. L. V. Mejino, and C. Rosse. The role of definitions in biomedical concept representation. In *American Medical Informatics Association Fall Symposium*.

[MSB+04]   H. Maul, A. Scharf, P. Baier, M. Wuestemann, H.H. Guenter, G. Gebauer, and C. Sohn. Ultrasound Simulators: Experience with the SonoTrainer and Comparative Review of other Training Systems. *Ultrasound in Obs. & Gyn.*, 24:581–585, 2004.

[NFM00]   N. Noy, R. Fergerson, and M. Musen. The knowledge model of protégé-2000: Combining interoperability and flexibility. In *EKAW*, pages 17–32, 2000.

[NM01]   N. Noy and D. McGuinness. Ontology development 101: A guide to creating your first ontology. Smi-report, Stanford Knowledge Systems Laboratory, March 2001.

[NPA04]   U. Ngah, C. Ping, and S. Aziz. *Knowledge-Based Intelligent Information and Engineering Systems*, volume 3213/2004 of *Lecture Notes in Computer Science*, chapter Mammographic Image and Breast Ultrasound Based Expert System for Breast Diseases, pages 599–607. Springer Berlin/Heidelberg, 2004.

[PGB99]   R. Prager, A. Gee, and L. Berman. Stradx: real-time acquisition and visualization of freehand three-dimensional ultrasound. *Medical Image Analysis*, 1999.

[PNJ92] D. Pretorius, T. Nelson, and J. Jaffe. 3-dimensional sonographic analysis based on color flow doppler and gray scale image data - a preliminary report. *Ultrasound in Medicine*, 11:225232, 1992.

[RBvLH04] G. Reis, M. Betram, R.H. van Lengen, and H. Hagen. Adaptive Volume Construction from Ultrasound Images of a Human Heart. In *Eurographics/IEEE TCVG Visualization Symposium Proceedings*, pages 321–330, 2004.

[Rei05] G. Reis. *Algorithmische Aspekte des 4dimensionalen Ultraschalls*. PhD thesis, Technische Universität Kaiserslautern, 2005.

[RN00] T. Roxborough and G. Nielson. Tetrahedron Based, Least Squares, Progressive Volume Models with Applications to Freehand Ultrasound Data. In *Proceedings of IEEE Visualization*, pages 93–100, 2000.

[SM00] O. Schweikart and F. Metzger. Standardisierte Befunderfassung in der Echokardiographie mittels WWW: EchoBefundSystem. *Z Kardiol 89*, pages 176–185, 2000.

[SRvL05] K. Schwenk, G. Reis, and R.H. van Lengen. Real-time Artificial Ultrasound, 2005. submitted to Elsevier Science.

[SWS+00] F. Sachse, C. Werner, M. Stenroos, R. Schulte, P. Zerfass, and O. Dössel. Modeling the anatomy of the human heart using the cyrosection images of the visible female dataset, 2000. Third Users Conference of the National Library of Medicine's Visible Human Project.

[TDCF96] S. Tong, D. Downey, H. Cardinal, and A. Fenster. A three-dimensional ultrasound prostate imaging system. *Ultrasound in Medicine and Biology*, 22:73546, 1996.

[TGP+02] G. Treece, A. Gee, R. Prager, C. Cash, and L. Berman. High Resolution Freehand 3D Ultrasound. Technical report, University of Cambridge, 2002.

[vLKB+04] R.H. van Lengen, S. Köhn, M. Bertram, B. Klein, and H. Hagen. Mit Ontologien visualisieren. *Informatik Spektrum Sonderheft - Computergraphik*, March 2004.

[WSSL83] R.F. Wagner, S.W. Smith, J.M. Sandrick, and H. Lopez. Statistics of Speckle in Ultrasound B-Scans. *IEEE Trans. Son. Ultrason*, 30:156–163, 1983.

[WWP+00] M. Weidenbach, C. Wicks, S. Pieper, K.J. Quast, T. Fox, G. Grunst, and D.A. Redel. Augmented Reality Simulator for Training in Twodimensional Echocardiography. *Computers and Biomedical Research*, 33:11–22, 2000.

[YLLL91] Z. You, M. Lusk, R. Ludwig, and W. Lord. Numerical Simulation of Ultrasonic Wave Propagation in Anisotropic and Attenuative Solid Materials. *IEEE Trans. Ultrasonics, Ferroelectrics and Frequency Control*, 38(5), 1991.

Supporting Depth and Motion Perception in Medical Volume Data

Jennis Meyer-Spradow, Timo Ropinski, and Klaus Hinrichs

Visualization and Computer Graphics Research Group, Department of Computer Science, University of Münster
{spradow, ropinski, khh}@math.uni-muenster.de

Summary. There are many application areas where dynamic visualization techniques cannot be used and the user can only view a still image. Perceiving depth and understanding spatio-temporal relations from a single still image are challenging tasks. We present visualization techniques which support the user in perceiving depth information from 3D angiography images, and techniques which depict motion inherent in time-varying medical volume datasets. In both cases no dynamic visualization is required.

1 Introduction

Volume rendering has become a mature field of interactive 3D computer graphics. It supports professionals from different domains when exploring volume datasets, representing for example medical or meteorologic structures and processes. Current medical scanners produce datasets having a high spatial resolution, and even dynamic datasets can be obtained. Especially Computer Tomography (CT) datasets are highly suitable for showing anatomical structures. However, when dynamic processes have to be explored a single 3D volume dataset is often insufficient. For example, Positron Emission Tomography (PET) imaging techniques are used for exploring the dynamics of metabolism. Successive scans produce a time series of images each showing the distribution of molecules at a certain point in time. Time-varying volume data is also produced when applying functional Magnetic Resonance (MR) imaging techniques to evaluate neurological reaction to certain stimuli.

Although 3D volumes can be displayed separately to view the data for specific points in time, the dynamics contained in a time-varying dataset can only be extracted when considering all of them. Therefore time-varying volume datasets are often visualized by displaying the 3D volumes sequentially in the order they have been acquired. When rendering these 3D volumes at a high frame rate, the viewer is able to construct a mental image of the dynamics. Besides the possibility to acquire time-dependent data also the spatial resolution of medical scanners has been increased leading to large and possibly more complex datasets. This complexity demands the development of interactive visualization techniques supporting an efficient and effective analysis. Since the medical datasets acquired through different imaging technologies differ in the way they are explored, specialized techniques have to be developed. We propose visualization techniques which have been developed to support the exploration of angiography datasets. We will show how the depth perception and thus the spatial cognition of these datasets can be improved and thus results in a more efficient as well as effective diagnosis.

Often dynamics are used to visualize both time-varying as well as high resolution datasets. In the latter case rotations or other interactive techniques are applied to explore the datasets. However, in some domains these techniques are inappropriate, since viewers prefer still images rather than dynamic image sequences. The following two problems arise when dealing with image sequences of 3D volumes. When several people watch an image sequence, a viewer usually points at a specific feature within an image to exchange findings with the other participants. Since the images are dynamic, it is hard to pinpoint such an item of interest, especially if it is moving, and the animation needs to be paused. Another problem is the exchange of visualization results. In medical departments diagnostic findings are often exchanged on *static media* such as film or paper. However, it is not possible to exchange time-varying volume datasets in this manner. Although the sequential viewing process could be simulated by showing all 3D volumes next to each other, this could lead to registration problems, i.e., it would be more difficult to identify a reference item across the 3D volumes.

We address these problems by proposing visualization techniques which represent the motion dynamics extracted from a time series of 3D volumes within a single static image and support the exploration of high resolution datasets. All presented techniques are implemented in our Volume Rendering Engine *Voreen* [Vor07].

2 Related Work

Depicting Motion

Mainly non-photorealistic rendering techniques are used for depicting motion in a single image. Masuch et al. [MSR99] present an approach to visualize motion extracted from polygonal data by speedlines, repeatedly drawn contours and arrows; different line styles and other stylization give the viewer the impression of a hand-drawing. Nienhaus and Döllner [ND05] present a technique to extract motion information from a behavior graph, which represents events and animation processes. Their system automatically creates cartoon-like graphical representations. Joshi and Rheingans [JR05] use speedlines and flow ribbons to visualize motion in volume datasets. However, their approach handles only feature extraction data and is mainly demonstrated on experimental datasets. Meyer-Spradow et al. [MRVH06] use speedlines to visualize motion dynamics extracted in a pre-processing step. Furthermore they use edge detection and transparency techniques to show several time steps in one still image.

A lot of research in the area of dynamic volume visualization has been dedicated to flow visualization. Motion arrows are widely used [LHD*04], but are often difficult to understand because information is lost when performing a 2D projection to the screen. Boring and Pang [BP96] tried to tackle this problem by highlighting arrows, which point in a direction specified by the user. Texture based approaches compute dense representations of flow and visualize it [MCD92]. First introduced for 2D flow, it has also been used for 3D flow visualization, for example with volume line integral convolution used by Interrante and Grosch [IG97]. Svakhine et al. [SJE05] emulate traditional flow illustration techniques and interactive simulation of Schlieren photography.

Hanson and Cross [HC93] present an approach to visualize surfaces and volumes embedded in four-dimensional space. Therefore they use 4D illuminated surface rendering with 4D shading and occlusion coding. Woodring et al. [WWS03] interpret time-varying datasets as four-dimensional data fields and provide an intuitive user interface to specify 4D hyperplanes, which are rendered with different techniques. Ji et al. [JSW03] extract time-varying isosurfaces and interval volumes considering 4D data directly.

Supporting Depth Perception

The most common technique used for visualizing vessel structures is the maximum intensity projection (MIP). In contrast to this direct rendering technique, model-based approaches generate and visualize a model of the vessel system to support generation of high quality images. The initial work on this topic has been done by Gerig et al. in 1993 [GKS*93]. Hahn et al. [HPSP01] describe an image processing pipeline to extract models of vascular structures in order to generate high quality visualizations. In 2002 Kanitsar et al. [KFW*02] have proposed a model-based visualization technique based on curved planar reformation. Oeltze and Preim [OP05] have introduced in 2005 the usage of convolution surfaces to further enhance visualization quality especially at vessel furcations. While all other techniques focus on visualizing the vessel structures without contextual information, the VesselGlyph provides also context information given by surrounding tissue [SCC*04].

In this paper we use monoscopic visualization techniques to enhance depth perception. We do not discuss monoscopic depth cues in detail, for an overview we refer to the work done by Lipton [Lip97] and Pfautz [Pfa00]. A detailed comparison of the influence of depth cues on depth perception is described by Wanger [WFG92]. Ropinski et al. [RSH06] propose and compare different kinds of depth cues. Different models have been proposed for evaluating the interplay of depth cues. These models consider combinations of different depth cues and postulate how these depth cues contribute to the overall depth perception. The models incorporate for instance the weighted linear sum [BC88], a geometric sum [DSW86], or the reliability of depth cues in the context of other cues and additional information [YLM93].

3 Illustrating Dynamics

We visualize the motion dynamics that is inherent in a volume dataset in a single static image which is constructed from the images of the dataset at succeeding points in time. We distinguish three groups of 3D volumes, depending on their points in time. The current volume is used as a reference. The second group contains the chronologically preceding volumes, and the third group the chronologically succeeding ones.

The current volume is rendered using a direct volume rendering (DVR) technique. The resulting image contributes most to the final image and is then enriched with information gained from the other two groups. For the visualization of these groups we propose two techniques: edge detection and color coding of differences between the volumes. The parameterization is based on the difference in time between the current and the preceding/succeeding volumes. In comparison to using motion blur these techniques have the advantage that they avoid blurring information contained in the images.

For the edge detection technique we first render the current time step with DVR. We also use DVR to render all preceding/succeeding volumes we are interested in. On the resulting images we use an image based edge detection by applying an appropriate filter kernel. Thus we receive the silhouettes of the preceding/succeeding volumes. We use an edge threshold (by considering the lengths of gradients) to show only edges exceeding a certain thickness. The color of the edges depends on the point in time of the corresponding volume. These parameters can be chosen freely. For the example in Figure 1 (*left*)

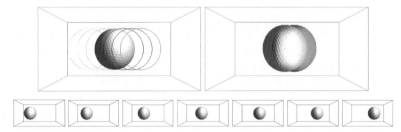

Fig. 1. Two visualization techniques to depict the motion of the golf ball volume dataset. The current 3D volume is rendered using isosurface shading. Edges of preceding and succeeding 3D volumes are shown (*left*); the combinations of the current and the succeeding 3D volumes are rendered with different colors (*right*). Renderings of seven time steps extracted from the dataset (*bottom*).

we used cold colors for the edges of the preceding and warm colors for the edges of the succeeding volumes. In addition, the edge threshold for preceding volumes depends on their point in time so that older volumes are depicted with thinner edges. In a final step the individual images need to be composed into a single image as final result. This can be done by either blending the images or masking them by considering the background color.

The color coding technique shows the differences between two time steps. Both time steps are rendered using DVR. Then three sets of pixels are distinguished: the pixels determined by

- the image of the current volume,
- the image of the succeeding volume, and
- the intersection of the images.

For the intersection the user has to decide in which time step he is more interested in. This set of pixels will then be shown as is. For the other two sets only the brightness information is used and multiplied with a color, for instance red for the current volume and green for the succeeding one. This ensures that the alpha-part of a transfer function can be used. Finally the three disjoint sets are blended into the final image as can be seen in Figure 1 (*right*).

4 Depth Enhancement

Because of the complex depth structure of angiography datasets spatial cognition is one of the most challenging tasks during their exploration. We introduce techniques to support the user in recognizing the depth structure of vessels even in static images while the user concentrates on regions he is interested in, and we try to minimize the visualization of unwanted information that may distract him.

4.1 Depth-Based Mouse Cursor Replacement

In most desktop applications a computer mouse together with its screen representation is used. Usually the screen representation is an arrow-shaped cursor that is used to point at certain locations of interest or to interact with objects. We alter this representation and provide additional depth information with the mouse cursor.

The user is not only interested in one depth value, but would like to compare depth values at different positions, e.g., which of two vessels is closer and which is farther away. For this purpose we use a so called cross-hair cursor instead of the standard cursor. A cross-hair cursor consists of two orthogonal lines, which are parallel to the x- resp. y-axis. Their intersection point corresponds to the original mouse cursor position, and the cross-hair moves with the mouse. To show different depth values simultaneously we use a depth-dependent color coding for every point of the cross-hair cursor. The color of such a point is chosen according to the depth value at its position. Since vessel visualization commonly uses gray shades only, color coding for visualizing additional information can be applied without any drawbacks.

The cross-hair is colored in real time while the user moves with the mouse over the angiography image. The remaining image stays untouched. As the cross-hair is moved by the mouse, the user can control the two perpendicular lines easily and exploit their color to compare the depth of different vessels. Once he finds an interesting constellation where the cross-hair cursor shows the relevant information, he can make, e.g., a printout for further use.

Coding Schemes

A color scheme maps the range of depth values in a one-to-one manner to different colors, for example a red to black gradient. Then parts of the vessels covered by the cursor which are closer to the viewer will be colored in bright red shades, when the user moves the cross-hair over them, and vessels which are farther away will be colored using a darker red or black. Such a mapping scheme has the advantage that the same color always represents the same depth value. But it also has a drawback: Angiography images may contain many depth values from a relatively large depth range. Mapping the depth values in a one-to-one manner to colors may result in very small and hard to detect color nuances (see Figure 2 (*left*)). This problem can be addressed by choosing not only a red to black gradient, but using the whole color spectrum. But it can be further improved.

Preliminary tests have indicated that users prefer to compare vessels lying on the cross-hair lines without moving the cursor. Hence it is possible to show the user not only small absolute differences, but also larger relative differences (see Figure 2 (*right*)). To do this we count the number of different depth values covered by both lines of the current cross-hair cursor and divide the available color interval into the same number of subintervals before mapping every depth value to a color. As a (simplified) example assume that the cross-hair cursor crosses three vessels with the (normalized) depth values 0.7, 0.9 and 0.8, and these values should be visualized with a red to black gradient. In this case the gradient will be divided into three intervals—bright red, dark red, and black—and the cross-hair will be colored with these different colors at the appropriate locations. This also works with many different depth values

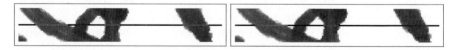

Fig. 2. Color coding of the depth structure. The detail images show the differences when using an absolute mapping (*left*) and a relative mapping (*right*). In the left image the right vessel seems to have the same depth value as the left ones. The right image show that actually the right vessel is further away from the observer.

in such an intervall, because values near 0.7 will still be visualized very dark and values near 0.9 will still be shown as a bright red. In case of an absolute mapping all values would have been mapped to slightly different shades of dark red.

Of course in some situations with a large amount of widespread depth values the proposed mapping yields only a small advantage, but in most cases it enhances the visualization. Especially when applied to the very sparse angiography datasets it is easy to find an appropriate mapping. However, the user has to be aware of the fact that the color coding is only valid within one position of the cross-hair cursor, because the current depth range may change when the cursor is moved. The use of a gradient with more than two colors may further improve the visualization.

4.2 Splitview Visualization

Another intuitive way to perceive depth information is to change the viewing position in order to look from the left or from the right. But when switching the viewing position the user may get disoriented within the vessel structure. Showing both views simultaneously solves this problem. Therefore we use a multi-view visualization with linked cameras and some extensions. Usually the user is not interested in a side view of the whole vessel structure but concentrates on some details in the image. Therefore we divide the vessel structure into two parts, and in the side view we visualize only the part being in focus (see Figure 3). Looking from the front we cut along a plane that is

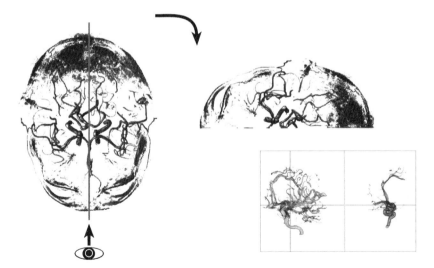

Fig. 3. The splitview seen from bird's eye view and an example what the user sees (small image bottom right). The right view shows the vessel structure in a side view. The current slice is marked red, and the front region is clipped.

determined by two vectors: the first vector points from the viewer's eye to the mouse cursor, the second is the up vector. This results in a vertical plane that divides the vessel structure in a left and a right part. In the side view only the left part is visualized, i.e., the parts of the vessel structure being behind the cutting plane. The cutting plane and the side view are calculated and visualized in real time, so the user can alter the view with the mouse and see the result immediately.

To illustrate vessels that cross the cutting plane, the intersections are marked with red color. To clarify the relation between both views we use a cross-hair cursor whose horizontal line is continued through the side view. Furthermore this line makes it easier for the user to orient herself. This technique needs a short familiarization, but has two big advantages: the user can perceive depth relations immediately, and also very small depth differences are clearly visible.

5 Application Examples

5.1 Heart Motion

We have used a segmented time-varying volume dataset of the human heart to test our visualization techniques. First we have applied our techniques to different time steps of the same slice. In Figure 4 (*left*) we use direct volume rendering to display one time step and silhouette rendering, i.e. edge detection, for the other time steps. In Figure 4 (*right*) the differences between two time steps are visualized with different colors. The *succeeding only* pixels are colored with red shades, and the *current only* pixels with green shades. In the *intersection* the pixels of the original image of the current dataset are shown.

Both techniques are suitable for visualizing motion of a complete volume by displaying all slices simultaneously; the slices can be arranged for instance in a checker-board pattern physicians are familiar with.

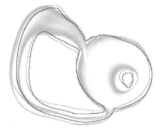 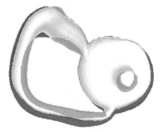

Fig. 4. Visualization of heart motion. Edges of preceding and succeeding volumes are shown (*left*). Differences between two time steps are color coded (*right*).

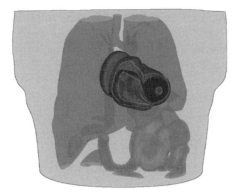

Fig. 5. Context information added to the visualization of heart motion.

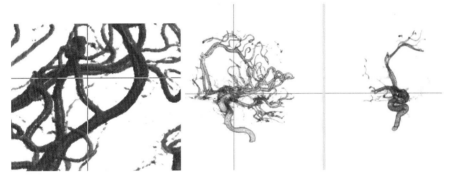

Fig. 6. The cross-hair cursor visualizes depth information (*left*). Here a mapping to a rainbow gradient is used. The splitview (*right*) shows a front view of the vessels in its left part and a sideview in the right part. Structures lying in the right half of the front view are cut away in the side view.

In Figure 5 we add some additional context information to the rendering. We have cut the dataset with a clipping plane. In addition we rendered some of the surrounding organs using tone shading. To achieve an impression similar to visualizations used in medical illustrations we added a rendering of the body's silhouette.

5.2 Vessel Visualization

We have applied both techniques for enhancing depth perception to an angiography dataset. Without any depth cues it is nearly impossible to figure out the exact depth structure of the vessels (see Figure 6 (*left*)). The color coding of depth values with the cross-hair allows spatial cognition. A gradient from red over green and yellow to blue is used for the depth coding, using red for near structures and blue for those farther away.

The cross-hair visualization is very suitable for depth structures of medium complexity but not sufficient for more complex ones, because for every pixel only the nearest structure is visible. In such cases the split view visualization works well (see Figure 6 (*right*)). In the left view a front perspective of the vessels is shown, in the right part a side view. All structures lying in the right half of the left view are cut away for the side view. Thus the precise depth structure of the plane determined by the vertical line of the cross-hair and the viewing direction is visible in the right view. To further improve the side view, structures cutting the plane are colored red.

User Study

The results of a survey evaluating the usage of interactive techniques and the techniques described above as well as further discussions with physicians have indicated that users prefer to view interactive visualizations over static ones. But as stated above, interactive exploration is often not possible. And especially medical visualization experts also appreciated the cross-hair cursor, which shows additional information even in a static view.

On a five-point Likert-scale, where 1 corresponds to not helpful, and 5 corresponds to very helpful, participants assessed the feature of being able to rotate the vessel structures with 4.73 on average. Medical experts rated the cross-hair cursor with 4.5 (color gradient) and 4.0 (red gradient); non-experts assessed these techniques as not beneficial (color gradient: 2.5, red gradient: 2.0). However, many participants especially used the horizontal line of the cross-hair in order to evaluate depth using a scan line analogue.

Independent of the experience with medical visualization, nearly all participating users have evaluated the splitview visualization as very helpful (on average 4.5). In particular the reddish cutting slice in the right view (see Figure 6 (*right*)) has been evaluated as very helpful (in average 4.73). Although we assumed that the usage as well as interpretation of the result of this technique requires some effort to get used to it, the users remarked that after a short period of practice the handling is intuitive instead of complex.

6 Conclusions and Future Work

In this paper we have presented visualization techniques for depicting dynamics and depth characteristics in still renderings of (time-varying) volume datasets. The techniques are applied automatically and do not need any user input. In order to obtain interactive frame rates they have been implemented as extensions to GPU-based ray-casting. To demonstrate the usability of our visualization techniques we have described the application to real world data.

With the techniques presented in the first part it is possible to present a reasonable subset of the motion information contained in a time-varying dataset in a single image. Thus this information can be communicated more easily since it can also be visualized on static media. The techniques to improve depth perception enable the user to build a precise mental image of depth structures even for very complex data.

To get further directions we have conducted informal interviews with physicians which work with angiography datasets. These interviews have revealed that in addition to the interactive rotation of a dataset in order to better perceive its depth relation, the possibility to switch between combinations of different perspectives is very welcome.

Another open issue is the development of an appropriate lighting model which improves spatial comprehension in angiography. While in general the phong lighting model was preferred to the tone-based color coding, the highlights were perceived as distracting. Therefore it would be necessary to evaluate in how far existing lighting models are able to give shape cues without tampering the depth perception.

7 Acknowledgments

This work was partially supported by grants from the Deutsche Forschungsgemeinschaft (DFG), SFB 656 MoBil Münster, Germany (project Z1). The golf ball dataset is available at voreen.uni-muenster.de. It has been generated with the voxelization library *vxt* developed by Milos Sramek. The medical dataset has been generated using the 4D NCAT phantom.

References

[BP96] Boring, E., Pang, A.: Directional Flow Visualization of Vector Fields. In: VIS '96: Proceedings of the conference on Visualization, 389–392, IEEE Computer Society (1996)

[BC88] Bruno, N., Cutting, J.: Minimodality and the perception of layout. In: Journal of Experimental Psychology, **117**, 161–170 (1988)

[DSW86] Dosher, B.A., Sperling, G., Wurst, S.A.: Tradeoffs between stereopsis and proximity luminance covariance as determinants of perceived 3d structure. In: Journal of Vision Research, **26(6)**, 973–990 (1986)

[GKS*93] Gerig, G., Koller, T., Szekely, G., Brechbühler, C., Kübler, O.: Symbolic description of 3-d structures applied to cerebral vessel tree obtained from mr angiography volume data. In: IPMI '93: Proceedings of the 13th International Conference on Information Processing in Medical Imaging, 94–111, Springer-Verlag (1993)

[HPSP01] Hahn, H.K., Preim, B., Selle, D., Peitgen, H.O.: Visualization and interaction techniques for the exploration of vascular structures. In: VIS '01: Proceedings of the conference on Visualization, 395–402, IEEE Computer Society (2001)

[HC93]   Hanson, A.J., Cross, R.A.: Interactive visualization methods for four dimensions. In: VIS '93: Proceedings of the conference on Visualization, 196–203, IEEE Computer Society (1993)
[IG97]   Interrante, V., Grosch, C.: Strategies for Effectively Visualizing 3D Flow with Volume LIC. In: VIS '97: Proceedings of the conference on Visualization, 421–424, IEEE Computer Society (1997)
[JSW03]   Ji, G., Shen, H.-W., Wenger, R.: Volume Tracking using Higher Dimensional Isocontouring. In: VIS '03: Proceedings of the conference on Visualization, 201–208, IEEE Computer Society (2003)
[JR05]   Joshi, A., Rheingans, P.: Illustration-inspired techniques for visualizing time-varying data. In: VIS '05: Proceedings of the conference on Visualization, 679–686, IEEE Computer Society (2005)
[KFW*02]   Kanitsar, A., Fleischmann, D., Wegenkittl, R., Felkel, P., Gröller, E.: Cpr: Curved planar reformation. In: VIS '02: Proceedings of the conference on Visualization, 37–44, IEEE Computer Society (2002)
[LHD*04]   Laramee, R.S., Hauser, H., Doleisch, H., Vrolijk, B., Post, F.H., Weiskopf, D.: The State of the Art in Flow Visualization: Dense and Texture-Based Techniques. Computer Graphics Forum, **23(2)**, 203–221 (2004)
[Lip97]   Lipton, L.: Stereographics developers handbook. StereoGraphics Corporation (1997)
[MSR99]   Masuch, M., Schlechtweg, S., Schulz, R.: Speedlines – Depicting Motion in Motionless Pictures. In: SIGGRAPH '99 Conference Abstracts and Applications, 277 (1999)
[MCD92]   Max, N., Crawfis, R., Williams, D.: Visualizing Wind Velocities by Advecting Cloud Textures. In: VIS '92: Proceedings of the conference on Visualization, 179–184, IEEE Computer Society (1992)
[MRVH06]   Meyer-Spradow, J., Ropinski, T., Vahrenhold, J., Hinrichs, K.H.: Illustrating Dynamics of Time-Varying Volume Datasets in Static Images. In: Proceedings of Vision, Modeling and Visualization 2006, 333–340 (2006)
[ND05]   Nienhaus, M., Döllner, J.: Depicting Dynamics Using Principles of Visual Art and Narrations. In: IEEE Computer Graphics & Applications, **25(3)**, 40–51 (2005)
[OP05]   Oeltze, S., Preim, B.: Visualization of vascular structures: method, validation and evaluation. In: IEEE Transactions on Medical Imaging, **24(4)**, 540–548 (2005)
[Pfa00]   Pfautz, J: Depth perception in computer graphics. Doctoral dissertation, University of Cambridge, UK (2000)
[RSH06]   Ropinski, T., Steinicke, F., Hinrichs, K.H.: Visually Supporting Depth Perception in Angiography Imaging. In: Proceedings of the 6th International Symposium on Smart Graphics (SG06), 93–104 (2006)
[SCC*04]   Straka, M., Cervenansky, M., Cruz, A.L., Kochl, A., Sramek, M., Gröller, E., Fleischmann, D.: (2004). The vesselglyph: focus & context visualization in ct-angiography. In: VIS '04: Proceedings of the conference on Visualization, 385–392, IEEE Computer Society (2004)
[SJE05]   Svakhine, N., Jang, Y., Ebert, D., Gaither, K.: Illustration and Photography Inspired Visualization of Flows and Volumes. In: VIS '05: Proceedings of the conference on Visualization, 687–694, IEEE Computer Society (2005)

[WFG92]   Wanger, L.C., Ferwerda, J.A., Greenberg, D.P.: Perceiving spatial relationships in computer-generated images. In: IEEE Computer Graphics and Applications, **12(3)**, 44–51, 54–58 (1992)
[WWS03]   Woodring, J., Wang, C., Shen, H.-W.: High Dimensional Direct Rendering of Time-Varying Volumetric Data. In: VIS 03': Proceedings of the conference on Visualization, 417–424, IEEE Computer Society (2003)
[YLM93]   Young, M.J., Landy, M.S., Maloney, L.T.: A perturbation analysis of depth perception from combinations of texture and motion cues. In: Journal of Vision Research, **33(18)**, 2685–2696 (1993)
[Vor07]   http://voreen.uni-muenster.de

Part III

Visualization of Multi-channel Medical Imaging Data

Multimodal Image Registration for Efficient Multi-resolution Visualization

Joerg Meyer

Department of Electrical Engineering and Computer Science
644E Engineering Tower, Irvine, CA 92697-2625
jmeyer@uci.edu

Summary. Arising from the clinical need for multimodal imaging, an integrated system for automated multimodal image registration and multi-source volume rendering has been developed, enabling simultaneous processing and rendering of image data from structural and functional medical imaging sources. The algorithms satisfy real-time data processing constraints, as required for clinal deployment.

The system represents an integrated pipeline for multimodal diagnostics comprising of multiple-source image acquisition; efficient, wavelet-based data storage; automated image registration based on mutual information and histogram transformations; and texture-based volume rendering for interactive rendering on multiple scales.

Efficient storage and processing of multimodal images as well as histogram transformation and registration will be discussed. It will be shown how the conflict of variable resolutions that occurs when using different modalities can be resolved efficiently by using a wavelet-based storage pattern, which also offers advantages for multi-resolution rendering.

1 Introduction

This chapter describes a set of algorithms that form an integrated system for automated image processing, multimodal image registration and multi-source volume rendering. The system has been developed to enable simultaneous processing and rendering of image data from multiple medical imaging sources.

A complete system (Figure 1) consists of an integrated pipeline for multimodal diagnostics. The components include multiple-source image acquisition; automated image registration based on histogram transformations and optimization of normalized cross-correlation; efficient, wavelet-based data storage and transmission; and 3D-texture-based, multi-resolution volume rendering for interactive visualization.

The focus of this chapter is on histogram transformation and registration. A short section in the end will discuss efficient storage, processing and rendering of multimodal images. It will be shown how different modalities, even

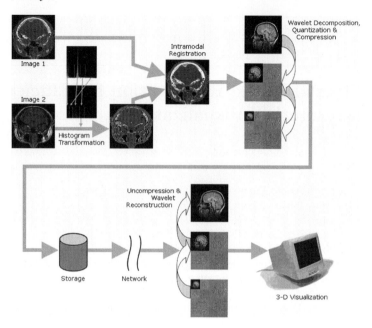

Fig. 1. Rendering pipeline for multimodal images.

when they express various tissue types differently or when their resolution differs, can be combined and stored efficiently by using a histogram transformation and a wavelet-based storage pattern. The latter also offers advantages for network-based data transmission, and for visualization on multiple scales. In the end, references will be given for further study.

This chapter is structured as follows. Section 2 describes the difference between intermodal and intramodal registration and provides examples of both techniques. Section 3 describes the pre-processing step of segmenting the images in each modality and correlating the scalar values for each cluster. Section 4 outlines the registration process using cross-correlation as a similarity measure, and Section 5 shows the results obtained using this method. Section 6 explains how multi-resolution storage patterns and three-dimensional, texture-based volume rendering can be used to display multimodal scans in real time. Section 7 summarizes the findings and outlines future research goals and potential directions.

2 Background

2.1 Intermodal vs. Intramodal Registration

This section addresses the problem of automatic registration of CT and MR images. The task is non-trivial due to the fact that particular tissue types are represented in dissimilar ways in different modalities and may appear in a different order in the gray level spectrum. For example, bone is visually represented as bright white (high X-ray absorption) in CT scans (Figure 9(a)), and as a black band (low proton density) in MRI scans (Figure 9(b)). Hence, the correlation between corresponding pixels in each scan is not linear (Figure 9(c)). This makes the task of aligning them, without the use of stereotactic frames or markers, difficult.

As an alternative approach to intermodality registration, a method that uses feature segmentation based on intensity histograms [Gon02] of the two images is proposed here. The idea is that if one of the scans is transformed to match the gray value distribution of the other scan, then the similarity measures used in *intramodality* registration techniques can be used for *intermodality* registration. Section 3 outlines techniques to achieve a reasonable segmentation of the various tissue types and correlation of the same to achieve a robust registration. The fact that the algorithm in the current implementation uses several different tissue types to control the alignment process and the correlation computations adds additional robustness to the approach.

Combining multiple scans of the same specimen into a single image or volume has numerous advantages in diagnostics, pre-surgical planning and radiation therapy planning, especially when multiple modalities depicting various physical properties of the scanned sample are involved. Some tissue types may be better visible in one type of scan, others may only appear in another scan.

Similar scans (e.g. multiple CT scans) can be registered by using a simple differential function as a similarity measure (*intramodal registration*). Combining images taken from different modalities (*intermodal registration*) is a much harder problem. The main reason is the fact that the various tissue types that occur in the images are represented in different ways in different modalities and may appear in a permutated order in the gray level spectrum. For example, a linear transfer function maps the high X-ray absorption values of bone to an intensity value that is visually represented as a bright white, whereas the same linear transfer function would produce a black pixel for the same location in an MRI scan due to the low proton density of bone matter. Hence, the correlation between corresponding pixels in each scan is not linear over the entire range of values. Moreover, the intensity ranges associated with a particular tissue type do not appear in the same order in a histogram. This makes the task of aligning two images from different modalities difficult.

2.2 Registration Methods

Three different, classical registration methods with different prerequisites and application areas are presented here. The *Global Difference* method is based on the assumption that intensities in the two participating images are linearly correlated. This assumption makes the method not suitable for intermodal registration, because absolute values, the spacing between these values in an intensity histogram, and the order of clusters in such a histogram vary due to the difference in the way material properties and the physical quantities measured by a particular scanning method differ.

The second method is called *Geometric Features*. It compares outlines of objects, which can be identified by extracting intensity gradients using a Laplacian or high-pass filter. This method is based on the assumption that tissue gradients are expressed similarly in both images. Unfortunately, this is not the case for the same reason as for the absolute values, i.e., because of the differences in the physical quantities being measured by the two incompatible image acquisition devices.

A third method is known as *Mutual Information*. This technique computes the statistical dependence or information redundancy between image intensities of corresponding pixels in both images. No specific assumptions about feature expression are being made. This method is suitable for intermodal registration and has been widely studied in the literature [Mae97].

Mutual information (MI) works for different modalities that contain the same information, but expressed differently. It is typically computed as illustrated in equation 1.

$$MI(I_1, I_2) = \sum_{i_2(x,y) \in I_2} \sum_{i_1(x,y) \in I_1} h(i_1(x,y), i_2(x,y)) \cdot \log \frac{h(i_1(x,y), i_2(x,y))}{h(i_1(x,y)) \cdot h(i_2(x,y))} \quad (1)$$

I_1 and I_2 describe the two images, and i_1 and i_2 the respective pixels in these images. $h(i)$ refers to the respective histograms, while $h(i_1, i_2)$ refers to the mutual histogram (products of pixel counts per intensity value). The goal is to maximize MI.

The idea of the method presented in this chapter is that methods suitable only for *intramodal registration* (same imaging device type) can be adapted to work also for *intermodal registration* (different imaging device types). The method can be efficiently implemented using a look-up table and, depending on the implementation of the intramodal registration algorithm, is therefore potentially faster than the computationally expensive MI method.

2.3 Advanced Methods

Numerous attempts to accelerate the alignment process, ranging from manual to semi-automatic to fully automatic methods [Els94, Els95, Stu96, Pen98, Hil01, Jen02, LoC03], have been proposed in the literature. These methods include histogram matching [Hua98], texture matching [Ash95], intensity cross-correlation [Mai98], kernel-based classification methods [Cri00], information divergence minimization [Sto98] and optical flow matching [Lef01]. A good survey can be found in [Bro92].

More recently developed fully automated methods essentially revolve around entropy [Mey97, But01, Zhu02] and mutual information (MI) [Vio95, Wel96, Plu03, Wel96]. Even though they are more convenient than manual or semi-automatic techniques, they are still time consuming and computationally expensive. The general consensus on MI-based techniques is that they provide the best results in terms of alignment precision and are generally consistent with manual, expert-based alignment procedures. However, the loss of spatial coherence during the registration, which is due to the different physical quantities being measured and the resulting variations in the feature boundaries and gradients, is a potential cause of inaccurate registration.

2.4 Using Intramodality Techniques for Intermodality Registration

As described before, we propose an alternate method for registering CT and MRI scans, usually requiring *intermodality* techniques. The method described here employs *intramodality* methods to accomplish the alignment task. Our approach is rather intuitive and inexpensive. We hypothesize that if we convert one of the scans (e.g. MRI) to match the other (e.g. CT) in terms of the scalar value representation of various tissue types that occur in the images, then by applying a similarity measure based on the correlation between corresponding features and globally optimizing this parameter, we can achieve alignment of the two different modalities at a relatively cheaper cost, providing results comparable in quality to those of intermodality registration techniques.

The task of converting the representation of the various tissue types present in one scan into the representation given by another scanning technology or modality involves the use of feature segmentation followed by histogram matching [Hua98]. By sorting the intensities of each image into clusters corresponding to the intensity ranges of characteristic features, we can approximately segment the scans into the various compounds or tissue types. These individual segmentation clusters can then be mapped onto relevant bins from the second scan. A histogram matching of these segmented images, which includes a permutation and adaptation of the bin widths in the histogram, generates a transfer function that defines a mapping between the original scan and the alternate scan.

3 Segmentation Using Histogram Binning

In order to correlate the individual representations of the various anatomical structures in each scan, we use intensity histograms. Figure 10(a) shows the intensity profile of a CT scan (data specification: 512 x 512 grayscale image). Figure 10(b) shows the intensity profile of an MRI scan (data specification: T1-weighted, 256 x 256 grayscale image) of a human head. All images and histograms were stored in an 8-bit format (value range [0...255]). The adjustment of the resolution (scaling factor: 2) is accomplished as a by-product of the wavelet reconstruction described in Section 6, which is not carried out to the highest detail level. Other scaling factors can be obtained using bi-linear interpolation of the next higher resolution step. The non-standard Haar wavelet decomposition [Mey03] provides resolution steps in multiples of two in each dimension.

Histogram binning is an approximation technique based on known expressions of particular tissue types in various modalities. Depending on the number of expected distinct features (i.e. tissue types), we create bins of intensity clusters. For example, we have one bin to store all the high intensity pixels in a CT image which are characteristic of skull bone (high X-ray absorption). This would map onto the bin containing the low intensities from the skull in the MRI scan (low hydrogen proton concentration in bone). Similarly, the gray matter, white matter and cerebral spinal fluid (CSF) can be stored in one bin corresponding to the uniform gray region of brain matter in the CT.

The bins thus correspond to a partitioning of the image into characteristic tissue expressions. Initial bins were obtained and refined using user interaction, and a table of mapping values was generated (Table 1).

Having thus created the desired bins, we then generate a transfer function to carry out the mapping between each segment (bin) of the MRI and the CT scan based on Table 1, using the principles of histogram matching [Hua98]. This process of selective histogram matching ensures a good approximation.

Figure 11 shows a graphical representation of the bins and a typical permutation. The arrows indicate the mapping shown in Table 1. It becomes obvious that bone, for instance, is represented differently in different modalities, and that the width of the bins varies. Therefore, the mapping is not linear. A piece-wise linear transfer function is designed to perform the permutation and scaling of the histogram bins.

After applying the transfer function, the MRI image now contains intensity values that correspond to those found in the CT scan. This means that the various tissue types are now represented by a similar intensity value. Figure 12(a) shows the original superimposed histograms of the CT and the MRI scan before application of the transfer function. The image shows that there is minimal overlap, which makes image registration of features that are expressed differently difficult. Figure 12(b) shows the superimposed histograms of the CT and the transformed MRI scan. Obviously, there is a much better match in the overall shape of the histogram, indicating that the two images, even though generated by two different modalities, will yield a much

Table 1. Mapping of MRI intensities to corresponding CT intensities for each feature space.

CT intensity	MR intensity	Feature
0 – 9	0 – 1	Background
10 – 60	175 – 255	Adipose
61 – 140	56 – 119	Soft tissue
141 – 255	2 – 55	Bone

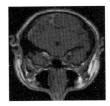 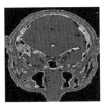

Fig. 2. (a) Original MRI image. (b) Transformed MRI image. (c) Original CT image. The transformed MRI (center) matches the geometry of the original MRI (left) and the intensity profile of the CT scan (right).

better registration of individual features. Figures 2(a) and (b) show the corresponding images, i.e., the original MRI image, and the new MRI image after application of the transfer function. Figure 2(b) is now similar to the original CT image, which is shown in Figure 2(c) for comparison.

The quality of the transformation can be evaluated by subtracting either the histograms or the images. Using a low-pass filter on the images or a quantization on the histogram is necessary to obtain a reasonable measure of the quality, because due to the difference in the image acquisition process, the individual pixels or bin distributions will still be very different, but the average distribution is similar.

Our registration algorithm takes as input a file that contains the segmentation information, i.e., the number of clusters or bins followed by the intensity range for each cluster. The algorithm then optimizes a global similarity measure based on correlation, as described in the following section.

4 Registration

The registration was done using an intensity-based similarity measure and an optimization algorithm for various rotation angles and translation vectors. Optimization of a normalized cross-correlation (NCC) function using Powell's multi-dimensional line maximization [Jac77] formed the basis of the similarity measure.

$$R = \frac{\sum_{k=0}^{n}[(r(k) - \mu(r)) - (f(k) - \mu(f))]}{\sqrt{\sum_{k=0}^{n}(r(k) - \mu(r))^2 - \sum_{k=0}^{n}((f(k) - \mu(f))^2}} \quad (2)$$

Equation 2 shows the NCC function. Here r and f represent the reference and the floating image, respectively. $\mu(r)$ and $\mu(f)$ represent the mean intensity values of the reference and the floating image. In our case, the CT scan is the reference image and the transformed MRI is the floating image. R represents the metric used to determine the similarity between the CT and the transformed MR image. $R = 1$ implies maximum correlation, $R = 0$ implies no correlation, and $R = -1$ implies an opposing relation between the intensities from the two modalities. Powell's multi-dimensional line maximization algorithm, which employs Brent's one-dimensional optimization algorithm, was used to compute different variants and to determine the optimum value for R in the X-Y plane.

A "sum of squared differences" technique may also be used for the registration process, since specific optimization algorithms work very efficiently for the same. However, the choice of NCC as a similarity measure was dictated by the fact that tissue-type-based clustering is not an exact science due to large variations in individual tissue expressions across patients. In addition, noise, if not removed prior to the transformation, is also transformed by the histogram matching process, has a strong effect on the "sum of squared differences" method and interferes with the registration, especially when the noise is not uniformly distributed or Gaussian. The NCC function together with a low-pass filter ensure the robustness of the algorithm.

5 Results

Figure 13(a) shows the superimposed images of the CT and the original MRI scans in their initial, unregistered position. Note that the bone (bright red or bright cyan) is not aligned (white, composition of red and cyan). Figure 13(b) shows the registered set, again with no good match (few white pixels). The hollow space (transparent) in the MRI is completely filled by the skull (cyan) from CT. This means that the bone is hardly visible in the given MRI scan.

Figure 13(c) shows the same results for a transformed MRI image. Note that the bone (white, composition of cyan and red) is much better visible now. Especially after registration (Figure 13(d)), the bone appears mostly white. This means that it is visible in both images (CT and transformed MRI). This property is most obvious for tissues that appear bright in the current representation (bone), but it also applies to other tissue types, as is shown in the following images. The assignment of intensity ranges (histogram bin mapping) is arbitrary and typically chosen to match one of the two modalities, in this case the CT scan.

Figures 3 and 4 were created using thresholding of the pixels that indicate the highest structural similarity (bone only). As stated before, not only the pixels with the highest values (typically bone) can be used for a correspondence analysis. Other tissue types can be used as well, as shown in Figures 5 and 6 (all tissue types).

The last image in each series (Figures 4(b) and 6(b)) shows the largest number of pixels usable for correspondence analyses. An analysis of the pixel count indicates that the transformation using transfer functions and histogram permutation increases the number of pixels that can be used for correspondence analyses, therefore increasing the accuracy and robustness of the method.

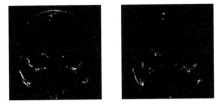

Fig. 3. CT and original MRI. (a) unregistered, (b) registered. (Bone only.)

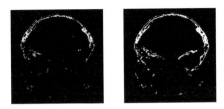

Fig. 4. CT and transformed MRI. (a) unregistered, (b) registered. (Bone only.) White regions indicate matched bone regions expressed similarly in the CT and the transformed MRI scan.

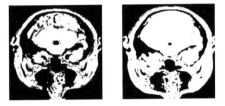

Fig. 5. CT and original MRI. (a) unregistered, (b) registered. (All tissue types.)

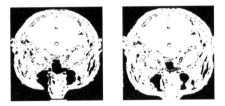

Fig. 6. CT and transformed MRI. (a) unregistered, (b) registered. (All tissue types.) White regions indicate matched regions of all tissue types expressed similarly in the CT and the transformed MRI scan. The right image shows an almost perfect match (all tissue types).

The algorithm can be very efficiently implemented using look-up tables so that it was possible to run it on a laptop computer. In comparison to other intensity-based techniques for intermodality registrations that use entropy or MI calculations, NCC computations are less expensive. The preprocessing step is also inexpensive and is independent of the alignment routine.

146   J. Meyer

Performance data of the various algorithms is difficult to compare due to large variations in the implementations and applications. Therefore, absolute timing measurements were taken, which may serve as a guideline for similar implementations.

The computation of the transformation of the original MRI to its CT-look-alike took approximately 10 milliseconds, and the alignment algorithm took 3.25 seconds for a single slice (approx. 6 minutes for a typical data set of 113 slices) on an IBM R40 Thinkpad with 256MB RAM and 1.4 GHz PentiumM processor. The graphics card is an ATI Mobility Radeon with 16MB memory.

6 Multimodal Image Acquisition and Multi-resolution Visualization

This section will give a brief overview of the image acquisition, storage, processing and rendering pipeline for multimodal images. The registration algorithm described in the previous sections will be used in this context.

When images are acquired using two different medical scanners, registration of the three-dimensional volume data sets is always required. Even if the scanners are perfectly calibrated, if the geometry of the scanned area is similar and the scanned specimen is the same, there is still a margin of error and deviation due to positioning of the specimen in the scanner, shifting of tissues (e.g., movements of a brain inside a skull), and geometric distortions of the scanner. Some of these deviations and deformations require linear, others non-linear transformations. Also, the resolution must be adjusted using an interpolation or multi-resolution technique (see Section 3).

The image registration described earlier in this book chapter applies to the two-dimensional case, which requires at least the body axes to be aligned in the scanners. This constraint can be dropped if the algorithm is translated into the three-dimensional domain, which is pretty straightforward. In this case, the normalized cross-correlation function (equation 2) must be computed for three-dimensional images r and f.

Since three-dimensional data sets, especially when obtained from two independent storage devices, can become very large, efficient data compression is required for data transmission over limited-bandwidth networks and archival storage. Since there is no correlation in the absolute data values obtained from two different modalities, the compression algorithm cannot take advantage of similarities in the features. Data compression usually becomes more efficient when smaller values are involved, increasing the likelyhood of finding similar or identical values. Quantization helps to increase the number of identical values and therefore improves the compression ratio by sacrificing some of the image quality.

Two-dimensional wavelet decomposition (Figure 7) works by separating an image into low-frequency and high-frequency coefficients [Mey03]. Three-dimensional wavelet compression algorithms have been shown to provide

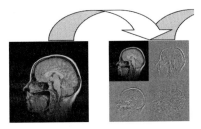

Fig. 7. Wavelet decomposition.

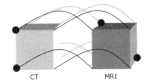

Fig. 8. Four-dimensional, non-standard Haar Wavelet decomposition. Detail coefficients (differences between adjacent voxel values) in the fourth dimension (modality number) are typically smaller for CT–transformed MRI pairs (right) than for CT–original MRI pairs (left) and therefore yield better compression ratios.

an efficient way to compress volume data, as they take advantage of the similarity of neighboring voxels in three spatial directions. Wavelet decomposition algorithms become more efficient in higher dimensional data sets if the data is correlated in each dimension. By extending the three-dimensional wavelet compression into the fourth dimension, where the fourth dimension represents the various modalities (transformed into a common representation of a reference modality), the compression algorithm that follows the wavelet decomposition can take advantage of the improved correlation between the two modalities after histogram transformation (Figure 8).

In this figure, the dots at the voxel corners indicate high values, while the other corners represent low values. Please note that the voxel corners in the CT–transformed MRI pair on the right are more similar to each other than the voxel corners in the CT–original MRI pair on the left. This is due to the histogram transformation and turns out to be an advantage when a non-standard Haar wavelet decomposition is applied. Since the two voxels are similar, the average of the two voxels (low-frequency component) is similar to both voxel values, while the difference between the two voxels (high-frequency component) is small. Detail coefficients (high-frequency components) can be quantized and stored with fewer bits, and small coefficients can be ignored.

The efficiency increases with the dimensionality of the data set, because the average (low-frequency component) is computed between two elements in 1-D, between $2*2$ (four) elements in 2-D, $2*2*2$ (eight) elements in 3-D, and so forth. If the modality number is considered the fourth dimension, an average of sixteen elements can be computed and represented as a single number.

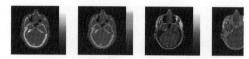

Fig. 9. Bone. (a) CT scan (green), (b) MRI scan (red), (c) superimposed (yellow).

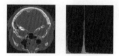

Fig. 10. (a) CT scan and (b) MRI scan with intensity histograms.

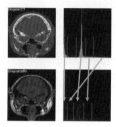

Fig. 11. Mapping of CT histogram (green) onto MRI histogram (red). Bins (purple).

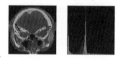

Fig. 12. CT (green) with MRI (a) original (b) transformed. Histogram in (b) shows both original (red) and transformed (yellow) MRI. Yellow is a better match.

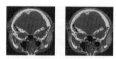

Fig. 13. (a), (c) unregistered; (b), (d) registered CT (cyan) and MRI (red). Left pair: Original MRI. Right pair: Transformed MRI. White: Matched bone.

The detail coefficients (high-frequency components) are typically small if the modalities are represented similarly and therefore compress well in a subsequent compression algorithm (run-length encoding with quantization, entropy encoding, Lempel-Ziv-Welch, etc.)

A wavelet decompression and reconstruction algorithm (the inverse of the wavelet decomposition algorithm) can be used to obtain volumetric images at multiple levels of detail (Figure 7). The advantage of such a multi-resolution reconstruction is that a low resolution image can be displayed instantly, while increasing levels of detail can be reconstructed sequentially as they become available (either as streaming data from a network or as high-resolution data from an archive). The details of the wavelet decomposition and reconstruction algorithm as well as the three-dimensional, texture-based rendering algorithm are described in [Mey03].

7 Summary

This chapter explained how histogram information of a grayscale image can be adapted based on another image using previous knowledge about the bins of the reference image histogram. By transforming one image into the value range of the reference image using a non-linear mapping, *intramodal* registration techniques, as opposed to computationally more expensive *intermodal* techniques, can be used to align the images. Optimization of a normalized cross-correlation function (NCC) is an efficient way to implement an intramodal image registration algorithm. Histogram transformation also benefits file storage and data compression without changing the information content in the image, allowing for simultaneous storage and rendering of multiple modalities.

8 Acknowledgements

This project was funded in part by the National Institute of Mental Health (NIMH) through a subcontract with the Center for Neuroscience at the University of California, Davis (award no. 5 P20 MH60975), and by the National Partnership for Advanced Computational Infrastructure (NPACI), Interaction Environments (IE) Thrust (award no. 10195430 00120410).

The author would like to thank Ruparani Chittineni for contributing some of the code, images and descriptions.

References

[Ash95] Ashley J., Barber R., Flickner M., Lee D., Niblack W., and Petkovic D.: Automatic and semiautomatic methods for image annotation and retrieval in qbic. In: Proc. SPIE Storage and Retrieval for Image and Video Databases III, 24–35 (1995)

[Bro92] Brown L. G.: A survey of image registration techniques. ACM Computing Surveys 24, 4, 325–376 (1992)

[But01] Butz T. and Thiran J.: Affine registration with feature space mututal information. In: Lecture Notes in Computer Science 2208: MICCAI 2001, Springer-Verlag Berlin Heidelberg, 549–556 (2001)

[Cri00]   Cristiani N. and Shaw-Taylor J.: Suport Vector Machines and other kernel-based learning methods. Cambridge U. Press (2000)

[Els94]   Van den Elsen P.A., Pol D. E., Sumanaweera S.T., Hemler P. F., Napel S., Adler J. R.: Grey value correlation techniques used for automatic matching of CT and MR brain and spine images. Proc. Visualization in Biomedical Computing, SPIE 2359, 227–237 (1994)

[Els95]   Van den Elsen P. A., Maintz J. B. A., Pol D. E., Viergever M. A.: Automatic registration of CT and MR brain images using correlation of geometrical features. IEEE Transactions on Medical Imaging 14, **2**, 384–396 (1995)

[Gon02]   Gonzalez R. C., Woods R. E.: Digital Image Processing. Prentice Hall (2002)

[Hil01]   Hill D., Batchelor P., Holden M., and Hawkes D.: Medical image registration. Phys. Med. Biol., **26**, R1–R45 (2001)

[Hua98]   Huang J., Kumar S., Mitra M., and Zhu W.: Spatial color indexing and applications. In: Proc. of IEEE International Conf. Computer Vision ICCV '98, Bombay, India, 602–608 (1998)

[Jac77]   Jacobs D. A. H.: The state of the art in numerical analysis. Academic Press, London (1977)

[Jen02]   Jenkinson M., Bannister P., Brady M., and Smith S.: Improved methods for the registration and motion correction of brain images. Technical report, Oxford University (2002)

[Lef01]   Lefébure M. and Cohen L.: Image registration, optical flow and local rigidity. J. Mathematical Imaging and Vision, (14) 2, 131–147 (2001)

[LoC03]   Lo C. H., Guo Y., Lu C. C.: A binarization approach to CT-MR registration using Normalized Mutual Information. Proc. IASTED Signal and Image Processing, 399 (2003)

[Mae97]   Maes F., Collignon A., Vandermeulen D., Marchal G., Suetens P.: Multimodality image registration by maximization of mutual information. IEEE Transactions on Medical Imaging 16, **2**, 187–198 (1997)

[Mai98]   Maintz J. B. and Viergever M.: A survey of medical image registration. Medical Image Analysis, (2) 1, 1–36 (1998)

[Mey97]   Meyer C. R., Boes J. L., Kim B., Bland P. H., Zasadny K. R., Kison P. V., Koral K. F., Frey K. A., and Wahl R. L.: Demonstration of accuracy and clinical versatility of mutual information for automatic multimodality image fusion using affine and thin-plate spline warped geometric deformations. Medical Image Analysis, (1) 2, 195–206 (1997)

[Mey03]   Meyer J., Borg R., Takanashi I., Lum E. B., and Hamann B.: Segmentation and Texture-based Hierarchical Rendering Techniques for Large-scale Real-color Biomedical Image Data. In: Post F. H., Nielson G. H., Bonneau G.-P., eds., Data Visualization - The State of the Art, Kluwer Academic Publishers, Boston, 169–182 (2003)

[Pen98]   Penney C., Weese J., Little J., Hill, D., and Hawkes, D.: A comparison of similarity measures for used in 2-D-3-D medical image registration. IEEE Trans. on Medical Imaging, (17) 4, 586–595 (1998)

[Plu03]   Pluim J. P. W., Maintz J. B. A., Viergever M. A.: Mutual Information based registration of medical images: a survey. IEEE Transactions on Medical Imaging 22, **8**, 896–1004 (2003)

[Sto98] Stoica R., Zerubia J., and Francos J. M.: The two-dimensional wold decomposition for segmentation and indexing in image libraries. In: Proc. IEEE Int. Conf. Acoust., Speech, and Sig. Proc., Seattle (1998)

[Stu96] Studholme C., Hill D. L. G., Hawkes D. J.: Automated 3-D registration of MR and CT images of the head. Medical Image Analysis 1, **2**, 163–175 (1996)

[Vio95] Viola P. and Wells III W. M.: Alignment by maximization of mutual information. In: Proceedings of IEEE International Conference on Computer Vision, Los Alamitos, CA, 16–23 (1995)

[Wel96] Wells W. M., Viola P., Atsumi H., Nakajima S., Kikinis R.: Multi-modal volume registration by maximization of mutual information. Medical Image Analysis 1, **1**, 35–51 (1996)

[Zhu02] Zhu Y. M.: Volume image registration by cross-entropy optimization. IEEE Transactions on Medical Imaging 21, 174–180 (2002)

A User-friendly Tool for Semi-automated Segmentation and Surface Extraction from Color Volume Data Using Geometric Feature-space Operations

Tetyana Ivanovska and Lars Linsen

Computational Science and Computer Science
School of Engineering and Science
Jacobs University
Bremen, Germany
{t.ivanovska,l.linsen}@jacobs-university.de

Summary. Segmentation and surface extraction from 3D imaging data is an important task in medical applications. When dealing with scalar data such as CT or MRI scans, a simple thresholding in form of isosurface extraction is an often a good choice. Isosurface extraction is a standard tool for visualizing scalar volume data. Its generalization to color data such as cryosections, however, is not straightforward. In particular, the user interaction in form of selection of the isovalue needs to be replaced by the selection of a three-dimensional region in feature space. We present a user-friendly tool for segmentation and surface extraction from color volume data. Our approach consists of several automated steps and an intuitive mechanism for user-guided feature selection. Instead of overburden the user with complicated operations in feature space, we perform an automated clustering of the occurring colors and suggest segmentations to the users. The suggestions are presented in a color table, from which the user can select the desired cluster. Simple and intuitive refinement methods are provided, in case the automated clustering algorithms did not immediately generate the desired solution exactly. Finally, a marching technique is presented to extract the boundary surface of the desired cluster in object space.

1 Introduction

Segmentation and surface extraction are an important task for many clinical observations. For example, it is important to measure the size of certain organs or tumors, which can only be done when having an explicit surface representation of the organs/tumors boundary at hand. Since the most commonly used in-vivo scanning techniques such as computed tomography (CT) or magnetic resonance imaging (MRI) produce 3D images representing a scalar field, most segmentation and surface extraction techniques have been developed for such applications.

For medical purposes it is also of relevance to relate the clinical measurements to higher-resolution ex-vivo measurements such as cryosections. Cryosections typically lead to 3D images of color RGB data. Hence, one would like to have a tool that allows similar segmentation and surface extraction mechanisms for color data. The selection mechanisms in 3D feature space, i. e. in color space, are obviously much more complex than the analogous ones in 1D feature space are. Still, all selection mechanisms have to be intuitive not only to visualization specialists but in particular to medical people. Complex operations in feature space should not be visible to the user, but should be replaced by some intuitive selection steps. Moreover, user interaction should be reduced to a minimum.

Segmentation of color data is an intensely researched field in the area of image analysis. The segmentation algorithms mainly concentrate on clustering similar colors to some kind of "average" color. These automated clustering procedures easily generalize to color volume data. However, when dealing with volume data, occlusion prohibits the visualization of the entire segmentation output. Further processing steps are required in addition to the clustering, which can chiefly be grouped into a selection mechanism and a subsequent extraction and rendering step. The selection step comprises the methods to determine regions of interests or "features". A final step extracts the features and displays them in a three-dimensional setting.

While the extraction and rendering step can typically be performed fully automated, there needs to be a certain portion of user interaction or, better, user guidance during the selection step. The user needs to communicate to the system, what parts of the data set he/she would like to visually explore. Ideally, the user interaction is limited to a minimal amount of intuitive selection choices.

The outcome of the entire feature extraction processing pipeline could be a volume-rendered image or a surface representation. In many applications and especially in medical ones when dealing with cryosection data, one favors surface representations, as one can precisely reveal the geometric structure of the underlying data field distribution. In particular, surface representations allow for quantitative analyses of the segmented objects.

We present a user-friendly tool for segmentation and surface extraction of color volume data. Our approach consists of several automated steps and an intuitive mechanism for user-guided feature selection. For clustering purposes we make use of and compare three of the most commonly used clusterization methods. Their output is presented to the user in a transparent fashion using a sorted color table. Simple and thus intuitive selection mechanisms are provided to select the desired region of interest based on the clusters.

The selection by the user is processed in feature space using geometric operations. The final surface extraction step iterates through all the cells in object space, while computing points on the surface using linear interpolation in feature space. All these processing steps are executed subsequent to the user selection are fully automated, as a user should not be faced with any incomprehensible user interaction in feature space. Medical people would have a hard time to understand such mechanisms. In particular, all geometric feature-space operations are hidden from the user.

2 Related Work

Displaying surfaces is an effective technique for visualizing 3D data fields. To extract surfaces from scalar volume data, intensive research effort has been undertaken in past decades since the pioneer work presenting the Marching Cubes algorithm for isosurface extraction [LC87]. Several improvements have been made leading to robust, fast, and flexible algorithms. To date, many approaches for surface extraction from scalar data exist. Unfortunately, they do not scale to color data in a straightforward manner.

In recent years a lot of effort has been taken in the segmentation of color images. They are based on clustering similar colors due to some metric in a certain color space and various grouping strategies. A survey of all these methods is beyond the scope of this paper, but three of the most commonly used techniques are discussed in this paper. The selection of these three clustering approaches was based on their efficacy and efficiency. Efficiency plays an important role when dealing with large volume data with non-synthetic colors. Thus, our decision was to use simple and fast approaches with a broad acceptance in the community with respect to the quality of their results.

The generalization of such segmentation approaches that operate on a 2D domain to 3D applications does not suffice to fulfill our goals. In 3D, one is also interested in segmenting the color images but, in addition, one needs to extract the relevant geometry in form of surface representations in a subsequent step.

Prominent approaches in the area of segmentation and surface rendering from color volume data were presented by Schiemann [STH97] and Pommert [PHP+00]. These approaches are most closely related to our work. Schiemann classifies objects based on ellipsoidal regions in RGB space. The author used a semi-automated method for segmentation. The procedure is based on thresholding followed by binary mathematical morphology and connected component labeling. The subsequent rendering step, however, is a mere volume rendering approach. No explicit surface representation is generated.

There are also several approaches for direct volume rendering of photographic volumes [KKH02, EMRY02], where the authors present sophisticated transfer function generations, but extracting surfaces from the color volume data has hardly been addressed. A hardware-assisted volume renderer for the visualization of cryosection data has been presented by Takanashi [TLM+02].

3 Semi-automated Segmentation and Surface Extraction of Color Volume Data: The Main Approach

Our processing pipeline for segmentation and surface extraction from color volume data comprises several automated steps and a user-guided feature selection mechanism. The entire pipeline reaching from reading the RGB color data to the rendering of the extracted surfaces is shown in Figure 1.

Since standard distance metrics applied to RGB color space are biased by the selection of the three coordinates of the color space, we start with converting the data into $L^*a^*b^*$ color space, where Euclidean distance is correlated with human perception. Afterwards, we apply the clustering algorithms, thus, operating in $L^*a^*b^*$ color space. For clusterization we provide the choice of three standard clustering algorithms, namely the median cut, the k-means and the c-means algorithm. These approaches are fast, yet produce decent results. If desired, they could be replaced by any other, possibly more sophisticated and more time-consuming, clustering technique. The output of the clustering algorithm is a set of clusters in $L^*a^*b^*$ color space each associated with a representative color.

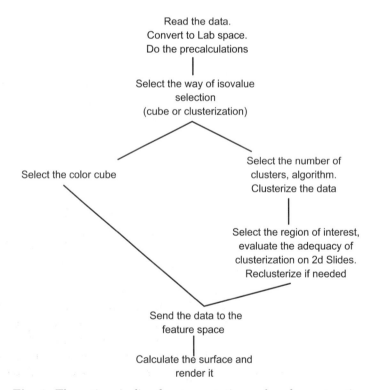

Fig. 1. The entire pipeline for segmentation and surface extraction.

The generated clusters are presented to the user by displaying their representative in a color list that is sorted by the frequency of occurrences of the initial colors belonging to the respective cluster. Some high-level selection mechanisms are provided to the user to pick the appropriate set of clusters. With the selection of clusters in the color table the respective 2D contours in object space are visualized overlaid with a $2D$ color slice. The user can go through all the original 2D color slices. If the user is not satisfied with the results of the clusterization, the result can be refined by simply selecting the representative colors of the desired clusters and applying a reclusterization of this reduced set of colors into a larger amount of clusters. This process can be repeated until the user gets the desired result. A few selections and iterations typically suffice. The output of the selection procedure is the collection of clusters, which correspond to a point cloud in feature space of all colors that belong to the clusters.

Given the point cloud in feature space, we would like to define a continuous region in feature space that covers all the selected points but excludes all not selected colors. Although this region does not have to be convex, in general, we obtain an approximate representation by computing the convex hull of the point cloud. Dealing with a convex region significantly speeds up computations.

This segmentation is fed into the surface-extraction method to compute a triangular mesh. A marching method is used for surface extraction, where the marching cubes table can be applied to determine the topology of the resulting surface. The points on the surface are computed by linear interpolation in feature space, which involves the computations of intersections of the convex hull with feature-space edges. Special care has to be taken in regions, where the convexity assumption violates the inside/outside-property.

The final step of the pipeline is the surface rendering step, which makes use of the actual color values given at the points on the surface.

The details on the individual processing steps are given in the subsequent sections.

4 Conversion to $L^*a^*b^*$ Color Space

We decided to use the $L^*a^*b^*$ color space for all feature space operations, as Euclidean distance in $L^*a^*b^*$ color space closely corresponds to perceptual dissimilarity [Pas03]. Roughly, this means that if two colors look to be different by about the same amount, then the numbers that describe those colors should differ in the same way. $CIE\ L^*a^*b^*$ is the most complete color model used conventionally to describe all the colors visible to the human eye. The three parameters in the model represent the luminance L of the color, where $L = 0$ yields black and $L = 100$ indicates white, its position a between red and green, where negative values indicate green while positive values indicate red, and its position b between yellow and blue, where negative values indicate blue and positive values indicate yellow.

The most commonly used color space for image acquisition and storage, on the other hand, is the RGB color space. Assuming that the original data is given in RGB color space, we need to convert the colors to a representation in the $L^*a^*b^*$ color space. We first transform the RGB data to the $CIE\ XYZ$ color space and, afterwards, convert the XYZ values to $L^*a^*b^*$ colors. Note that the matrix of transformation from RGB data to XYZ depends on the chosen RGB standard. We consider the $R709\ RGB$ standard. Hence, the three channels of the $L^*a^*b^*$ colors are computed by

$$L^* = \begin{cases} 116 \cdot \left(\frac{Y}{Y_n}\right)^{\frac{1}{3}} - 16, & \text{if } \frac{Y}{Y_n} > 0.008856 \\ 903.3 \cdot \left(\frac{Y}{Y_n}\right), & \text{otherwise} \end{cases},$$

$$a^* = 500 \cdot \left(\left(\frac{X}{X_n}\right)^{\frac{1}{3}} - \left(\frac{Y}{Y_n}\right)^{\frac{1}{3}}\right), \text{ and}$$

$$b^* = 200 \cdot \left(\left(\frac{Y}{Y_n}\right)^{\frac{1}{3}} - \left(\frac{Z}{Z_n}\right)^{\frac{1}{3}}\right),$$

where X_n, Y_n, and Z_n are the values of X, Y, and Z, respectively, for a specified reference of white, i. e. illuminant, color, and X, Y, and Z are computed by

$$\begin{bmatrix} X \\ Y \\ Z \end{bmatrix} = \begin{pmatrix} 0.412453 & 0.357580 & 0.180423 \\ 0.212671 & 0.715160 & 0.072169 \\ 0.019334 & 0.119193 & 0.950227 \end{pmatrix} \cdot \begin{bmatrix} R \\ G \\ B \end{bmatrix}.$$

The main characteristics of the $L^*a^*b^*$ color space is very important for us, because on the clusterization step of the pipeline the user will be proposed to choose the regions of interest via the colors in the dataset. The chosen colors should be close in the color space to form a cluster. We rely on the perceptual choice of the user. Although conversion from RGB to $L^*a^*b^*$ is a time-consuming step, it can be done as a precomputation. Hence all subsequent feature-space processing steps are executed in the $L^*a^*b^*$ space.

5 Clusterization

Color clustering or color image clustering (quantization) is a process that reduces the number of distinct colors occurring in an image to a chosen smaller number of distinct colors. Usually, the reduction is performed with the intention that the new image should be as visually similar as possible to the original image [Sch97]. Since the 1970s, many algorithms have been introduced. In this work we use three popular unsupervised learning algorithms that are known as being simple, fast, and effective in solving the clustering problem [DHS00]. The used approaches are known as the median cut [Hec82], k-means [Mac67], and c-Means [Bez81] approaches. We tested them for our purposes, as they represented a good trade-off between efficacy and efficiency. Certainly, any other clustering algorithm could be used instead.

The median-cut method partitions the color space with the goal to balance the number of pixels in each color cell. The idea behind the median cut algorithm is to generate a synthesized color look-up table with each color covering an equal number of pixels of the original image. The algorithm partitions the color space iteratively into subspaces of decreasing size. It starts off with an axes-aligned bounding box in feature space that encloses all the different color values present in the original image. The box is given by the minimum and maximum color component in each of the three coordinate directions (color channels). For splitting the box one determines the dimension in which the box will be (further) subdivided. The splitting is executed by sorting the points by increasing values in the dimension where the current box has its largest edge and by partitioning the box into two sub-boxes at the position of the median. Approximately equal numbers of points are generated on each side of the cutting plane. Splitting is applied iteratively and continued until k boxes are generated. The number k may be chosen to be the maximum number of color entries in the color map used for the output. The color assigned to each of the k boxes is calculated by averaging the colors of each box. The median-cut method performs well for pixels/voxels, whose colors lie in a high-density region of the color space, where repeated divisions result in cells of small size and, hence, small color errors. However, colors that fall in low-density regions of the color space are within large cells, where occurrences of large color errors may be observed.

The main idea of the k-means algorithm is to define k centroids, one for each cluster. These centroids should be placed as far from each other as possible. Typically, this is approximated by using randomly distributed centroids. Each point from the initial data set is associated with the nearest centroid. When all the points have been assigned a centroid, the k centroids are recalculated as the average centers of each cluster. Thereafter, the assignment of colors to centroids is updated using the original color points but new centroids. Again, one assigns to each color the nearest centroid. This procedure is iterated until the assignments stabilize. The main advantages of this algorithm are its simplicity and speed which allows it to run on large datasets. Its drawback is its sensitivity to the initial choice of the centroids, i.e. that it does not yield the same result with each run, since the resulting clusters depend on the initial random assignments. The k-means algorithm maximizes inter-cluster (or minimizes intra-cluster) variance, but does not ensure that the computed solution actually represents a global optimum.

The idea of the fuzzy c-means method is similar to the k-means approach, but it allows one data point to belong to two or more clusters. Fuzzy partitioning is carried out through an iterative optimization of the data points membership in the clusters and the corresponding update of the cluster centers. The iteration terminates when no further changes with respect to the membership results occur with some given tolerance. The algorithm minimizes intra-cluster variance as well, but has the same problems as k-means: The minimum is a local minimum, and the results depend on the initial choice of weights.

As the results of the k-means and c-means approaches happen to be rather sensitive to the choice of the initial centroids, we thought of a good seeding strategy for our task: We take a random data point as the first initial centroid and then iteratively find other centroids among the rest of the points by taking points that are located as far as possible from the already chosen ones.

The user can select the quantity of clusters in all three methods. Unfortunately, there is no general theoretical solution to find the optimal number of clusters for any given data set. Different approaches for recommending a preferable number of clusters exist, but such general heuristic approaches always have to be adjusted to the given application. For our purposes, a too large number of clusters can lead to too many selection steps during the user interaction to define the region of interest. On the other hand, a too small number of clusters can lead to clusters that unite colors that the user would like to see separated. For both cases we provide intuitive solutions. A single slider can be used to adaptively combine closest clusters in case of too many clusters, and an adaptive cluster refinement can be used in case of too few clusters.

6 Intuitive User Interaction for Cluster Selection, Combination, and Refinement

To cluster the colors a user is supposed to choose the appropriate clustering algorithm and to choose the number of clusters. Figure 2 shows the user interface. The resulting clusters are shown in the sorted cluster color table. Each cluster is represented by its average color, see Figure 3. The clusters are sorted in descending order according to the number of voxels that are covered by the colors that belong to the respective cluster.

Ideally, the generated clusters represent the desired regions of interest and the user can just pick the desired feature (such as an organ, part of an organ, or a tumor) by a single mouse click on the cluster's representative. Obviously, the clustering algorithms do not always generate perfect solutions. The desired feature may be represented by several clusters or it may be contained in a cluster together with undesired features. We provide mechanisms to quickly and intuitively resolve such situations.

If a feature is represented by by several clusters, the user may just click on the clusters' representatives in the cluster list and they are combined to larger clusters. If many clusters are to be united, a semi-automated method can be used by clicking at one of them and moving a slider until all the desired

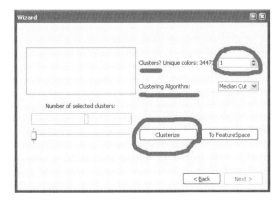

Fig. 2. For the clusterization step, the user can choose between three clustering algorithms and can set the number of generated clusters.

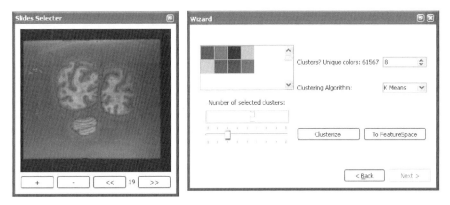

Fig. 3. The clusterization result is a sorted cluster list, which is drawn using cluster representatives (right). Clusters can be selected by clicking at the representatives and increased by dragging the slider. Any selection can be adaptively refined by reclusterization. The segmentation result is shown in a preview in object space by overlaying a 2D contour (pink) with any 2D slice ((left)

clusters are included. When moving the cluster, the next closest clusters are subsequently included with respect to the Euclidean metric in $L^*a^*b^*$ color space. As mentioned above, this distance is related to the color difference that a human eye can perceive.

If one or several clusters contain both desired and undesired features, user can select these clusters as before and apply a refinement of the clustering. For the refinement, only the selected clusters are taken into account. The user can, again, choose the clusterization method and the desired number of new clusters.

These interactions can be iterated. Typically, a few iteration steps suffice to retrieve the desired features. This iterative procedure with few clusters at a time is not only intuitive for the user due to a low complexity of the displayed items, it also reduces the computation times during clusterization.

While selecting, combining, and refining clusters, the user is provided with a "preview" of the current selection in object space. The preview contains a 2D contour rendered on top of a 2D slice. Any of the original 2D color slices can be chosen. In Figure 3, the pink contour encloses the white matter of the cryosection slice through a human brain. When selecting other or additional clusters, the contour is immediately updated. A slice-based preview is desirable, as the user can simultaneously observe the contour, the interior, and the exterior of the current selection, which allows for a good judgment of the accuracy of the feature selection.

Once the user is satisfied with his/her selection, the result is handed to the automatic feature-space segmentation algorithm.

7 Feature-space Segmentation

The feature space is represented by an $L^*a^*b^*$ color space, where the colors of the initial dataset are shown as points using a scatter plot, see Figure 4. The user selection left us with a combination of certain clusters that define the region of interest. These clusters contain a set of colors, which in feature space represent a point cloud that is a subset of the points shown in Figure 5. We need to build a hull around the point cloud in feature space. The hull must include all the points selected by the user and exclude all the other points. In general, the hull's shape is not convex. As the subsequent surface extraction algorithm can be designed significantly more efficiently when assuming a convex shape, we approximate it by constructing the convex hull over the selected points.

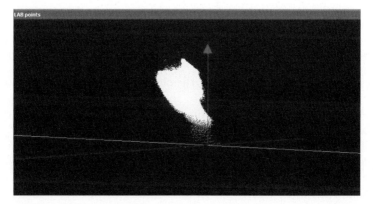

Fig. 4. The feature space with the colors of the initial dataset presented as a scatter plot.

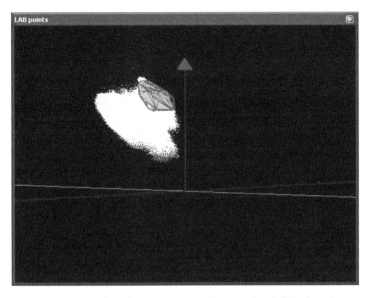

Fig. 5. The convex hull (pink) of the selected point cloud (blue) in feature space.

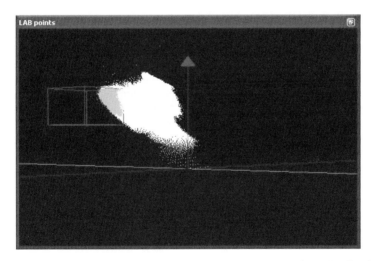

Fig. 6. Feature selection is feature space using an axes-aligned cuboid.

For the computation of the convex hull we chose the Quick Hull algorithm [BDH96]. The process of a hull computation is automated and does not require user interaction. Figure 5 shows the result for a selected point cloud, where the point cloud is highlighted in blue and the convex hull is shown in pink.

We compare our results with the results one would obtain using an axes-aligned cuboid for selection of colors in the feature space. The cuboid selection as shown in Figure 6 does not require any clusterization and allows a fast surface extraction, but obviously it is not capable of extracting features as precisely as we can do with our method.

8 Object-space Segmentation and Surface Extraction

Having segmented the feature space, we can segment the object space and extract the boundary surface of the desired feature. Since we are dealing with a stack of registered 2D color slices, we operate on a structured rectilinear grid. Thus, we can apply a marching technique that steps through all the cells and extracts the surface components within each cell independently.

For each cell we observe the colors at the eight corners. Since we know exactly, which colors belong to the desired feature, we do not have to perform any inside/outside-test but can just look up the inside/outside-property from the feature-space collection. Based on the inside/outside-property we can determine the topology of the surface within the cell. This is the same decision that is made when operating on scalar data. Thus, we can just use the standard marching cubes look-up Table [LC87].

For each edge of the cell that is intersected by the surface, we have to compute the exact intersection point. Let $\mathbf{p}_1$ and $\mathbf{p}_2$ be the two endpoints of the edge. Then, we look up the colors $\mathbf{p}_1.color$ and $\mathbf{p}_2.color$ at these vertices. In feature space, the edge between $\mathbf{p}_1.color$ and $\mathbf{p}_2.color$ cuts the convex hull, as shown in Figure 7.

First we have to determine, which face of the convex hull is the intersected one. For a general polyhedron such a test is rather expensive. The computation

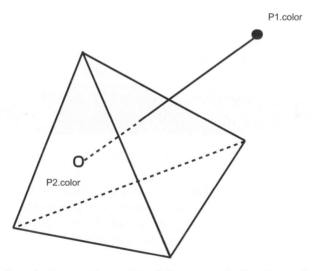

Fig. 7. Finding the intersection point of the convex hull and an edge in feature space.

of the intersection point of an edge with a plane is simple. The expensive part is to find out, whether the intersection point actually lies on the face. However, we can use the convexity of the generated hull. Hence, we can just compute all intersection points and take the one that is closest to $\mathbf{p}_2.color$, assuming that $\mathbf{p}_2.color$ is the color within the convex hull.

Having found the intersection point $\mathbf{p}_1.color + \lambda \cdot \mathbf{p}_2.color$ in feature space, the surface point in object space is given by $\mathbf{p}_1 + \lambda \cdot \mathbf{p}_2$. Figure 8 shows the correspondence between feature and object space.

As the convex hull is only an approximation of a "real" hull of the point cloud given by the user selection, some colors may lie within the convex hull, although they do not belong to the chosen point cloud. Since we know which colors are supposed to be inside and outside, respectively, we always choose the correct marching cubes case. The only problem is the computation of the surface point, as our intersection point computation would fail. For this case we just pick a surface point close to the endpoint of the edge labeled as being inside. This simple solution, of course, inserts some inaccuracy, but for the sake of a major speed-up in computation we have chosen to go this way. This choice was motivated by the fact that we are typically dealing with high-resolution data, where speed becomes an issue and the introduced inaccuracy become negligible.

If the data sets are of very high resolution, we even propose to use an alternative solution for "previewing" the segmentation results. This alternative solution circumvents the computation of the intersection point in feature space

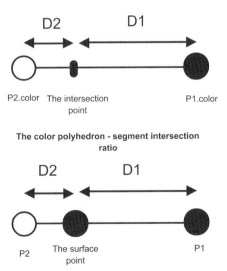

Fig. 8. Correspondence between computation on surface point along an edge in object space and the computation of the intersection point between a color edge and the convex hull in feature space.

by always picking the center point of the cells' edges as surface point. We refer to this method as constant division.

9 Results and Discussion

We have applied our methods to two types of data sets. The first data set presented in this paper is a cryosection data set of a human brain.[1] The resolution of the data set is $512 \times 512 \times 275$. The second data set we present is a fluorescence microscopy data set of a cancer cell.[2] The resolution of the data set is $100 \times 100 \times 40$. To the data we have applied all of the three introduced clustering algorithms and the segmentation and surface extraction methods using hull intersection, constant division, and even the application of the axes-aligned cuboid.

Concerning the clustering algorithms we observed that all of them produced decent results but none of them prefect ones. Which clustering algorithm to pick depends on the properties of the data set. More detailed explanations have been given in the Section 5. However, we observed that misclassification during clustering can easily be fixed using our reclusterization option. The shown pictures are produced using median cut and k-means clustering.

Concerning the segmentation and surface extraction methods, the hull intersection method produced the best results, as expected. For high-resolution data set, the difference to the constant division method decreases. Obviously, the error is in the subvoxel range. The axes-aligned cuboid method proved to be hardly applicable for real-world examples. The cuboid was just not flexible enough to extract the features precisely. The shown pictures are produced using hull intersection and constant division.

Figures 9 and 10 show the application of our algorithms to the cryosection data set. For the generation of the results shown in Figure 9 we applied a few iterations of clusterization with k-means to separate all brain tissue from the surrounding background. In Figure 10, we show the result when extracting the white matter of the brain. In particular, the white matter needed to be separated from the gray matter. In both pictures, we also depict the segmentation and surface extraction applied to a few consecutive slices such that the contours can be compared to the shown 2D slice. As main memory becomes an issue for large data sets, we only used part of the brain data set for the generation of these images. In Figure 11, the result when applying our techniques to a downsampled version of the entire brain data set can be observed. The rendering of the brain cortex, obviously, does not exhibit as crisp contours anymore. A new data management system is left for future work.

---

[1] Data set courtesy of A. Toga, University of California, Los Angeles.
[2] Data set courtesy of the Department of Medicine, University of California, San Diego.

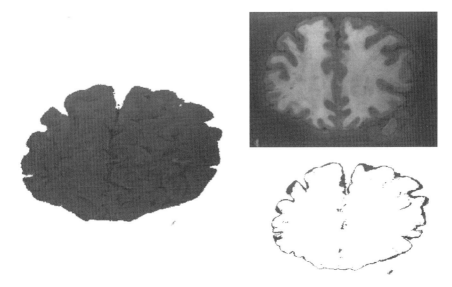

Fig. 9. Segmentation and surface extraction from cryosection data representing a human brain. We separated the brain tissues from the surrounding background for part of the brain (left). We also show one of the original slices (upper right) and a band of the surface when applying our algorithm to a few consecutive slices (lower right).

Fig. 10. Segmentation and surface extraction to extract white matter of part of the brain (right). We also show a band of the surface when applying our algorithm to a few consecutive slices (left).

Figure 12 shows that when applying our techniques to the fluorescence microscopy data set, red regions in the cancer cell are extracted.

The computation times are governed by the computation time for the clustering and the convex hull generation, where the clustering time depends on the methods used. For both the clustering and the convex hull generation, we have been using standard state-of-the art algorithms. The timings can be found in the respective referenced articles.

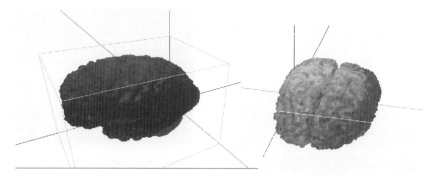

Fig. 11. Segmentation and surface extraction applied to a downsampled version of the entire brain.

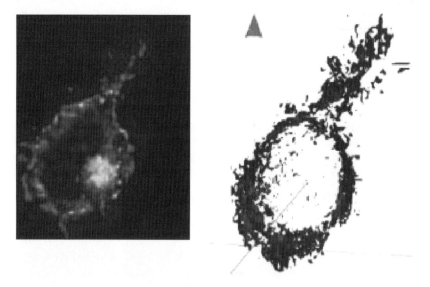

Fig. 12. Segmentation and surface extraction from fluorescence microscopy data representing a cancer cell. We extracted the regions that show up as red (right) and compare the result to one of the 2D slices (left).

10 Conclusions and Future Work

We have presented a tool for the segmentation and surface extraction from color volume data. Our tool incorporates many automated features such as clusterization, feature space segmentation, and object space segmentation and surface extraction. It leaves the guidance to the user where appropriate. Thus, the selection what feature to extract is left to the user by clicking at the respective cluster. Intuitive mechanisms for selecting, adaptively combining, and adaptively refining clusters have been introduced.

We showed that our tool can fulfill the demanded tasks with few user interaction. The user interaction is intuitive, which makes our tool appropriate for handing it to people with a medical background. No direct manipulations in feature space are required.

Apart from making our system more efficient in terms of speed and memory consumption, we want to develop a new clustering algorithm that is targeted towards our goals in future work. If two voxels happen to have exactly the same color assigned to it, although they belong to different tissues, a merely feature-based clustering algorithm cannot separate the voxels. Hence, we must apply a clustering technique that takes into account feature-space distances as well as object-space distances or some derived values.

References

[BDH96] C. Bradford Barber, David P. Dobkin, and Hannu Huhdanpaa. The quickhull algorithm for convex hulls. *ACM Transactions on Mathematical Software*, 1996.

[Bez81] J. C. Bezdek. *Pattern Recognition with Fuzzy Objective Function Algorithms*. Plenum Press - New York, 1981.

[DHS00] R. O. Duda, P. E. Hart, and D. G. Stork. *Pattern classification*. Wiley, 2000.

[EMRY02] D. S. Ebert, C. J. Morris, P. Rheingans, and T. S. Yoo. Designing effective transfer functions for volume rendering from photographic volumes. *IEEE Transactions on Visualization and Computer Graphics*, 8(2):183–197, 2002.

[Hec82] P. Heckbert. Color image quantization for frame buffer display. In *Computer Graphics (Proceedings of ACM SIGGRAPH 82)*, pages 297–307, 1982.

[KKH02] Joe Kniss, Gordon Kindlmann, and Charles Hansen. Multidimensional transfer functions for interactive volume rendering. *IEEE Transactions on Visualization and Computer Graphics*, 8(3):270–285, 2002.

[LC87] William E. Lorensen and Harvey E. Cline. Marching cubes: A high resolution 3d surface construction algorithm. In *Computer Graphics (Proceedings of SIGGRAPH 87)*, volume 21, pages 163–169, July 1987.

[Mac67] J. B. MacQueen. Some methods for classification and analysis of multivariate observations. In *Proceedings of 5-th Berkeley Symposium on Mathematical Statistics and Probability*, volume 1, pages 281–297, 1967.

[Pas03] D. Pascale. *A Review of RGB Color Spaces...from xyY to R'G'B'*. Babel Color, 2003.

[PHP+00] Andreas Pommert, Karl Heinz Hhne, Bernhard Pflesser, Martin Riemer, Thomas Schiemann, Rainer Schubert, Ulf Tiede, and Udo Schumacher. A highly realistic volume model derived from the visible human male. *The Third Visible Human Project Conference Proceedings*, 2000.

[Sch97] P. Scheunders. A comparison of clustering algorithms applied to color image quantization. *Pattern Recognition Letters*, 18, 1997.

[STH97] T. Schiemann, U. Tiede, and K. H. Hhne. Segmentation of the visible human for high quality volume based visualization. *Med. Image Analysis*, 1997.

[TLM+02] Ikuko Takanashi, Eric Lum, Kwan-Liu Ma, Joerg Meyer, Bernd Hamann, and Arthur J. Olson. Segmentation and 3d visualization of high-resolution human brain cryosections. In Robert F. Erbacher, Philip C. Chen, Matti Groehn, Jonathan C. Roberts, and Craig M. Wittenbrink, editors, *Visualization and Data Analysis 2002*, volume 4665, pages 55–61, Bellingham, Washington, 2002. SPIE- The International Society for Optical Engineering, SPIE.

Part IV

Vector and Tensor Visualization in Medical Applications

Global Illumination of White Matter Fibers from DT-MRI Data

David C. Banks[1] and Carl-Fredrik Westin[2]

[1] UT/ORNL Joint Institute for Computational Sciences
  and Harvard Medical School dbanks@cs.utk.edu
[2] Harvard Medical School westin@bwh.harvard.edu

Summary. We describe our recent work in applying physically-based global illumination to fiber tractography. The geometry of the fiber tracts is derived from diffusion tensor magnetic resonance imaging (DT-MRI) datasets acquired from the white matter in the human brain. Most visualization systems display such fiber tracts using local illumination, a rendering technology provided by the video card on a typical desktop computer. There is indirect evidence that the human visual system perceives the shape of a fiber more quickly and more accurately when physically-based illumination is employed.

1 Introduction

This paper describes current work at the Laboratory for Mathematics in Imaging (LMI) at Harvard Medical School to incorporate global illumination into the visual data analysis pipeline for displaying 3D imagery of the human brain. The LMI works within the Department of Radiology at Brigham and Women's Hospital and has a particular emphasis on developing tools and techniques to analyze diffusion-tensor magnetic resonance imaging (DT-MRI). These tools are used by our collaborators to identify clusters of white-matter fiber tracts and to distinguish between fibers that infiltrate a tumor and fibers that pass around a tumor. In these tasks, an expert user examines a static image or a dynamic 3D display of geometry reconstructed from brain scans. The fiber tracts in these displays are long thin tubes, suggestive of the actual fiber bundles that innervate the brain's cortical surface (the gray matter). Water diffuses preferentially along the direction of the fibers, yielding orientation information in the DT-MRI data from which polygonal geometry is synthesized to create a 3D scene. The scene may contain many thousands of tubes representing fiber tracts. In the case of a patient with a brain tumor, surgical resection of the affected tissue destroys both the tumor and any fibers within it. As a result, the cortical regions connected through these fibers will lose their function. The treatment outcome may therefore depend on the relative

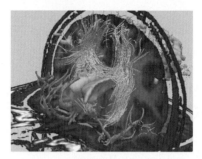

Fig. 1. Brain tumor (ganglioglioma) surrounded by blood vessels and infiltrated by fiber tracts. Visualization produced by *3D Slicer* using local illumination. Tumor represented in green; fiber tracts represented in yellow.

pose of the fibers and the tumor; this information may be of great importance to the patient.

The geometric complexity of the fibers surrounding, and perhaps infiltrating, a tumor produces a 3D scene that is sometimes difficult to comprehend. The visual determination of whether a fiber passes through, or only near, the tumor surface can be a difficult task. Figure 1 illustrates the complex geometry of a tumor (green) surrounded by blood vessels (pink) and infiltrated by fiber tracts (yellow). The current state of the practice in DT-MRI visualization is to render the white matter fiber tracts as colored curves or tubes with either no illumination or with local illumination as provided by the underlying graphics hardware. We are investigating how to make physically-based global illumination of fiber tracts available to surgeons and patients.

2 Illumination

Illumination can be understood in different ways. It can be considered as a scattering process in which photons interact at the quantum level with molecules near the surface of an object. It can be considered as a mathematical equation for light transport. It can be considered as an algorithm implemented in hardware or software to produce an image. And it can be considered as the resulting perceptual phenomena that the human observer responds to. These latter two approaches (thinking of illumination as an algorithm or as a perceptual effect) are discussed below.

2.1 Algorithms for Illumination: Local and Global

Many 3D rendering techniques have been devised to improve the visual quality of rendered scenes via incremental changes within the basic polygon-rendering pipeline, where a polygon is illuminated without consulting other geometry in the scene; this technique is called local illumination. The one-polygon-at-a-time rendering approach is well suited for implementation in hardware, thus allowing a user to interactively manipulate a 3D scene containing thousands of locally-illuminated fibers.

Since the mid-1980's, efforts have been made to formulate the rendering process as a solution to the equation for light transport. In the 1980's, work at Cornell [CG85] [GCT86] produced the first images in which radiosity techniques generated realistic images of scenes containing diffusely reflecting surfaces, while other work approximated the solution to the governing integral equation for light transport by using Monte Carlo techniques [CPC84] [Kaj86]. Interpreting renderers as light-transport solvers has led to increasingly realistic images that are sometimes indistinguishable from photographs. This strategy, which requires that the illumination at one point in a scene must account for illumination at other points in the scene, is called global illumination. Generally speaking, global illumination is performed in software rather than via graphics hardware and is therefore too slow to be used in an interactive visualization system.

The difference between local illumination and global illumination can be appreciated in Figure 2, which shows scenes containing a red tube and a white tube. In the scene shown on the top row, the tubes are close together, within a tube-radius of each other. In the scene shown on the bottom row, the tubes are far from each other, separated by several radii. When the tubes are viewed from above, the two poses are nearly indistinguishable with illumination (left images). But from the same viewpoint one can distinguish the two poses when the images are rendered using physically-based illumination (middle images): shadows and inter-reflections indicate when one tube is close to the other. When viewed from the side (right images), of course, the two poses can be distinguished immediately. Figure 3 shows a more complex scene containing many cylinders; note the shadows cast on and cast by the cylinder near the middle when global illumination is computed.

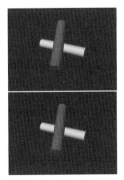 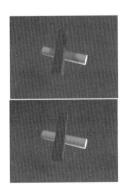 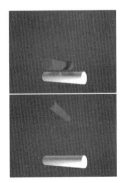

Fig. 2. Ambiguous poses of two tubes. Top row: tubes are closely spaced. Bottom row: tubes are far apart.

Fig. 3. Synthetic scene containing randomly placed cylinders. Left: local illumination. Right: global illumination.

Two distinct threads have thus emerged in 3D graphics. One thread includes hardware-accelerated real-time video games and real-time visualization systems that use local illumination, where physical realism is sacrificed in favor of interactivity. The other thread includes studio films where interactivity is sacrificed in favor of realistic appearance of globally-illuminated 3D characters. The visual analysis of 3D scientific and medical data has generally followed the hardware-based thread. Visualization software tools have historically relied on the default capability of the graphics card to render images on the screen via local illumination (typically by implementing the OpenGL standard). The statement is sometimes repeated within the visualization community that "scientists don't use global illumination." But the scientific community may not have access to global illumination within the visualization tools they use. Unless the scientists are motivated to re-implement a physically-based light-transport code for data analysis, they are unlikely to develop an opinion regarding the value of global illumination or to demand it in their data analysis software. Thus a stable state currently exists in which visualization software generally relies on local illumination for its rendering since global illumination is not demanded by users; meanwhile, users generally do not render their scenes with global illumination since it is not available from their visualization tools. The absence of demand for global illumination in DT-MRI visualization software tools is both a cause for and the result of an absence of its supply, as illustrated below.

A basic unresolved question for those who design 3D visualization tools for displaying the complex geometry arising from brain data is: in order to make 3D shapes of lesions, fibers, blood vessels, and cortex perceptually evident to the human viewer (*e.g.* the surgeon), does physically-based rendering offer an improvement over the standard "local illumination" provided by graphics

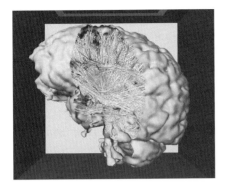 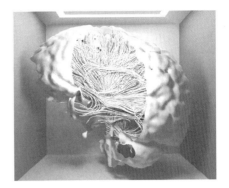

Fig. 4. White matter fibers within a cut-away approximation of the cortex. Left: local illumination. Right: global illumination. Brain dataset courtesy of Gordon Kindlmann at the Scientific Computing and Imaging Institute, University of Utah, and Andrew Alexander, W. M. Keck Laboratory for Functional Brain Imaging and Behavior, University of Wisconsin-Madison.

hardware? If so, then surgeons would demand this improvement and developers of visualization tools would be motivated to offer global illumination as a rendering option in their software systems.

Our work inserts global illumination into the visualization pipeline for analyzing DT-MRI data, giving neurosurgeons and radiologists the opportunity for the first time to analyze complex 3D scenes rendered with physically realistic illumination. Figure 4 illustrates the difference between local illumination and global illumination for a scene composed of several thousand fibers, represented as polygonal tubes, surrounded by a cut-away of a surface approximating the cortex. The brain geometry is situated within a rectangular box, exhibiting shadows and inter-reflections as a result of physically-based global illumination. Informal comments from colleagues (including radiologists and surgeons) indicate that the globally illuminated display of fibers "looks more 3D," "looks more real," "looks prettier," or "looks like you could touch it." Although there may be intrinsic value in the esthetic appeal of global illumination, it has not yet been shown whether global illumination of fibers provides measurable improvements in shape inference. We are currently engaged in designing a user study to measure differences in recognition and performance that may arise as a result of local versus global illumination of fibers.

2.2 Perception of Illumination: Shadows

Although there is evidence that global illumination enhances 3D shape perception in general, this hypothesis is the subject of debate. The issue of whether global illumination adds perceptual value to a scene is sometimes characterized by whether shadows enhance perception of 3D shape. From a physics standpoint, this characterization is awkward: when we are situated in an unlit room, every object is "in shadow" from sunlight, being illuminated indirectly by skylight through the window, and yet fully shadowed 3D shapes are perceived all around us within the room.

While the physics of light transport proceeds without regard to the phenomenological qualities a human observer may attribute to the perceived scene, an artist may "add" specular highlights to shiny surfaces, "add" shading to rounded objects, and "add" shadows onto the painted canvas, and therefore conceive categories (highlights, texture, shadow) and subcategories (shadow is attached, grounding, or cast) of perceptual phenomena. The distinction is important when considering how physically-based illumination may improve the perceptual quality of a rendering of fiber tracts. A 3D visualization system can be designed to numerically solve the equation of light transport to greater or lesser degrees of accuracy, or else it can be designed to add perceptual phenomena to a greater or lesser degrees of completeness within artistically conceived categories. Historically, the latter artistic approach has led to disputes about whether or not shadows should be "added" to a 3D visualization.

There is conflicting evidence about the role that shadowing (as a phenomenological effect) plays in the perception of geometric attributes of a 3D scene. Madison [MTK+01] and Mamassian [MKK98], for example, found that cast shadows provide a positive depth cue when present in a scene. But shadows can be problematic in a complex 3D scene. For example, shadows from tree branches on a sunny day can ruin the appearance of human faces in an outdoor photograph, in part because these shadows act as camoflage and in part because they interfere with the face's natural shading that would be exhibited were the tree not interposed between the face and the sun. This a problem may be eliminated by using area lights rather than a single point light; Figure 5 illustrates the difference in appearance between a small-area and a large-area light source. The small-area luminaire produces sharp shadows case by the fiber tracts and blood vessels onto the surface of the tumor; these shadows may hinder visual determination of which fibers penetrate the tumor.

Fig. 5. Shadows in globally illuminated ganglioglioma dataset (used in Figure 1) resulting from luminaires of different sizes. Tumor represented in green; fibers represented in yellow. Left: Small-area luminaire produces sharp shadows. Right: Large-area luminaire produces soft shadows.

The human visual system is somewhat insensitive to inaccuracies of shadows in a synthetic rendering: Ostrovsky [OCS01] found that inconsistent shading of objects is slow and/or difficult to detect, while Jacobson [JW04] found subjects could not distinguish between a stripe-textured cube shadowed by leaves or a leaf-textured cube shadowed by stripes. But if 3D shadow and shape perception occurs early in visual processing, then the cognitive task of inferring a globally consistent light source or determining the source of a shadow may not correlate with the early mental formation of a geometric model influenced by the presence of shadows. The issue of whether shadow-processing is performed by an early-level system has not been resolved, but experiments by Rensink [RC04] suggest that shadow-processing is indeed performed during the early stages of visual search. Since visual search is an essential component of the surgeon's task during visual data analysis, display technologies that exploit early-level visual processing may improve the surgeon's spatial perception of the complex geometric interplay between lesions, fibers, blood vessels, and cortex.

3 Visual Integration of Glyphs and Curves

Figure 2 illustrates the difference between local and global illumination for only a single pair of straight tubes. Our actual concern is to apply global illumination to many long curvilinear tubes, each constructed explicitly as a polygonal mesh as in Figure 4, or else implicitly as a set of disconnected glyphs as in Figure 6.

The direction of a fiber is generally assumed to align with the principal eigenvector of the diffusion tensor acquired by DT-MRI (at least when the principal eigenvalue is much larger in magnitude than the other two

Fig. 6. Superquadric glyphs derived from DT-MRI data. The extent and orientation of the glyphs invite the visual system to integrate their small shapes into larger coherent structures. This visual integration process may be enhanced by global illumination. Left: local illumination. Right: global illumination.

eigenvalues); an overview of the process can be found in the summary by Westin [WMM+02]. The centerline of the tube follows the integral curve of the principal eigenvector. Vector integration of the principal eigenvector $v_1(x)$ at point $x = (x[0], x[1], x[2])$ produces a streamline curve $c(t)$ defined by the equation

$$c(t) = \int_0^T v_1(c(t))dt$$

which can be approximated by standard numerical techniques for vector integration. Fiber tracts are expected to align, subject to some confidence bounds, with these glyphs. Two basic strategies are employed to visualize the fiber tracts: either (1) display discrete glyphs representing either the principal eigenvector or the entire tensor, or (2) display the actual streamlines, whose geometry is expected to follow the tracts.

In strategy (1), the glyphs may be uniformly placed at gridpoints of the 3D volume, or may be packed at nonuniform densities [KW06]. A glyph may take the shape of an ellipsoid or superquadric to incorporate tensor information about all three diffusion directions.

Glyphs and color. In examining the display of individual 3D glyphs, the surgeon is required to visually integrate the underlying vector field and infer fiber pathways, presumably by estimating streamlines through the glyphs. It is an open question whether illumination contributes to this visual-integration task. Beaudot [BM01] recently investigated the role of color and curvature on the visual task of linking oriented glyphs across a display in order to infer the presence of a 2D path. This 2D task is simpler than, but similar to, the task of visually linking 3D tensor glyphs to infer fiber tracts. Beaudot measured reaction time where subjects pressed a mouse button after determining whether a path did or did not exist within a field of densely packed 2D oriented glyphs. The study found reaction times increased (worsened) by 100ms when the path was curved rather than straight, and increased by 100ms when the glyph/background combinations possessed saturated values of red/green or blue/yellow rather than achromatic shades of gray varying in luminance.

The current state of 3D visualization of DT-MRI data exhibits, generally speaking, the worst case scenario for path following. Most glyph-based visualization tools display fully saturated glyphs representing curved paths. Figure 6 (left) shows an example image representative of glyphs rendered by a typical visualization tool. An informal sampling of presentations at the recent International Society of Magnetic Resonance in Medicine meeting (ISMRM 2006) revealed that nearly all of the 3D images of diffusion MRI data used fully saturated glyphs/streamlines to represent curving paths, a coloring scheme that Beaudot found to increase visual processing time, or else employed luminance changes resulting from local illumination only; Figure 7 illustrates these two widely-used techniques. Although Beaudot did not measure spatial error in visual path integration, it is plausible that increased visual processing time indicates increased complexity for perception and thus correlates with

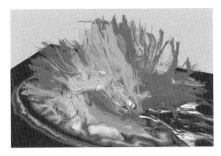

Fig. 7. Representative examples of current state-of-the-art visualization of fiber tracts derived from DT-MRI data. Most of the fibers are occluded by other fibers, requiring the visual system to perform amodal completion of the hidden fiber trajectories. Left: saturated constant-hue curves. Right: polygonal tubes with local illumination (image by H. J. Park).

increased spatial error by the visual system. The presence of shadows resulting from physically based illumination may provide information through the luminance channel to assist the visual system in path-following.

Shape completion. Whether DT-MRI data is displayed using 3D glyphs or 3D curves, the complexity of the resulting geometry practically ensures that most of the glyphs/curves are occluded by others. The visual system is adept at "amodal completion" [MTC91], allowing us to perceive a partially hidden object as complete. Completion seems to result from both local cues (e.g., extrapolating shape direction across the discontinuous edges of the occluder) and global cues [SPF94], and may require 75–200 ms of processing time. What factors influence the visual task of perceiving a 3D curve amidst the clutter of other fragments of geometry in the scene, and how can they be measured?

In experiments by Plomp *et al.* [PNBvL04], an eye-tracking system was used to avoid the extra processing time involved in a motor task (*e.g.* pressing a mouse button). Subjects were asked to visually search for target shapes that were partly occluded while their eye movements were being recorded. The premise of the study was that completion of an occluded figure correlates with longer gaze duration when the eye fixates on the occluded object. They found that subjects demonstrated longer gaze durations looking at targets than non-targets, about 450–600 ms versus 350–450 ms on average. They found further that a less familiar target (hexagon versus circle) added to the gaze duration, by an increment of about 50 ms. They propose that familiar objects are completed more quickly when occluded. Although no experiments have yet been conducted to determine the effect of illumination on amodal completion of fiber tracts surrounded by occluders, it is plausible that 3D tubes subjected to physically-based lighting (as opposed to local illumination or no illumination) are more familiar from daily experience and might therefore be completed more quickly and accurately by the visual system.

Fig. 8. Fibers automatically clustered into fascicles. Left: local illumination. Right: global illumination.

Amodal completion of fibers operates at a coarse spatial scale when the surgeon or radiologist analyzes clusters (fascicles) of tracts. In this case, not only do individual fibers occlude each other, but entire functional groups of fibers hide each other. Figure 8 illustrates the situation, where several fascicles have been automatically grouped (by a technique described by O'Donnell [OKS+06] [OW05] and Brun [BKP+04]) and then color coded.

4 Results and Acknowledgments

The process of upgrading a visualization software system from local illumination to global illumination can be attacked in various ways. In our case, we wished to introduce global illumination into the *Slicer* visualization tool developed by Harvard and MIT [GNK+01] (*Slicer* can be downloaded from http://slicer.org). *Slicer* was designed to use VTK, the Visualization Toolkit [SML06], which manages geometry and renders it using OpenGL (and therefore local illumination). We used VTK's ability to export geometry in the Open Inventor [WtOIAG04] scene-graph file format so that the 3D scene could be fed into a stand-alone global illumination tool that implements photon mapping [Jen01]. This pipeline of tools allows the 3D scenes from *Slicer* to pass through a post-processing step that generates 2D images of the globally illuminated scene [BB06]. On a typical desktop computer, the renderer requires 1-100 minutes to generate an image, depending on the number of photons used, the density estimation algorithm applied, and the size of the final image. The initial results of incorporating global illumination into the *Slicer* pipeline can be seen in Figures 1 and 8.

The National Alliance for Medical Image Computing (NA-MIC), which supports and distributes *Slicer*, uses the *Teem* library. *Teem* a group of tools for manipulating and displaying scientific and medical data; it can be found on the Web at http://teem.sourceforge.net. We converted *Teem*'s superquadric glyphs into Open Inventor format and rendered the resulting scene graph to produce the images in Figure 6.

Within the computer graphics research community, a considerable amount of effort is being applied to develop real-time systems for global illumination. Our proof-of-concept demonstration offers a preview of what may one day become the standard for 3D visualization systems for analyzing DT-MRI datasets acquired from the human brain.

This work was supported by NSF award CCF 0432133 and NIH NIBIB NAMIC U54-EB005149, NIH NCRR NAC P41-RR13218, NIH NCRR mBIRN U24-RR021382, and R01 MH 50747. The authors express their appreciation to Kevin Beason (Rhythm and Hues Studios), Yoshihito Yagi (Florida State University), and Israel Huff (University of Tennessee) for assistance with rendering and generation of test scenes; Gordon Kindlmann (Laboratory of Mathematics in Imaging, Harvard Medical School) for his DT-MRI dataset and his glyph dataset; Lauren O'Donnell for her expertise with *Slicer* and VTK, and for use of the clustered-fiber dataset; Ion-Florin Talos for use of the ganglioglioma dataset; and Susumu Mori (JHU) for diffusion MRI data (R01 AG20012-01, P41 RR15241-01A1).

References

[BB06] David C. Banks and Kevin Beason. Pre-computed global illumination of mr and dti data. In *International Society for Magnetic Resonance in Medicine 14th Scientific Meeting (ISMRM '06)*, page 531, 2006. Seattle, WA, 6-12 May 2006, Abstract 2755.

[BKP+04] A. Brun, H. Knutsson, H. J. Park, M. E. Shenton, and C.-F. Westin. Clustering fiber tracts using normalized cuts. In *Seventh International Conference on Medical Image Computing and Computer-Assisted Intervention (MICCAI '04)*, pages 368–375, 2004. Rennes - Saint Malo, France.

[BM01] William H. A. Beaudot and Kathy T. Mullen. Processing time of contour integration: the role of colour, contrast, and curvature. *Perception*, 30:833–853, 2001.

[CG85] Michael F. Cohen and Donald P. Greenberg. The hemi-cube: A radiosity solution for complex environments. In *Computer Graphics (Proceedings of SIGGRAPH 85)*, pages 31–40, August 1985.

[CPC84] Robert L. Cook, Thomas Porter, and Loren Carpenter. Distributed ray tracing. In *Computer Graphics (Proceedings of SIGGRAPH 84)*, pages 137–145, July 1984.

[GCT86] Donald P. Greenberg, Michael Cohen, and Kenneth E. Torrance. Radiosity: A method for computing global illumination. *The Visual Computer*, 2(5):291–297, September 1986.

[GNK+01] D. Gering, A. Nabavi, R. Kikinis, N. Hata, L. Odonnell, W. Eric L. Grimson, F. Jolesz, P. Black, and W. Wells III. An integrated visualization system for surgical planning and guidance using image fusion and an open mr. *Journal of Magnetic Resonance Imaging*, 13:967–975, 2001.

[Jen01] Henrik Wann Jensen. *Realistic Image Synthesis Using Photon Mapping*. AK Peters, 2001.

[JW04] Jayme Jacobson and Steffen Werner. Why cast shadows are expendable: Insensitivity of human observers and the inherent ambiguity of cast shadows in pictorial art. *Perception*, 33:1369–1383, 2004.

[Kaj86] James T. Kajiya. The rendering equation. In *Computer Graphics (Proceedings of SIGGRAPH 86)*, pages 143–150, August 1986.

[KW06] Gordon Kindlmann and Carl-Fredrik Westin. Diffusion tensor visualization with glyph packing. *IEEE Transactions on Visualization and Computer Graphics (Proceedings Visualization / Information Visualization 2006)*, 12(5), September-October 2006.

[MKK98] P. Mamassian, D. C. Knill, and D. Kersten. The perception of cast shadows. *Trends in Cognitive Sciences*, 2:288–295, 1998.

[MTC91] A. Michotte, G. Thines, and G. Crabbe. Les complements amodaux des structures perceptives. In G. Thines, A. Costall, and G. Butterworth, editors, *Michotte's Experimental Phenomenology of Perception*, pages 140–167, Hillsdale, NJ, 1991. Lawrence Erlbaum Associates.

[MTK+01] C. Madison, W. Thompson, D. Kersten, P. Shirley, and B. Smits. Use of interreflection and shadow for surface contact. *Perception and Psychophysics*, 63:187–194, 2001.

[OCS01] Y. Ostrovsky, P. Cavanagh, and P. Sinha. Perceiving illumination inconsistencies in scenes. Technical report, Massachusetts Institute of Technology, Cambridge, MA, USA, November 2001.

[OKS+06] Lauren O'Donnell, Marek Kubicki, Martha E. Shenton, Mark E. Dreusicke, W. Eric L. Grimson, and Carl-Fredrik Westin. A method for clustering white matter fiber tracts. *American Journal of Neuroradiology (AJNR)*, 27(5):1032–1036, 2006.

[OW05] Lauren O'Donnell and Carl-Fredrik Westin. White matter tract clustering and correspondence in populations. In *Eighth International Conference on Medical Image Computing and Computer-Assisted Intervention (MICCAI '05)*, Lecture Notes in Computer Science, pages 140–147, 2005. Palm Springs, CA, USA.

[PNBvL04] Gijs Plomp, Chie Nakatani, Valérie Bonnardel, and Cees van Leeuwen. Amodal completion as reflected by gaze durations. *Perception*, 33:1185–1200, 2004.

[RC04] Ronald A. Rensink and Patrick Cavanagh. The influence of cast shadows on visual search. *Perception*, 33:1339–1358, 2004.

[SML06] Will Schroeder, Ken Martin, and Bill Lorensen. *Visualization Toolkit: An Object-Oriented Approach to 3D Graphics*. Kitware, 2006. 4th edition.

[SPF94] A. B. Sekuler, S. E. Palmer, and C. Flynn. Local and global processes in visual completion. *Psychological Science*, 5:260–267, 1994.

[WMM+02] C.-F. Westin, S. E. Maier, H. Mamata, A. Nabavi, F. A. Jolesz, and R. Kikinis. Processing and visualization of diffusion tensor MRI. *Medical Image Analysis*, 6(2):93–108, 2002.

[WtOIAG04] Josie Wernecke and the Open Inventor Architecture Group. *The Inventor Mentor: Programming Object-Oriented 3D Graphics with Open Inventor*. Addison-Wesley Professional, 2004.

Direct Glyph-based Visualization of Diffusion MR Data Using Deformed Spheres

Martin Domin[1], Sönke Langner[1], Norbert Hosten[1], and Lars Linsen[2]

[1] Department of Diagnostic Radiology and Neuroradiology
  Ernst-Moritz-Arndt-Universität Greifswald
  Greifswald, Germany
  {martin.domin,soenke.langner,hosten}@uni-greifswald.de
[2] Computational Science and Computer Science
  School of Engineering and Science
  Jacobs University
  Bremen, Germany
  l.linsen@jacobs-university.de

Summary. For visualization of medical diffusion data one typically computes a tensor field from a set of diffusion volume images scanned with different gradient directions. The resulting diffusion tensor field is visualized using glyph- or tracking-based approaches. The derivation of the tensor, in general, involves a loss in information, as the $n > 6$ diffusion values for the n gradient directions are reduced to six diverse entries of the symmetric 3×3 tensor matrix. We propose a direct diffusion visualization approach that does not operate on the diffusion tensor. Instead, we assemble the gradient vectors on a unit sphere and deform the sphere by the measured diffusion values in the respective gradient directions. We compute a continuous deformation model from the few discrete directions by applying several processing steps. First, we compute a triangulation of the spherical domain using a convex hull algorithm. The triangulation leads to neighborhood information for the position vectors of the discrete directions. Using a parameterization over the sphere we perform a Powell-Sabin interpolation, where the surface gradients are computed using least-squares fitting. The resulting triangular mesh is subdivided using a few Loop subdivision steps. The rendering of this subdivided triangular mesh directly leads to a glyph-based visualization of the directional diffusion measured in the respective voxel. In a natural and intuitive fashion, our deformed sphere visualization can exhibit additional, possibly valuable information in comparison to the classical tensor glyph visualization.

1 Introduction

Magnetic Resonance Imaging (MRI) is a well-known method for in-vivo observation of human brains. One can obtain structural informations of a patient's brain and spot tumors. Nevertheless, normal MRI does not tell a neurosurgeon what happened to the neuronal fibers in the region covering or surrounding the tumor. When the tumor forms and grows, fibers may be deflected, destroyed, or surrounded by the tumor. Insight on how fibers penetrate or bypass the tumor influence the decision of the treatment. In particular, a surgeon needs to know, how he can access the tumor and what regions he can cut without damaging certain fibers.

Such information on the fibers' structures can be derived from Magnetic Resonance (MR) Diffusion Imaging. Assuming that diffusion is higher along the fibers' direction, the brain is scanned several times using various diffusion gradients. The fibers' structure is often derived using tracking approaches. Starting with appropriate seeding locations, fibers are typically tracked by going in the direction of highest diffusion. Since, in practice, diffusion is only measured in a few directions, the direction of highest diffusion is approximated or interpolated from the given measurements, for example, by applying statistical methods.

The most common technique to analyze MR diffusion imaging data is to derive a diffusion tensor that describes the diffusivity in the directions of the principal diffusion axes. Using principal component analysis on the gradient directions one derives a symmetric 3×3 matrix describing the diffusion tensor. The eigenvectors of the matrix are the principal diffusion axes. The eigenvector to the largest eigenvalue is the direction of highest diffusion. Most visualization techniques operate on this diffusion tensor, although it obviously does not contain all originally measured information. The symmetric matrix only holds six diverse entries, while using a gradient field with substantially more than six directions, i.e. one obtains more than six measured values, is very common.

We present visualization techniques that directly operate on the measured data, i.e. the diffusion gradient vectors and the respective measured diffusion signal attenuation. The data is processed in a way that directly leeds to a glyph-based visualization. The underlying model is that of a deformed sphere, where the deformation describes the diffusion anisotropy within a voxel of the given data set. Thus, the sphere should be stretched most in the direction of highest diffusion, and proportionally less in the other directions. The model is described in detail in Section 4.

Since diffusion signal attenuation is only measured in some discrete gradient directions, we have to interpolate the diffusion values in between. For interpolation purposes, we first establish neighborhood information between the sample locations. Operating on a spherical domain, we can triangulate the domain easily by applying a convex hull computation, see Section 4. Then, we apply the Powell-Sabin interpolation scheme over the triangulated domain. We have chosen this interpolation scheme, as it produces C^1-transitions between two adjacent triangles. The known Powell-Sabin scheme described over a planar domain needs to be adjusted to operate on a sphere. The details are given in Section 6. The Powell-Sabin scheme interpolates both function values and gradients. For the computation of the function's gradients we use a least-squares fitting method. As an interpolation result we obtain six Bézier patches

Fig. 1. Processing pipeline from diffusion MR measurements to glyph-based visualization.

for each triangle. Instead of evaluating the Bézier polynomials we apply a few steps of the Loop-subdivision scheme, see Section 7.

The resulting geometrical shape is a discrete approximation of the deformed sphere model. The precision of the approximation depends on the number of applied subdivision steps. Obviously, the discretized version of the deformed sphere model can directly serve as a 3D glyph representing the diffusion of the respective voxel, see Section 8. We present results for different cases and compare our results to tensor-based glyphs. We observe that our visualizations can convey additional information when compared to diffusion tensor imaging visualizations.

Figure 1 illustrates the entire processing pipeline from diffusion MR measurements to the rendering of the generated glyphs.

2 Related Work

For the visualization of MR diffusion imaging data two major approaches are commonly used, namely the visualization of fibers tracked through the brain and the visualization of local diffusion behavior via glyphs.

Tractography, a term first applied to DTI analysis by Basser [BPP+00], yields curves of neural pathways, which are continuous and hard to represent using discrete glyphs. Streamlines and their derivatives are widely used for tractography results. Zhang et al. used streamtubes and streamsurfaces to visualize the diffusion tensor field [ZDL03]. Streamtubes visualize fiber pathways tracked in regions of linear anisotropy: the trajectories of the streamtubes follow the major eigenvectors in the diffusion tensor field; the color along the streamtubes represent the magnitude of the linear anisotropy; the cross-section shape represent the medium and minor eigenvector. Streamsurfaces visualize regions of planar anisotropy: the streamsurface follows the expansion of major and medium eigenvectors in the diffusion tensor field; the color is mapped to magnitude of planar anisotropy.

Due to the possibility of mixed diffusion directions in one voxel, most tensor-based approaches have difficulties to resolve these different directions. Furthermore tractography is also sensitive to noise, whereby a small amount of noise can lead to significantly different results. Some groups have tried to address these problems by regularizing diffusion data set [BBKW02a, WKL99, ZB02] or direction maps [OAS02, PCF+00]. Brun et al. use sequential importance sampling to generate a set of curves, labeled with probabilities, from each seed point [BBKW02b]. Batcherlor et al. generate an isosurface of solution by solving a diffusion-convection equation [PDD+02]. Parker et al. use front propagation in fast marching tractography [PSB+02]. High angular resolution diffusion imaging is reported to ameliorate ambiguities in regions of complex anisotropy [TWBW99, TRW+02].

For representing the data with its shape, color, texture, location, etc. often a parameterized icon, a glyph, is used. Diffusion ellipsoids are a natural glyph to summarize the information contained in a diffusion tensor [PB96]. The three principal radii are proportional to the eigenvalues and the axes of the ellipsoid aligned with the three orthogonal eigenvectors of the diffusion tensor. The preferred direction of diffusion is indicated by the orientation of the diffusion ellipsoid. Arrays of ellipsoids can be arranged together in the same order as the data points to show a 2D slice of DTI data. In a still image it is often hard to tell the shape of an ellipsoid with only surface shading information. Westin et al. used a composite shape of linear, planar and spherical components to emphasize the shape of the diffusion ellipsoids [WMM+02]. The components are scaled to the eigenvalues. Kindlmann's approach adapted superquadrics, a traditional surface modeling technique [Bar81], as tensor glyphs. He created a class of shapes that includes spheres in the isotropic case, while emphasizing the differences among the eigenvalues in the anisotropic cases. Cylinders are used for linear and planar anisotropy, while intermediate forms of anisotropy are represented by approximations to boxes. As with ellipsoid glyphs, a circular cross-section accompanies equal eigenvalues, for which distinct eigenvectors are not defined.

All these glyph-based methods have in common that they operate on the computed diffusion tensor and its eigenanalysis. Thus, they are based on diffusion values in three orthogonal directions, whereas the original data may have contained much more subtle information about the diffusion distribution in various directions. Tuch et al. [Tuc04b] introduced the q-ball imaging method to calculate glyphs and one or more preferred diffusion directions without relying on the diffusion tensor. Their method, however assumes high angular resolution diffusion imaging (HARDI) data, which takes too long to acquire for clinical purposes. We present a method based on the deformed sphere model that considers all measured diffusion values and is applicable to clinically relevant data.

Recently, Hlawitschka and Scheuermann [HS05] presented an approach that represents diffusion MR data using higher-order tensors. Although their visualization approach seeks for tracking lines they also make use of glyph-based renderings of the tensors.

An approach to visualize tensor data by applying direct volume rendering has been presented by Kindlmann and Weinstein [KW99]. They use two-dimensional color maps and a tensor-based lighting model to render the second-order tensor field.

3 Diffusion Imaging Background

The diffusion in biological tissue, based on the Brownian motion of molecules in fluids, is the basis for MR diffusion imaging. In general, water molecules tend to move along the gradient of concentration differences, from the highest to the lowest concentration. Using Magnetic Resonance Imaging (MRI) techniques this movement becomes measurable. The common diffusion-weighted MRI measures a single scalar value, the apparent diffusion coefficient (ADC). This scalar coefficient is sufficient to represent the movement of water in homogeneous (i.e. isotropic) tissues where the diffusion is largely orientation-independent.

In contrast to the isotropic diffusion, anisotropic diffusion is heavily depending on the direction of the surrounding tissue. Anisotropic media such as muscle or white matter in a human brain create boundaries that force the moving water into an ordained direction.

In diffusion-weighted imaging (DWI) one applies one magnetic gradient with an ordained orientation at a time. The resulting ADC represents the projection of all molecular displacements along this direction at a time. Hence, the technique is one-dimensional. To resolve multiple displacement directions one has to apply multiple magnetic diffusion gradients from different directions. To obtain as much information as possible, the directions should be non-collinear and non-coplanar.

For each direction, one measures the diffusion-weighted signal $S(b)$ and compares it to a standard MRI signal $S(0)$, i.e. the measured non-diffusion-weighted signal, using the formula $\ln\left(\frac{S(b)}{S(0)}\right)$. The formula describes the anisotropic diffusion. The weighting factor b can vary within an experimental set-up, but typically is chosen to be constant for a scanning sequence.

Obviously, the quality of the measurements increases with increasing number of scans. However, for clinical purposes one cannot expect a patient to be scanned for hours. Thus, the number of scans is limited. A typical number of gradient directions within a sequence would be in the range of 12.

If n is the number of gradient directions, the resulting data set consists of $n+1$ 3D scalar fields that are reconstructed from $n+1$ stacks of 2D grayscale images. One scalar field stores the non-diffusion-weighted scan, each of the other n scalar fields store a field of diffusion-weighted signals. In other words, for each voxel we store one non-diffusion-weighted signal and n diffusion-weighted signal, from which we compute n diffusion values, one for each gradient direction. It is desired that the gradient directions are distributed as uniformly as possible. For further details on the concept and application of diffusion MR imaging we refer to the survey by LeBihan [LeB01].

4 Deformed Sphere Model

Let V be a voxel of our volume data set. The diffusion at voxel V is represented by n gradient directions and the respective diffusion values. When considering the isotropic case, diffusion is independent of the gradient direction. Thus, all diffusion values at V are equal. If one wants to display this information geometrically in Euclidean space, one could draw a sphere surrounding the center of voxel V. The sphere would be the unit sphere whose radius is scaled by the diffusion value. The sphere illustrates that no direction is preferred over another direction.

For the anisotropic case, on the other hand, diffusions vary for different directions. When we apply these diffusion values to our geometric model of a sphere, the sphere gets deformed. It gets stretched more in regions of high diffusion and less in regions of low diffusion. Thus, the degree of deformation represents the diffusion in the respective gradient direction. Figure 2 illustrates the idea of the deformed sphere model in a two-dimensional set-up. The deformation is governed by the diffusion-scaled gradient vectors.

The figure shows that the sphere can be deformed in any direction. In particular, the main deformation directions do not have to be orthogonal to each other. This property distinguishes our model from the common diffusion tensor model, as the eigenvectors of the diffusion tensor are always orthogonal to each other.

Since the gradient directions used are only unique up to its sign, the geometric representation is chosen to be point-symmetric. Due to the expected symmetry of the diffusion, measurements are taken with directions that lie in one hemisphere [Tuc04a]. Thus, for each gradient direction we use two diffusion-scaled gradient vectors, one in the gradient direction and one in the opposite direction. Both vectors are scaled with the same diffusion, see Figure 2.

From the measured data we only obtain deformations in the gradient directions. We do not know the precise directions of maximum or minimum

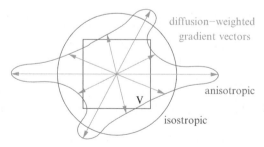

Fig. 2. Deformed sphere model: Unit sphere gets scaled by diffusion values in gradient directions (green). Sphere remains a sphere in isotropic cases (blue), but gets deformed in anisotropic cases (red).

deformation (or diffusion, respectively). Thus, we need to interpolate the given values to obtain a sufficiently precise approximation of the deformed sphere model.

5 Spherical Domain Triangulation

Our sphere model leaves us with a scattered data set over a spherical domain. Figure 3(a) displays the scattered data for $n = 6$ gradient directions. In order to apply interpolation techniques to the scattered data, we need to establish some neighborhood information between the scattered data points. We compute the neighborhood information by triangulating the domain, where the spherical domain is represented by the unit sphere.

Geometrically spoken, the triangulation algorithm needs to connect points that are closest to each other with respect to the metric on the unit sphere. As a sphere has distinct properties in terms of symmetry and convexity, it suffices to approximate the triangulation by a convex hull triangulation.

The convex hull of a set of points is the smallest convex set that contains the points [Aur91]. In our case, all points lie on the unit sphere and thus contribute to the convex hull. Hence, when triangulating the convex hull, we obtain a triangulation with the desired properties. The result of the convex hull triangulation is a tessellated sphere. For the computation of the convex hull a Java implementation of the Quickhull algorithm [Bar96] is used. Figure 3(b) shows the result of the convex hull triangulation applied to the scattered data in Figure 3(a).

After the triangulation of the spherical domain, we can scale the geometric construction by the ordinates at the respective abscissae. The ordinates are the diffusion values, the abscissae the coordinate vector of the respective diffusion gradient directions. The result is a triangular mesh that represents a crude approximation of our deformed sphere model. Figure 4 shows an example.

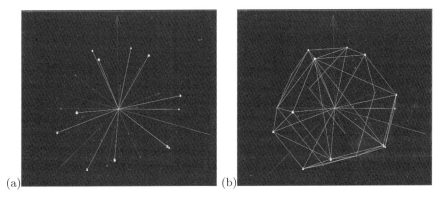

Fig. 3. (a) Abscissae of scattered data over spherical domain. (b) Triangulation of spherical domain in form of unit sphere.

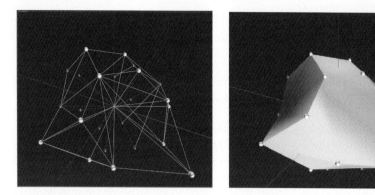

Fig. 4. Piecewise linear approximation of a deformed sphere (wireframe and shaded rendering).

6 Powell-Sabin Interpolation on a Sphere

The generated triangulation is a piecewise linear approximation of our deformed sphere model with $2n$ sample points, where n denotes the number of gradient directions in the diffusion MR measurements. We need to replace this linear approximation by a higher-order representation with continuous transitions at the triangles' borders. The Powell-Sabin interpolant is a piecewise quadratic bivariate function that interpolates function values and gradients. A quadratic function is defined over each triangle such that a C^1-continuity is obtained across the common edge of adjacent triangles. The use of a quadratic function is suitable, as a sphere is a quadric.

The Powell-Sabin interpolant subdivides each triangle into 24 smaller triangles by inserting new points, as shown in Figure 5. Let $\mathbf{q}_0$, $\mathbf{q}_1$, and $\mathbf{q}_2$ be the abscissae of the vertices of the triangle to be subdivided, and $f(\mathbf{q}_0)$, $f(\mathbf{q}_1)$, and $f(\mathbf{q}_2)$ the respective ordinates. Moreover, let $\mathbf{p}_{ijk}$ denote the abscissae of the vertices of the new triangles and $\mathbf{b}_{ijk}$ denote the respective ordinates. The indexing follows the scheme shown in Figure 5. Thus, we can set $\mathbf{p}_{400} = \mathbf{q}_0$, $\mathbf{p}_{040} = \mathbf{q}_1$, and $\mathbf{p}_{004} = \mathbf{q}_2$, while the remaining $\mathbf{p}_{ijk}$ are newly inserted vertices.

Since we are operating on a spherical domain, all $\mathbf{p}_{ijk}$ lie on the unit sphere with origin $\mathbf{o}$, as depicted in Figure 6.

In order to establish C^1-continuity across adjacent triangles, the ordinates and abscissae must fulfill certain coplanarity and collinearity conditions. Using the notation from Figures 5 and 6, the interpolation needs to fulfill the following requirements:

1. The abscissae $\mathbf{p}_{222}$, $\mathbf{p}_{220}$, and $\mathbf{p}'_{222}$, must be collinear, where $\mathbf{p}'_{222}$ denotes the center vertex for the respective adjacent triangle.
2. The ordinates $\mathbf{b}_{400}$, $\mathbf{b}_{310}$, $\mathbf{b}_{311}$, and $\mathbf{b}_{301}$ must be coplanar. They all must lie in the tangent plane at vertex $\mathbf{b}_{400}$.

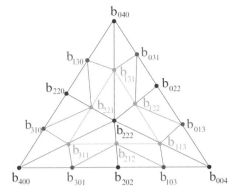

Fig. 5. The Powell-Sabin interpolant. $\mathbf{b}_{ijk}$ denote the ordinates.

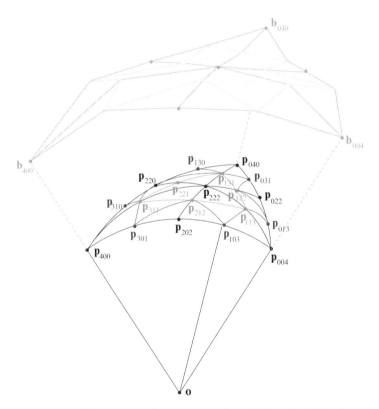

Fig. 6. The Powell-Sabin interpolant over a spherical domain. The abscissae $\mathbf{p}_{ijk}$ lie on the unit sphere with origin $\mathbf{o}$.

3. The ordinates $\mathbf{b}_{222}$, $\mathbf{b}_{221}$, $\mathbf{b}_{212}$, and $\mathbf{b}_{122}$ must be coplanar.
4. The ordinates $\mathbf{b}_{310}$, $\mathbf{b}_{220}$, and $\mathbf{b}_{130}$ must be collinear.
5. The ordinates $\mathbf{b}_{311}$, $\mathbf{b}_{221}$, and $\mathbf{b}_{131}$ must be collinear.

For the mathematical derivation of the conditions we refer to the original work by Powell and Sabin [PS77]. For the sake of legibility we formulated the conditions for some representative vertices. Symmetric conditions apply to the other vertices. Figure 7 summarizes all the coplanarity and collinearity requirements, where the coplanarity requirements are depicted by the four hatched areas and the collinearity requirements by the nine highlighted edges.

Since we are operating on a spherical domain, we cannot apply the standard construction for a Powell-Sabin interpolant, but have to modify the individual steps in accordance with the spherical domain property and the coplanarity/collinearity constraints. In the following, we describe the construction of a Powell-Sabin interpolant over a spherical domain. Again, we cover all cases pick by describing the procedure for a representative vertex of each case.

First, we need to place the center vertex $\mathbf{p}_{222}$. In the Powell-Sabin interpolation over a planar domain, $\mathbf{p}_{222}$ is chosen to be the barycenter. Replacing the Euclidean metric $\|.\|_2$ with the metric $d(\mathbf{x}, \mathbf{y}) = \arccos(\mathbf{x} \cdot \mathbf{y})$ that gives a measure of the distance on the unit sphere, we obtain a vertex on the sphere:

$$\mathbf{p}_{222} = \frac{\mathbf{p}_{400} + \mathbf{p}_{040} + \mathbf{p}_{004}}{\|\mathbf{p}_{400} + \mathbf{p}_{040} + \mathbf{p}_{004}\|_2} .$$

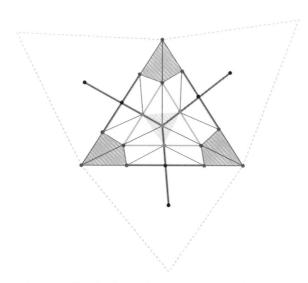

Fig. 7. The coplanarity (hatched areas) and collinearity (highlighted edges) conditions for the Powell-Sabin interpolant.

Using similar considerations, we set

$$\mathbf{p}_{220} = \frac{\mathbf{p}_{400} + \mathbf{p}_{040}}{\|\mathbf{p}_{400} + \mathbf{p}_{040}\|_2},$$

$$\mathbf{p}_{310} = \frac{\mathbf{p}_{400} + \mathbf{p}_{220}}{\|\mathbf{p}_{400} + \mathbf{p}_{220}\|_2},$$

$$\mathbf{p}_{311} = \frac{\mathbf{p}_{400} + \mathbf{p}_{222}}{\|\mathbf{p}_{400} + \mathbf{p}_{222}\|_2}, \text{ and}$$

$$\mathbf{p}_{221} = \frac{\mathbf{p}_{220} + \mathbf{p}_{222}}{\|\mathbf{p}_{220} + \mathbf{p}_{222}\|_2}.$$

To define the ordinates, we first set

$$\mathbf{b}_{400} = f(\mathbf{q}_0) \ .$$

Following coplanarity condition 2, the points $\mathbf{b}_{310}$, $\mathbf{b}_{311}$, and $\mathbf{b}_{301}$ must lie in the tangent plane at vertex $\mathbf{b}_{400}$. We approximate the tangent plane by fitting a plane through $f(\mathbf{q}_0)$ and its neighbored vertices, i.e. the vertices that belong to the 1-ring of the triangulation described in the previous section, in the least-squares sense. The least-fitting plane is translated to obtain a plane P_t that is parallel to the least-squares fitting plane and contains $f(\mathbf{q}_0)$. The position of point $\mathbf{b}_{310}$ is computed as the intersection point of the line $\mathbf{o} + \lambda \mathbf{p}_{310}$, $\lambda \in \mathbf{R}$ with plane P_t, see Figure 8. Analogously, we compute $\mathbf{b}_{311}$ and $\mathbf{b}_{301}$.

Next, we need to compute $\mathbf{b}_{222}$. From collinearity condition 1 and the spherical domain property we derive that $\mathbf{b}_{222}$ must lie in the plane P_n through the origin $\mathbf{o}$ and the center vertices $\mathbf{p}_{222}$ and $\mathbf{p}'_{222}$. Moreover, from collinearity

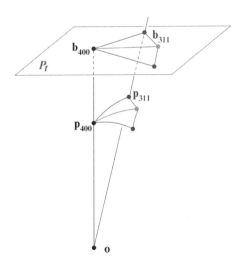

Fig. 8. Computation of $\mathbf{b}_{310}$ by intersection of line $\mathbf{o} + \lambda \mathbf{p}_{310}$, $\lambda \in \mathbf{R}$ with tangent plane P_t at vertex $\mathbf{b}_{400}$.

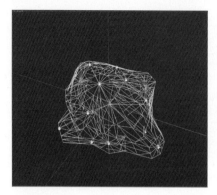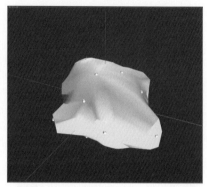

Fig. 9. Refined triangular mesh of Figure 4 with respect to the presented Powell-Sabin scheme over a spherical domain (wireframe and shaded rendering).

condition 4 we derive that $\mathbf{b}_{222}$ must lie on the edge between $\mathbf{b}_{310}$ and $\mathbf{b}_{130}$. Thus, we compute $\mathbf{b}_{222}$ as the intersection point of the line $\mathbf{b}_{310} + \lambda \mathbf{b}_{130}$, $\lambda \in \mathbf{R}$, and P_n.

The position of point $\mathbf{b}_{221}$ is determined, again, by collinearity condition 1 and the spherical domain property plus collinearity condition 5. Hence, $\mathbf{b}_{221}$ is computed as the intersection point of the line $\mathbf{b}_{311} + \lambda \mathbf{b}_{131}$, $\lambda \in \mathbf{R}$, and P_n.

Finally, the estimation of point $\mathbf{b}_{222}$ is ruled by coplanarity condition 3. We set $\mathbf{b}_{222}$ to the intersection point of line $\mathbf{o} + \lambda \mathbf{p}_{222}$, $\lambda \in \mathbf{R}$, and the plane that is spanned by $\mathbf{b}_{221}$, $\mathbf{b}_{212}$ and $\mathbf{b}_{122}$.

Figure 9 shows the example of a deformed sphere model of Figure 4, where each triangle has been refined with respect to the described Powell-Sabin scheme over a spherical domain. This is not the resulting Powell-Sabin interpolation yet, but just a rendering of the subdivided triangles. Hence, this is still a piecewise linear representation.

7 Loop Subdivision

Having established Bézier patches, we can evaluate the Bézier polynomials at any point of the underlying domain. In particular, for visualization purposes we would have to evaluate the polynomials over a dense discrete grid. These computations are rather complex. Therefore, we decided to not evaluate the polynomials analytically, but to use subdivision methods, instead.

The subdivision scheme by Loop [Loo87] is a well-known and commonly used subdivision scheme for triangular meshes. For regular triangular meshes, i.e. when all vertices have valence six, applying the subdivision scheme iteratively converges to a surface with the property of being C^2-continuous.

We are dealing with patches with regular triangular structure within the patch. We are not applying subdivision masks across the border of patches. In particular, the original vertices (before applying the Powell-Sabin scheme)

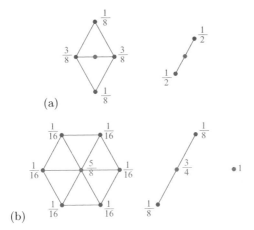

Fig. 10. Loop subdivision masks for (a) edges and (b) vertices. (a) Masks that insert new vertices are applied to interior edges (left) and boundary edges (right). (b) Masks that adjust old vertices are applied to interior vertices (left), boundary vertices except original vertices (middle), and original vertices (right).

should not be modified. They should always maintain their heights representing the measured diffusion (scattered data values). Thus, the subdivision scheme is not approximating but interpolating the original ordinates. This property is important to not fudge with the original data. To other vertices and edges on the patches' boundary we apply a one-dimensional subdivision scheme along the direction of the original edges.

We apply the Loop-subdivision masks shown in Figure 10. Figure 10(a) shows the masks for inserting new vertices, while Figure 10(b) shows the masks for adjusting the values of the old vertices. Figure 10(a) shows the mask applied to any interior edge of the patch on the left-hand side and the mask applied to any boundary edge of the patch on the right-hand side. Figure 10(b) shows (from left to right) the masks applied to any interior vertex of the patch, applied to any vertex on the boundary of the patch (excluding the original vertices), and applied to the original vertices.

In one subdivision step, all masks are applied once to all edges and vertices of all patches. The masks are applied simultaneously. For our purposes, we typically apply up to two subdivision steps. Figure 11 shows the Bézier mesh of Figure 9 after one (top) and two (bottom) subdivision steps. The resulting mesh is a sufficiently precise representation of our deformed-sphere glyph.

8 Results and Discussion

A healthy female volunteer was studied using a standard twelve-direction gradient scheme on a Siemens Symphony MRI. 50 slices were measured with an

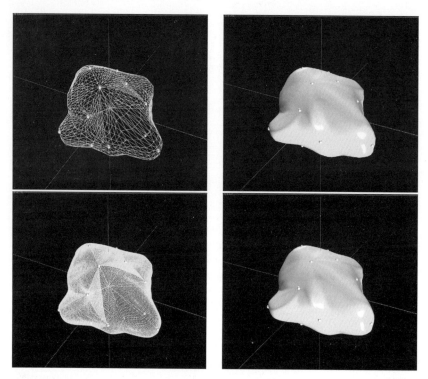

Fig. 11. Triangular mesh after one (top row) and two (bottom row) steps of Loop subdivision (wireframe and shaded rendering). Subdivision scheme is applied such that it interpolates the original heights.

image size of 128×128 pixel. The scanned voxel size was $1.64 \times 1.64 \times 2.2 mm^3$, which was resampled to isometric voxel size ($1 \times 1 \times 1 mm^3$) using a trilinear interpolation scheme. The acquisition for each slice was repeated four times for magnitude averaging and noise reduction. Scanning parameters for the measured sequence were given by the repetition time $TR = 10000 ms$ and the echo time $TE = 93$, and the b-value that describes the parameters of the diffusion-weighted gradients of the diffusion MR sequence was set to $1000 s/mm^2$.

We applied our visualization with deformed-sphere glyphs to the individual voxels of the given data set. Figure 12 shows our results for three distinguished cases and the comparison with ellipsoidal tensor glyphs. When using tensor glyphs, three main cases are distinguished, namely the prolate case, where one eigenvalue is significantly larger than the others and dominates the shape of the glyph, the oblate case, where the two largest eigenvalues are in the same range while the third one is significantly smaller, and the isotropic case, where the three eigenvalues are all in the same range. Figure 12 depicts these three tensor glyph cases and contrasts the results against the results we obtained

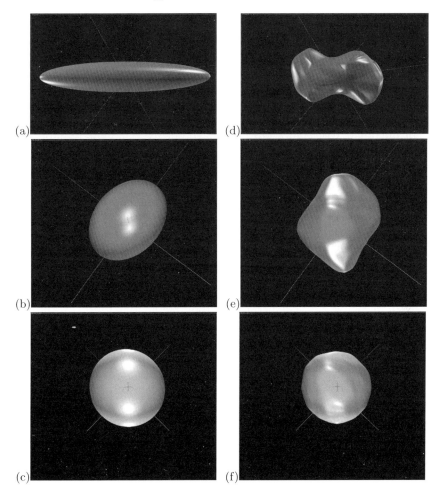

Fig. 12. Comparison of ellipsoidal tensor glyphs (left column) and deformed-sphere glyphs (right column) for the prolate (top row), oblate (middle row), and isotropic case (bottom row). The pictures exhibit a loss of information during tensor generation. This additional information is used for the deformed-sphere glyphs to visually extract the main diffusion directions.

using deformed-sphere glyphs. The tensor glyphs are rendered by an ellipsoid, where the axes correspond to the eigenvector directions of the tensor and the eigenvalues are used to scale the ellipsoid in the respective eigenvector direction.

In the prolate case shown in Figure 12(a) we obtain a cigar-shaped tensor glyph. As expected, the maximum deformation of our glyph shown in Figure 12(d) is approximately in the direction of the principal eigenvector. In many prolate cases the tensor glyph and the deformed-sphere glyph are similar. However, for the chosen example, our glyph illustrates that there are two directions with high diffusion. This may hint to a crossing or kissing of fiber bundles, or even a bifurcation. This additional information gets lost when operating with the tensor only. Since the angle between the two main directions is small, thus far from being orthogonal, the tensor glyph becomes a prolate ellipsoid, which is typically considered as a case with one main direction only.

In the oblate case shown in Figure 12(b) one expects more than one main direction. However, the flat ellipsoid we obtain when using tensor glyphs does not exhibit the main directions visually. One can only observe the plane they lie in. Certainly, one could use a different tensor-glyph representation that exhibits the two major eigenvector directions, but those would be orthogonal to each other, while the main directions are, in general, not orthogonal. Our deformed-sphere glyph shown in Figure 12(e) is also flat, but clearly shows the main diffusion directions.

Finally, in the isotropic case shown in Figures 12(c) and (f), one expects that our model has low deformations and is close to the spherical tensor glyph. Indeed, the pictures illustrate this behavior. The isotropic case is emerging either from highly mixed structures or from free floating fluids.

The images in Figure 12 display results when using twelve diffusion gradient directions. Similar results are obtained when using six directions only, i. e. when there should be no loss of information when using tensor glyphs.

The color scheme we are using to render the glyphs is based on the principle direction, i. e. the direction of the principal eigenvector for tensor glyphs and the direction of maximum deformation for the deformed-sphere glyphs. The direction of maximum deformation is estimated by evaluating the discrete vertex positions after having applied the Loop subdivision steps. Using this principal direction we map the respective vector to an RGB value following the mapping described by Pajevic and Pierpaoli [PP99]. The glyph is rendered with the respective color. The color scheme is related to the ideas by Kindlmann and Weinstein [KW99] and enhances the visual perception of the directions of main deformation.

As can be seen from Figure 12, the colors of the tensor and the deformed-sphere glyphs may differ. In the prolate case shown here (and to some extent in the oblate case), the principle direction of the tensor glyph is the average of the two main directions of the deformed-sphere glyph. Hence, instead of finding the global maximum, as done when using the tensor, for certain cases it would make sense to find two (or more) independent directions.

Figure 13 shows multiple views of a section of one slice of a data set with a tumor. The data set has been scanned using the same parameters as mentioned above. Overlaid with the slice we render an array of deformed-sphere glyphs. The colors indicate the principal directions. An FA (fractional anisotropy) threshold of 0.2 was used to exclude most of the isotropic voxels of the data for glyph generation. We obtain a good overview of the diffusion measurements within this slice. Moreover, one can observe how the tumor led to destruction

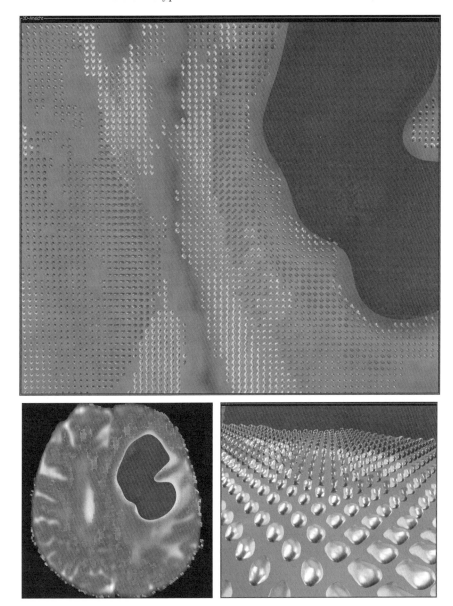

Fig. 13. Different views of a glyph-based visualization of a slice through a data set with a tumor using deformed spheres. The colors indicate the direction of maximum diffusion. Isotropic voxels with a fractional anisotropy of less than 0.2 are omitted.

and displacement of nerve fiber bundles. in particular, the glyphs surrounding the tumor indicate how the fibers pass by the tumor.

Our results document an approach to visualize diffusion MR data for estimating the organizational patterns of white matter structures. The tensor-based methods encounter several problems that can mostly be ascribed to the reduction of information in the course of the tensor calculation. The diffusion in highly anisotropic media such as white matter in the human brain is well-explained by means of the tensor matrix. In this case the off-diagonal components of the matrix tend to be near zero and the main eigenvector of the tensor represents, with a given uncertainty, the main direction of the diffusion. Regions with mixed structures, i.e. the crossing and kissing of nerve fiber bundles, result in tensors with non-negligible off-diagonal components. An eigenvalue analysis leads to two or even three eigenvalues that are nearly equal. Some attempts were made to solve the problem of oblate and spheroidal shaped tensors, as shown in Section 2. An important conclusion is that the tensor-based visualization methods are only capable of pointing out the three principle directions of the data, but not the principle directions of the underlying real-world structures. Our geometric approach demonstrated the ability to visually resolve mixed diffusion directions.

9 Conclusion and Future Work

We have presented a glyph-based visualization approach for diffusion MR data. Instead of operating on a symmetric 3 × 3 tensor that represents the data in each voxel by six distinct matrix entries, we use all measurements in the typically more than six diffusion gradient directions. This method allows us to visually extract more information concerning main diffusion directions. More than one diffusion directions are of interest with respect to bifurcation, crossing, or kissing of fiber bundles.

Our method is based on a deformed sphere model, where the deformation of a unit sphere directly visualizes the diffusion in the respective directions. The measurements are scattered over a spherical domain. To establish a continuous representation of the deformed sphere model, we apply a domain triangulation using a convex hull algorithm and a Powell-Sabin interpolation on a sphere. The resulting representation is evaluated by using some few Loop subdivision steps, which then is rendered using an intuitive color scheme.

For future work, we want to adopt our ideas and results for the generation of a fiber tracking algorithm. We want to evaluate the resulting fiber structures using a phantom. For the presented interpolation scheme, we want to evaluate how sensitive our method is with respect to the choice of the interpolation scheme. We also plan on comparing our results to the results when using higher-order tensors [HS05] and on evaluating the clinical relevance of our methods.

References

[Aur91]   F. Aurenhammer. Voronoi diagrams - a survey of a fundamental geometric data structure. *ACM Computing Surveys*, 23:345–405, 1991.

[Bar81]   A. Barr. Superquadrics and angle-preserving transformations. *IEEE Computer Graphics and Applications*, 18(1):11–23, 1981.

[Bar96]   C.B. Barber. The quickhull algorithm for convex hulls. *ACM Transactions on Mathematical Software*, 22(4):469–483, 1996.

[BBKW02a] Mats Bjornemo, Anders Brun, Ron Kikinis, and Carl-Fredrik Westin. Regularized stochastic white matter tractography using diffusion tensor mri. In *MICCAI 2002*, 2002.

[BBKW02b] A. Brun, M. Bjornemo, R. Kikinis, and C.-F. Westin. White matter tractography using sequential importance sampling. In *ISMRM 2002*, 2002.

[BPP+00] Peter J. Basser, Sinisa Pajevic, Carlo Pierpaoli, Jeffrey Duda, and Akram Aldroubi. In vivo fiber tractography using dt-mri data. *Magnetic Resonance in Medicine*, 44:625–632, 2000.

[HS05]   Mario Hlawitschka and Gerik Scheuermann. Hot-lines - tracking lines in higher order tensor fields. In Cláudio T. Silva, Eduard Gröller, and Holly Rushmeier, editors, *Proceedings of IEEE Conference on Visualization 2005*, pages 27–34, 2005.

[KW99]   Gordon Kindlmann and David Weinstein. Hue-balls and lit-tensors for direct volume rendering of diffusion tensor fields. In *Proceedings of IEEE Conference on Visualization '99*, pages 183–189, Los Alamitos, CA, USA, 1999. IEEE Computer Society Press.

[LeB01]   D. LeBihan. Diffusion tensor imaging: Concepts and applications. *Journal of Magnetic Resonance Imaging*, 13:534–546, 2001.

[Loo87]   C.T. Loop. Smooth subdivision surfaces based on triangles. Master's thesis, Department of Mathematics, University of Utah, 1987.

[OAS02]   O.Coulon, D.C. Alex, and S.R.Arridge. Tensor field regularisation for dt-mr images. In *Proceedings of British Conference on Medical Image Understanding and Analysis 2002*, 2002.

[PB96]   C. Pierpaoli and P. J. Basser. Toward a quantitative assessment of diffusion anisotropy. *Magn Reson Med.*, 36(6):893–906, 1996.

[PCF+00] C. Poupon, C. A. Clark, V. Frouin, J. Regis, I. Block, D. Le Behan, and J.-F.Mangin. Regularization of diffusion-based direction maps for the tracking of brain white matter fascicles. *NeuroImage*, 12:184–195, 2000.

[PDD+02] P.G.Batchelor, D.L.G.Hill, D.Atkinson, F.Calamanten, and A.Connellyn. Fibre-tracking by solving the diffusion-convection equation. In *ISMRM 2002*, 2002.

[PP99]   Sinisa Pajevic and Carl Pierpaoli. Color schemes to represent the orientation of anisotropic tissues from diffusion tensor data: Application to white matter fiber tract mapping in the human brain. *Magnetic Resonance in Medicine*, 42:526–540, 1999.

[PS77]   M.J.D. Powell and M.A. Sabin. Piecewise quadratic approximation on triangles. *ACM Transactions on Mathematical Software*, 3(4):316–325, 1977.

[PSB+02]   Geoffrey J.M. Parker, Klaas E. Stephan, Gareth J. Barker, James B. Rowe, David G.MacManus, Claudia A. M. Wheeler-Kingshott, Olga Ciccarelli, Richard E. Passingham, Rachel L. Spinks, Roger N. Lemon, and Robert Turner. Initial demonstration of in vivo tracing of axonal projections in the macaque brain and comparison with the human brain using diffusion tensor imaging and fast marching tractography. *NeuroImage*, 15:797–809, 2002.

[TRW+02]   D. S. Tuch, T. G. Reese, M. R. Wiegell, N. Makris, J. W. Belliveau, and V. J. Wedeen. High angular resolution diffusion imaging reveals intravoxel white matter fiber heterogeneity. *Magn. Reson. Med.*, 48(4):577–582, 2002.

[Tuc04a]   David S. Tuch. High angular resolution diffusion imaging reveals intravoxel white matter fiber heterogeneity. *Magnetic Resonance in Medicine*, 48:577–582, 2004.

[Tuc04b]   David S. Tuch. Q-ball imaging. *Magnetic Resonance in Medicine*, 52:1358–1372, 2004.

[TWBW99]   D. S. Tuch, R. M. Weisskoff, J. W. Belliveau, and V. J. Wedeen. High angular resolution diffusion imaging of the human brain. In *Proceedings of the 7th Annual Meeting of ISMRM*, page 321, 1999.

[WKL99]   David M. Weinstein, Gordon L. Kindlmann, and Eric C. Lundberg. Tensorlines: Advection-diffusion based propagation through diffusion tensor fields. In *IEEE Visualization '99*, pages 249–254, 1999.

[WMM+02]   C.-F. Westin, S. E. Maier, H. Mamata, A. Nabavi, F. A. Jolesz, and R. Kikinis. Processing and visualization for diffusion tensor mri. *Medical Image Analysis*, 6:93–108, 2002.

[ZB02]   L. Zhukov and A. Barr. Oriented tensor reconstruction: tracing neural pathways from diffusion tensor mri. In *Proceedings of the conference on Visualization 2002*, pages 387–394, 2002.

[ZDL03]   Song Zhang, Cagatay Demiralp, and David H. Laidlaw. Visualizing diffusion tensor mr images using streamtubes and streamsurfaces. *IEEE Transactions on Visualization and Computer Graphics*, 2003.

Visual Analysis of Bioelectric Fields

Xavier Tricoche, Rob MacLeod, and Chris R. Johnson

Scientific Computing and Imaging Institute
University of Utah
tricoche@sci.utah.edu, macleod@cvrti.utah.edu, crj@sci.utah.edu
http://www.sci.utah.edu

Summary. The bioelectric activity that takes place throughout the human body enables the action of muscles and the transmission of information in nerves. A variety of imaging modalities have therefore been developed to assess this activity. In particular, electrocardiography for the heart and electrocardiography and magnetoencephalography for the brain permit to measure non-invasively the resulting electric signal. Beyond their obvious clinical applications these techniques also open the door to a computational reconstruction of the physiological activity at the origin of this signal through the numerical solution of so-called inverse problems. In this case as well as in basic bioengineering research effective postprocessing tools are necessary to facilitate the interpretation of measured and simulated bioelectric data and permit to derive anatomical and functional insight from it. In contrast to scalar quantities for which effective depictions generally exist and are routinely used, the vector-valued nature of this information bears specific challenges that are insufficiently addressed by the visualization tools typically available to biomedical practitioners. This paper reports on the application of advanced vector visualization techniques to the postprocessing analysis of bioelectric fields as they arise in cardiovascular and inverse source reconstruction research. Our work demonstrates the ability of the corresponding visual representations to improve the interpretation of the data and support new insight into the underlying physiology.

1 Introduction

The bioelectric activity that takes place throughout the human body is essential to life. Electric, magnetic, and electromagnetic fields are produced by living cells that enable the action of muscles and the transmission of information in nerves. A variety of imaging modalities have therefore been developed to assess this activity. In particular, electrocardiography for the heart and electrocardiography and magnetoencephalography for the brain allow one to measure non-invasively the resulting electric signal. Beyond their clinical applications these techniques also provides the basis for a computational reconstruction of the physiological activity at the origin of this signal through the numerical solution of so-called inverse problems [9, 1].

To support the multiple bioengineering and biomedical applications that investigate the link between bioelectricity and physiology powerful post-processing tools are necessary that facilitate the interpretation of measured and simulated bioelectric field data and permit one to derive anatomical and functional insight. In contrast to scalar quantities, which have been the traditional focus of biomedical visualization and for which effective depictions generally exist, the vector-valued bioelectric field information bears specific challenges that are insufficiently addressed by the visualization tools typically available to biomedical practitioners. In particular, depictions restricted to scalar attributes derived from bioelectric vector fields essentially deprive the analyst of the wealth of structural contents that is encoded by a vector field and its associated mathematical flow. In our case, flows naturally arise as the interpretation of an arbitrary vector field as the right-hand side of a differential equation. This standard operation permits to define integral curves (or *streamlines*) and it provides a vector field with a continuous geometric description. The latter is essential from a visualization standpoint because it lends itself to expressive representations that better exploit the ability of the human visual system to perceive structures and patterns than glyph-based depictions of the discrete vector data itself. It is important to emphasize that the type of flow that we are considering in the context of bioelectric fields is obviously different from the fluid flows encountered in engineering applications of fluid dynamics, which have been the focus of most of the vector visualization methods developed within the scientific visualization community [14, 8]. In particular, the vector-valued data found in bioelectric problems does not correspond to a motion descriptor. Yet, a common mathematical formalism offers an interesting opportunity to explore the applicability and potential extension of existing flow visualization techniques to bioelectric field problems.

In this context, the present paper describes the mathematical and algorithmic aspects involved in the application of advanced vector visualization techniques to the visual analysis of bioelectric fields. The focus is on electric fields and electric current densities associated with the cardiovascular and cerebral activity in humans. The data considered in the examples shown hereafter were acquired through the solution of so-called forward problems, in which 3D bioelectric fields are computed for a known bioelectric source [6]. The solution of these numerical problems is instrumental in the solution of inverse problems, through which researchers try to reconstruct the source activity at the origin of the signals measured non-invasively on the body surface.

The contents of the paper are organized as follows. Section 2 introduces some basic notions of bioelectricity and explains the relationship between the corresponding scalar, vector, and tensor quantities. Section 3 provides an overview of advanced flow visualization techniques. Section 4 demonstrates the potential of these methods for cardiovascular research while section 5 comments on corresponding results in the context of brain research. We conclude the presentation by summarizing the main results introduced in the paper in section 6.

2 Basic Notions of Bioelectricity

For completeness, we introduce elementary notions of bioelectricity. The equations hereafter clarify the interconnections between scalar, vector, and tensor quantities that appear in the corresponding equations. Given the complexity of this topic, our ambition here is obviously limited and we restrict the presentation to an informal summary of basic notions that underlie the existence of bioelectric potentials and currents. For a more rigorous treatment of the subject we refer the reader to [11, 15].

2.1 Physical Laws

The Maxwell's equations are fundamental relations that link the electric field $\mathbf{E}$, the electric flux density (or electric displacement) $\mathbf{D}$, the electric current density $\mathbf{J}$, the magnetic field $\mathbf{H}$, the magnetic flux density (or magnetic induction) $\mathbf{B}$, and the electric free charge density ρ (bold letters denote vector quantities). Their differential form is as follows [15].

$$\nabla \times \mathbf{E} = -\frac{\partial \mathbf{B}}{\partial t} \tag{1}$$

$$\nabla \times \mathbf{H} = \mathbf{J} + \frac{\partial \mathbf{D}}{\partial t} \tag{2}$$

$$\nabla \cdot \mathbf{D} = \rho \tag{3}$$

$$\nabla \cdot \mathbf{B} = 0 \tag{4}$$

Additionally, current density and electric field are related through following equation.

$$\mathbf{J} = \sigma \mathbf{E}, \tag{4}$$

where σ corresponds to the electric conductivity, which is a second-order tensor field in general and reduces to a scalar field if the conductivity is homogeneously distributed. A number of simplifications can be made to these general formulae that take into account specific properties of the considered electromagnetic phenomenon. In particular, the bioelectric fields related to the cardiac and cerebral function are quasistatic. In this case, the above equations reduce to following expressions.

$$\nabla \times \mathbf{E} = \mathbf{0} \tag{5}$$

$$\nabla \times \mathbf{H} = \mathbf{J} \tag{6}$$

$$\nabla \cdot \mathbf{D} = \rho \tag{7}$$

$$\nabla \cdot \mathbf{B} = 0 \tag{8}$$

Hence, the electric field can be expressed as the gradient of an electric potential Φ:

$$\mathbf{E} = -\nabla \Phi \tag{8}$$

The relationship between the electric field and the resulting current flow of charged particles $\mathbf{J}$ is then defined by Ohm's law.

$$\mathbf{J} = \sigma \mathbf{E} = -\sigma \nabla \Phi, \tag{8}$$

If electric sources are present in the medium, they can be expressed in terms of a source density I_ν. Since the divergence operator measures the net outflow per unit volume, this source density is linked to the current density by following equation.

$$\nabla \cdot \mathbf{J} = I_\nu \tag{8}$$

Combining Equation (2.1) and Equation (2.1) we finally obtain the fundamental Poisson equation, which links scalar, vector, and tensor quantities in a single expression.

$$\nabla \cdot \mathbf{J} = I_\nu = -\nabla \cdot (\sigma \nabla \Phi) \tag{8}$$

2.2 Bioelectric Potentials and Impulse Propagation

Biological tissues behave as an electrolyte. In other words, they are conductive media in which ions play the role of charge carriers. Sodium, potassium, calcium, and chloride are present in concentrations that vary significantly between the intracellular and extracellular media of *excitable cells*, most notably nerves and muscles. The corresponding concentration gradient generates a diffusion flow of ions across the cell membrane. Another flow exists that is induced by the action of an electric field on the potential created by the charges associated with the ions.

The thin cell membrane controls the exchange of ions between intracellular and extracellular domains. One building block of the membrane are proteins called *pumps* that consume energy to maintain the concentrations of sodium and potassium on both sides of the membrane, and that against the flow caused by concentration gradient. On the opposite, *channels* use energy to permit the flow of ions down the concentration gradient in a very controlled way. The resulting *selective permeability* is modified in response to the presence of an electric field and allows for fast changes of the transmembrane potential.

At rest, excitable cells have a negative potential, which is called *polarized* state. When a stimulus reaches the cell, the behavior of the channels is modified. The flow of positive charges into the cell is suddenly facilitated, leading to a rapid potential increase. This is called an *action potential* and corresponds to a *depolarization* of the cell. Once the potential reaches a threshold value the behavior of the channels changes and the inflow of positive charges is interrupted. The pumps are then used to transfer ions in and out of the cell, which eventually returns to its polarized state. For long thin fibers, the action potential affects only a limited patch of the membrane at a time. This impulse propagates by subsequently affecting the potential of adjacent patches. This results in their depolarization and regenerates another action potential cycle down the fiber. This propagation mechanism applies to nerves and individual muscle fibers, but also to more complex structures like nerve trunk and cardiac muscle.

2.3 Data Acquisition

Non-invasive measurement techniques are essential in clinical practice. *Electrocardiography* (ECG) monitors cardiac activity by recording the evolution of the electric potential on the chest induced by successive excitation and repolarization waves occurring in heart. At a cellular level, depolarization and repolarization are linked to a phenomenon called *action potential* under which the varying permeability of the cell membrane entails a rapid increase of the cell potential (excitation or depolarization) that is followed by a decrease back to its original value (repolarization or recovery). The corresponding impulse propagates along muscular bundles, generating waves that are responsible for the electric signals measured on the body surface. ECG is used to detect cardiac arrhythmia (e.g. a skipped heart beat). It is also instrumental in the diagnosis of ischemia, a condition corresponding to a shortfall of the blood supply of the heart which reduces its pump function and can eventually lead to a heart attack. Similarly, the activity of the brain can be monitored through *Electroencephalography* (EEG), which records brain waves through the resulting variations of the electric potential on the scalp. In this case the electric signal is much weaker than the one produced by the heart and individual action potentials cannot be detected. Instead EEG measures the signal that arises form the synchronized bioelectric activity of a large number of neurons. This signal exhibits patterns of different frequencies that are related to different mental activities as well as pathologies. The latter includes *epilepsy* and *Alzheimer's disease* among others.

Computationally, a precise anatomical modeling combined with the use of realistic values for the conductivity of individual anatomical structures can be used to simulate bioelectric activity. Two types of computation exist. The forward problem consists in solving for the potential on the body surface when bioelectric sources are known. Conversely, inverse problems correspond to the localization of the source for which the boundary signal is known. Inverse problems are typically ill-posed and assumptions regarding the source model, the approximate location of the sources, and their number have to be made to obtain a unique solution. Source reconstruction by solution of an inverse problem is a prominent goal of the non-invasive techniques mentioned previously. Observe that solving the forward problem is instrumental in the solution of the corresponding inverse problem. As a matter of fact, identifying the parameters underlying observed data requires the repeated simulation of the bioelectric field distribution for a given source activity, as the space of possible parameters is being explored.

3 Visualization of Vector-Valued Data

The depiction of the vector data that is ubiquitous in scientific problems has always been a major research focus topic for the scientific visualization community. The primary focus of the techniques created for that purpose has traditionally been on problems generally related to the engineering applications of fluid dynamics, e.g. automotive industry, aircraft design, and turbomachinery. In contrast, the vector-valued contents of biomedical datasets has received, comparatively, very little attention from visualization practitioners. As a consequence, a striking disconnect exists between the advanced vector visualization methods devised after several decades of dedicated research and the tools typically available to biomedical researchers. While these experts are mostly concentrating their visual investigation on scalar quantities, it is important to observe that a quantity as fundamental as the bioelectric current is derived from the bioelectric potential through multiplication with the conductivity tensor, see Eq. 5. As such, it is not the mere gradient of a scalar quantity but a quantity whose full description requires a vector field.

This section offers a brief introduction to three major categories of such advanced techniques, deferring their application to bioelectric fields to subsequent sections. Observe that streamline techniques, which have been proposed early on to visualize bioelectric fields, are not specifically considered in the following since they have already benefited from a significant attention [10].

3.1 Texture Representations

Texture-based techniques permit to generate intuitive visualizations of steady as well as unsteady vector fields. In contrast to depictions based on a set of discrete geometric primitives, the resulting dense representation is a powerful way to convey essential patterns of the vector field while avoiding the tedious task of seeding individual streamlines to capture all the structures of interest. The basic idea of most of the corresponding methods consists in applying a one-dimensional low-pass filter to an underlying white noise image. The filter kernel is here defined over individual streamline trajectories at each pixel of the image, thus creating patterns of strongly coherent colors that are aligned with the underlying continuous flow. At the same time, a strong contrast is preserved along directions orthogonal to the vector field orientation, which results in intuitive pictures that mimic a dense coverage of the domain with thick streamlines. A wide variety of techniques have been proposed following this basic principle [8], including a recent application to electromagnetic fields [17]. Originally applied to planar flows, these algorithms were later extended to vector fields defined over curved surfaces [7, 20].

3.2 Topological Methods

Topological methods apply to the visualization of vector fields a mathematical framework developed for the qualitative study of dynamical systems [12].

This approach was introduced in the fluid dynamics community in an effort to provide a rational language to describe complex three-dimensional flow patterns observed in experimental flow visualizations. So-called critical points (where the magnitude of a vector field vanishes) are the nodes of a graph that segments the domain into regions of qualitatively different behaviors. The boundaries between neighboring regions called separatrices are streamlines or stream surfaces.

This technique was introduced in flow visualization [4, 3] in the form of methods that construct the topological graph for two-dimensional flows and laid the basis of an extension to three-dimensional flows. This seminal work stimulated numerous subsequent publications [16]. Of particular interest for the applications considered in the following is the extraction of the topological graph of a vector field defined over a curved boundary. The challenging aspect of this setting lies in the fact that the corresponding geometry is not smooth and the resulting discontinuities of the tangent plane lead to some specific topological features that must be accounted for in the visualization. In particular, the integration of separatrices can be done using the technique proposed by Polthier and Schmies [13], which is built upon the notion of geodesic on polygonal surfaces.

3.3 Stream Surfaces

Stream surfaces are formed by the advection of a (seed) curve along the flow. Hence, they embed in a continuous depiction the information conveyed by an infinite number of streamlines and enhance the understanding of depth and spatial relationship. They can be computed either implicitly [19] as isosurfaces of a pre-computed scalar function, or explicitly through numerical integration [5], which offers more flexibility in the choice of the initial curve and yields a geometric representation of better quality. The basic principle of this latter approach consists in propagating a polygonal front by integrating the trajectories followed by its discrete vertices. A divergent behavior of neighboring vertices can be dynamically accounted for by adaptively inserting additional vertices. The triangulation of the surface spanned by this motion is constructed on the fly. We recently introduced a method that allows a more precise control over the tessellation of the surface and creates stream surfaces of high visual quality, even in the challenging case of turbulent flows [18, 2].

4 Visualization of Cardiac Bioelectricity

The dataset that we consider in this section consists of a forward Finite Element computation of the electric potential Φ over the volume of a torso. The general context of this computation is a research project aimed at gaining new insight into the mechanisms and consequences of myocardial ischemia through the analysis of experiments and simulations. Myocardial ischemia corresponds to a shortfall in blood supply, which arises when blood supply to the heart does not meet demand, leading to chest pain (angina) and reduced pump function. In the extreme, a complete blockage of blood leads to a heart attack. The FEM method is chosen for the flexibility it provides in the choice of the torso geometry (the heart surface – the epicardium – constitutes here an internal boundary) and the distribution of tissue conductivity. The source information for the forward problem on the epicardium stems from experimental measurements carried out by Dr. Bruno Taccardi at the Cardiovascular Research and Training Institute (CVRTI) at the University of Utah. The elements used are linear tetrahedra. The resulting current density (obtained by derivation of the electric potential) is piecewise constant over the mesh and an averaging over the control volume (i.e. the direct neighborhood) of each node combined with piecewise linear interpolation yields the projected value of the current density field in a C^0 space suitable for flow integration methods.

4.1 Electric Field over Boundary Surfaces

The visualization techniques applied to the epicardium in cardiovascular research are typically restricted to representations of the electric potential, a scalar quantity that can be depicted using a combination of color coding and isocontours on the geometry of the heart. This picture can be enhanced by adding glyphs that show both the direction and the magnitude of the three-dimensional electric field crossing the epicardium as shown in Fig. 1.

Additional information can be conveyed by the use of a dense, texture-based representation of the electric field defined over the surface of the epicardium. Visualizing the associated topological graph permits to indicate the electric connectivity of the epicardium and highlight coherent structures of the current, which naturally complements the texture-based visualization. An interesting question in this context corresponds to the potential interpretation of the topological features exhibited by the electric field in terms of what is known in cardiac electrophysiology as epicardial breakthrough. This phenomenon occurs when an activation wave in the cardiac tissue breaks through the

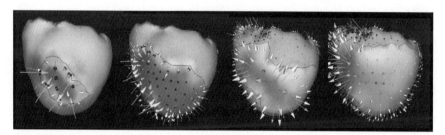

Fig. 1. Electric potential (color coding) and electric field (glyphs) on the epicardium. *Left and middle left:* excitation phase. *Middle right and right:* repolarization phase.

Fig. 2. Topology of bioelectric field on epicardium. The images show a texture representation of the potential gradient on the surface enhanced by the depiction of the associated topological graph. Green points correspond to potential minima, blue points mark potentials maxima. Left and middle images show an anterior view of the epicardium. Right image is seen from the left of the torso.

surface, generating a local potential minimum on the epicardial surface, as shown in Fig. 2.

Similar pictures can be achieved for the torso surface. In that case the topological structure provides little additional information in the absence of clearly defined separatrices, which is due to the overall dipolar nature of the return current. However, significant lines can be extracted, along which the surrounding flow tends to converge as it moves toward the chest boundary or away from it. These feature lines are shown on the left image in Fig. 3. These lines indicate where the electric current created by a dipolar source is divided in two parts along the domain boundary, i.e. the torso, since no current can escape the conducting volume. Texture representations can also be combined with a color coding of the electric potential to the show the interrelation between quantitative and structural aspects. They can also be computed over cutting planes as a means to unveil the patterns of the bio-electric return current in the chest volume. The use of transparency permits to mitigate the occlusion issues inherently associated with this type of depiction. In this case, the texture, through careful orientation of the corresponding plane, is able to exploit the inherent symmetry of the flow structure created by the dipolar source to capture the geometric signature of its associated three-dimensional pattern, as shown in Fig. 3. Specifically, a first plane is positioned so as to contain the dipole axis which shows most of the structure and a second, transparent plane is introduced orthogonal to the first one to reveal the flow behavior along the missing dimension.

4.2 Electric Current in the Torso

Previous visualizations are limited to planes and curved surfaces. To offer further insight into the structural properties of the current inside the torso volume, stream surfaces can be used to probe the three-dimensional geometry

Fig. 3. Texture representations of the return current in the torso. *Left:* Visualization of the torso surface combined with lines along which the volumetric current attaches or detaches from the outer boundary. *Middle:* Texture combined with color coding of electric potential (negative values mapped to blue, positive values to red). *Right:* Two textures computed over cutting planes combined by transparency. The geometry of the heart remains opaque for context.

of the current and inspect the resulting connectivity between distant locations distributed over the epicardium. The primary difficulty encountered when applying stream surfaces to explore 3D flow volumes lies in the choice of an appropriate seed curve. This curve must be chosen to capture interesting patterns. Because of the non-local nature of stream surfaces (their overall shape is determined by the numerical integration of the underlying vector field) finding a "good" seed curve is typically a difficult task that requires repeated tries to achieve satisfactory results.

To address this problem we leverage the dipolar nature of the cardiac source. The dipolar shape is namely a characteristic pattern of bioelectricity and surfaces can be used to visualize the resulting arches spanned by the return current across the volume. Specifically, we are interested in the interconnections between regions of inflow and outflow distributed over the epicardium. A way to investigate these connections consists in computing the dot product of the electric current with the surface normal on the epicardium and to select isocontours of the resulting flux (where positive values correspond to a source behavior while negative values indicate a sink) as seeding curves. The results are presented in Fig. 4. The arch-type connections induced by multiple dipoles as well as their intertwinement are clearly visible.

As presented here, stream surfaces are a powerful tool in the qualitative rather than quantitative assessment of three-dimensional bioelectric fields. Yet, an interesting extension of that type of representation would be to extract the full 3D (and transient) electric connectivity of the epicardium. In this case, objective results would be obtained that support comparisons and statistical analysis.

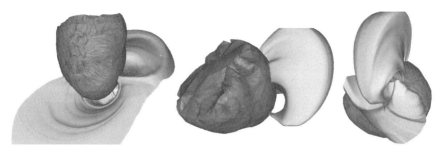

Fig. 4. Visualization of bioelectric field in the direct vicinity of epicardium with stream surfaces. The surfaces capture the geometry of the current induced by the dipole equivalent cardiac source. They also provide an effective representation of the interconnections that exist between different regions on the heart surface. The seeding curves correspond to isocontours of the electric flux selected close to local extrema. A rainbow color map is used along each seeding curve to visualize the stretching of the return current as it propagates through the torso.

5 Visualization of Cerebral Bioelectricity

The datasets considered in this section are all related to brain research and more specifically to the challenging problem of EEG source localization. The goal of the corresponding investigation is to assess the impact of different source localizations and different models of white matter conductivity anisotropy on the solution of the forward problem. In particular, this modeling strategy impacts the accuracy of the solution to the inverse problem required for source reconstruction since forward problems are building blocks of the corresponding computation. Practically, the forward problem is solved over a high-resolution FEM mesh composed of tetrahedral elements. The modeling of tissue conductivity is based on a segmentation that comprises skin, skull, cerebrospinal fluid (CSF), gray matter, and white matter [21]. A number of computations were carried out for different positions of a dipole source, yielding the values of the resulting electric potential on the head surface, which corresponds to the signal that can be acquired non-invasively using EEG. Moreover, the white matter conductivity was modeled as anisotropic in some simulations and compared to the solution of the forward problem obtained under the assumption of an isotropic conductivity of the white matter.

The task of the visualization in that context is to help elucidate the intricate structure of the three-dimensional return current in the brain and more specifically the impact of source localization and conductivity modeling on the corresponding patterns. It is important to emphasize that the non-uniformity of the conductivity and its anisotropy (in other words its tensor nature) result in a non-trivial relationship between electric potential – which is the focus of most visualizations in this type of application – and return current. The contribution of vector visualization techniques is therefore to shed some light on this relationship. Practically, as in the case of the torso dataset considered previously, a piecewise linear interpolation of the electric potential over the mesh yields piecewise constant electric field and return current (the conductivity tensor is cell-wise constant as well), which are transformed in continuous vector fields by averaging the values over the control volume of each vertex and applying piecewise linear interpolation to the resulting point-wise quantities.

5.1 Visualization of Bioelectricity on the Head Surface

As in the study of the heart, texture representations offer a natural means to visualize the return current on the boundary of the dataset, that is the head surface. Again, the natural complementarity between the directional information conveyed by the texture and the quantitative information (e.g. electric potential or current magnitude) encoded by a color map permits to offer a comprehensive picture of the bioelectricity on manifolds. Observe that the homogeneous Neumann boundary condition imposed on the simulation (i.e. the electric flux is zero across the head surface) justifies an analysis of the current restricted to boundaries since the vector field is tangent to the surface. The images shown in Fig. 5 reveal that the geometry of the return current is not affected by the conductivity model while its magnitude strongly varies between an isotropic and an anisotropic model.

5.2 Visualization of Bioelectricity in the Brain

As seen previously, texture representations can be used on cutting planes to help visualize the three-dimensional return current defined inside the brain. When applied to a dataset computed with an anisotropic model of white matter conductivity, this technique shows the asymmetry of the electric patterns, in strong contrast with the smooth and symmetric images obtained for

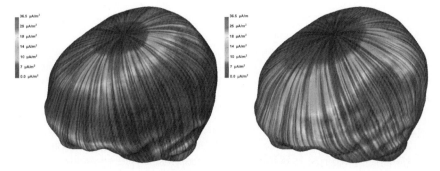

Fig. 5. Patterns of the surface return current corresponding to a source localized in the left thalamus. *Left:* an isotropic model has been used for the white matter. *Right:* a 1:10 anisotropic model has been applied to the white matter. The ratio corresponds to major vs. minor diffusion direction as inferred from a DTI scan. The color coding corresponds to the magnitude of the return current

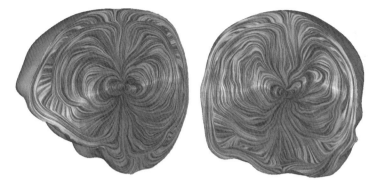

Fig. 6. Textures computed on sagittal and coronal clipping planes reveal details of the dipolar source and its interaction with the surrounding anisotropic tissue. The electric current is clearly diverted by the presence of white matter tracts that lie close to the source. The field also changes direction very rapidly as it approaches the skull just beneath the surface of the head.

isotropic models, as seen in Fig. 6. Notice that the canonical orientation of the planes corresponds to what is typically used in a clinical setting to visually assess the spatial distribution of scalar quantities. Moreover the planes are positioned so as to both contain the source position.

The complex patterns exhibited by these textures suggest that the return current has a very convoluted structure that cannot be fully captured by planar representations. Stream surfaces can therefore be employed to gain deeper insight into this three-dimensional structure. Yet, the intricacy of the return current makes the seeding task even more challenging than it was previously in the case of the heart dataset. Moreover, the natural seeding strategy provided by the epicardium and isocontours defined over it has no straightforward equivalent in this context. A possible solution consists in introducing an artificial surface, e.g. a sphere, around the known position of the source and to apply the flux-based seeding strategy mentioned in the previous section to this sphere. Alternatively, isopotential lines can be computed on the head surface and isolines located in the direct vicinity of potential extrema can play the role of seed curves. We remark that the homogeneous Neumann boundary conditions would impose that any field line started on the head surface remains embedded on that surface. Yet, an arbitrarily small offset from the head surface guarantees that the return current leaves the surface in one direction. These different strategies are illustrated in Fig. 7.

Because of the structural complexity of the return current and the lack of clearly defined symmetries (especially in the anisotropic case), visualizations composed of a few stream surfaces offer an intrinsically partial picture. Increasing the number of stream surfaces in an effort to achieve a dense filling of the space is faced with strong occlusion and visual clutter. A different solution therefore consists in using parameterized seed curves. In other words,

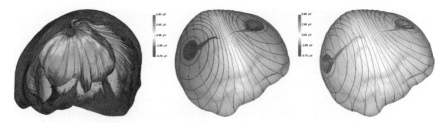

Fig. 7. Stream surface visualization of return current in the brain induced by dipolar source. *Left:* Seed curve corresponds to isocontour of electric flux on sphere enclosing the source (source located in left thalamus). *Middle:* Seed curves chosen as isocontours of the electric potential on the head surface (isotropic conductivity model, source in sematosensory cortex). *Right:* Same with 1:10 anisotropic conductivity model.

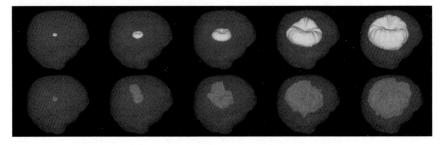

Fig. 8. Evolution of a stream surface integrated along the return current under the parameterization induced by the increasing radius of its seeding circle. *Top row:* Frames from an animation corresponding to isotropic white matter model. *Bottom row:* Frames of the animation obtained for anisotropic white matter model.

the seed curve becomes a dynamic curve and as the underlying parameter changes continuously, the resulting transformation of the associated stream surface can be visualized as an animation. To apply this general principle to the brain dataset, a circle of increasing radius can be used as a parametric curve, centered around the dipolar source and lying in the horizontal plane orthogonal to the dipole axis. An analysis of the differences between isotropic and anisotropic conductivity of the white matter can then be made in a side by side comparison of the corresponding animations. This method is illustrated in Fig. 8.

6 Conclusion

Electric and magnetic fields are vector fields that represent bioelectric activity in the human body. Yet, their interpretation has been hampered by the challenges of visualizing them in meaningful ways. One result has been that visualization and analysis have focused on the scalar potentials (electric and magnetic), for which mature visualization strategies exist. In this paper we have presented a variety of results from our ongoing research, which concentrates on the deployment and targeted extension of vector visualization methods that offer a global and expressive depiction of the continuous flow fields underlying the discrete information in numerical data sets. Our preliminary results show that visualizations combining a representation of the bioelectric current with related scalar attributes like potential and current amplitude permit a deeper understanding of the three-dimensional shape of the bioelectric sources and their fields. They also offer new insight into the impact of tissue characteristics, e.g., anisotropy, on the resulting bioelectric fields.

Acknowledgments

The authors are greatly indebted to Drs. Bruno Taccardi and Frank Sachse from CVRTI at the University of Utah, as well as to Dr. Carsten Wolters from the University of Muenster, Germany, who have provided their close collaboration, fruitful discussions, and the data sets that were used in the examples studied in the paper. This work was supported by the Center for Integrative Biomedical Computing, NIH NCRR Project 2-P41-RR12553-07.

References

1. A.M. Dale and M.I. Sereno. Improved localization of cortical activity by combining EEG and MEG with MRI cortical surface reconstruction. *J. Cognit. Neurosci.*, 1999.
2. C. Garth, X. Tricoche, T. Salzbrunn, T. Bobach, and G. Scheuermann. Surface techniques for vortex visualization. In *Data Visualization 2004 (Proceedings of Eurographics/IEEE TCVG Symposium on Visualization 2004)*, pages 155–164, 2004.
3. A. Globus, C. Levit, and T. Lasinski. A tool for visualizing the topology if three-dimensional vector fields. In *IEEE Visualization Proceedings*, pages 33 – 40, October 1991.
4. J.L. Helman and L. Hesselink. Representation and display of vector field topology in fluid flow data sets. *Computer*, 22(8):27–36, 1989.
5. J.P.M. Hultquist. Constructing stream surfaces in steady 3D vector fields. In *Proceedings of IEEE Visualization 1992*, pages 171–178, October 1992.
6. C.R. Johnson. Direct and inverse bioelectric field problems. In E. Oliver, M. Strayer, V. Meiser, D. Zachmann, and R. Giles, editors, *Computational Science Education Project*. DOE, Washington, D.C., 1995. http://csep1.phy.ornl.gov/CSEP/CSEP.html.
7. B. Laramee, J.J. van Wijk, B. Jobard, and H. Hauser. ISA and IBFVS: Image space based visualization of flow on surfaces. *IEEE Transactions on Visualization and Computer Graphics*, 10(6):637–648, nov 2004.

8. R.S. Laramee, H. Hauser, H. Doleisch, B. Vrolijk, F.H. Post, and D. Weiskopf. The state of the art in visualization: Dense and texture-based techniques. *Computer Graphics Forum*, 23(2):143–161, 2004.
9. R.S. MacLeod and D.H. Brooks. Recent progress in inverse problems in electrocardiology. *IEEE Engineering in Medicine and Biology Magazine*, 17(1):73–83, Jan/Feb 1998.
10. R.S. MacLeod, C.R. Johnson, and M.A. Matheson. Visualization of bioelectric fields. *IEEE Computer Graphics and Applications*, 14:10–12, Jul 1993.
11. R. Plonsey and R. C. Barr. *Bioelectricity, A Quantitative Approach. Second Edition.* Kluwer Academic / Plenum Publisher, 2000.
12. H. Poincaré. Sur les courbes définies par une équation différentielle. *J. Math. 1:167–244, 1875. J. Math. 2:151–217, 1876. J. Math. 7:375–422, 1881. J. Math. 8:251–296, 1882.*
13. K. Polthier and M. Schmies. Straightest geodesics on polyhedral surfaces. pages 391–408, 1998.
14. F.H. Post, B. Vrolijk, H. Hauser, R.S. Laramee, and H. Doleisch. The state of the art in flow visualization: Feature extraction and tracking. *Computer Graphics Forum*, 22(4):775–792, 2003.
15. F.B. Sachse. *Computational Cardiology: Modeling of Anatomy, Electrophysiology, and Mechanics.* Springer-Verlag New York, May 2004.
16. G. Scheuermann and X. Tricoche. Topological methods in flow visualization. In C.R. Johnson and C. Hansen, editors, *Visualization Handbook*, pages 341–356. Academic Press, 2004.
17. A. Sundquist. Dynamic line integral convolution for visualizing streamline evolution. *IEEE Transactions on Visualization and Computer Graphics*, 9(3):273–282, 2003.
18. X. Tricoche, C. Garth, T. Bobach, G. Scheuermann, and M. Ruetten. Accurate and efficient visualization of flow structures in a delta wing simulation. In *Proceedings of 34th AIAA Fluid Dynamics Conference and Exhibit, AIAA Paper 2004-2153*, 2004.
19. J.J. van Wijk. Implicit stream surfaces. In *Proceedings of IEEE Visualization '93 Conference*, pages 245–252, 1993.
20. D. Weiskopf and T. Ertl. A hybrid physical/device space approach for spatio-temporally coherent interactive texture advection on curved surfaces. In *Proceedings of Graphics Interface 2004*, pages 263–270, 2004.
21. C.H. Wolters, A. Anwander, X. Tricoche, D. Weinstein, and R.S. MacLeod. Influence of tissue conductivity anisotropy on EEG/MEG field and return current computation in a realistic head model: A simulation and visualization study using high-resolution finite element modeling. *NeuroImage*, 30(3):813–826, 2006.

MRI-based Visualisation of Orbital Fat Deformation During Eye Motion

Charl P. Botha[1], Thijs de Graaf[2], Sander Schutte[2], Ronald Root[2], Piotr Wielopolski[3], Frans C.T. van der Helm[2], Huibert J. Simonsz[4], and Frits H. Post[1]

[1] Data Visualisation, Delft University of Technology, Delft
[2] Biomechanical Engineering, Delft University of Technology, Delft
[3] Department of Radiology, Erasmus Medical Centre, Rotterdam
[4] Department of Ophthalmology, Erasmus Medical Centre, Rotterdam
   The Netherlands

Summary. Orbital fat, or the fat behind the eye, plays an important role in eye movements. In order to gain a better understanding of orbital fat mobility during eye motion, MRI datasets of the eyes of two healthy subjects were acquired respectively in seven and fourteen different directions of gaze. After semi-automatic rigid registration, the Demons deformable registration algorithm was used to derive time-dependent three-dimensional deformation vector fields from these datasets. Visualisation techniques were applied to these datasets in order to investigate fat mobility in specific regions of interest in the first subject. A qualitative analysis of the first subject showed that in two of the three regions of interest, fat moved half as much as the embedded structures. In other words, when the muscles and the optic nerve that are embedded in the fat move, the fat partly moves along with these structures and partly flows around them. In the second subject, a quantitative analysis was performed which showed a relation between the distance behind the sclera and the extent to which fat moves along with the optic nerve.

1 Introduction

The human eye is able to rotate at up to 1000° per second and has an angular range of 100° horizontally and 90° vertically. It is able to do all of this with almost no translation of its centre. Eye movement is driven by six eye muscles. The left image in Figure 1 shows the medial and lateral rectus muscles that are responsible for horizontal motion. The right image shows the superior and inferior rectus muscles that are responsible for vertical motion. The two oblique eye muscles are able to perform torsionary eye motion, i.e. rotation around the direction of gaze.

The exact nature and parameters of the mechanics supporting eye movement is still not entirely clear. For example, as part of the "Active Pulley

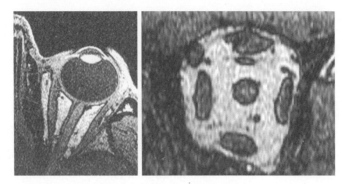

Fig. 1. The left image is of an axial slice of an MRI dataset of an eye showing the medial and lateral rectus muscles and the optic nerve. The white material surrounding the eye-muscles is orbital fat. On the left of the image, part of the nose is visible. The right image shows a slice orthogonal to that, intersecting at the diagonal line shown on the axial slice. This orthogonal slice shows the superior and inferior rectus muscles as well as the superior oblique muscle. The inferior oblique muscle is not visible in this image, but is diagonally opposite to the superior oblique.

Hypothesis", it was proposed that the connective tissue bands connecting the horizontal rectus muscles to the inside wall of the orbit (eye socket), first described by Tenon in 1816 and later called "pulleys" by Miller in 1989 [Mil89], are responsible for the fact that these muscles show a specific inflection or bending point during vertical eye motion [Dem02].

However, in recent work it was demonstrated with a finite element analysis model of the eye as well as clinical observation that these muscles show this inflection point without any connective tissue structures [SvdBK+03, Sch05]. It has also become clear that the orbital fat, i.e. the fat behind the eye, plays an important role in the mechanics of eye movement. Currently, relatively little is known about the mobility of this fat during eye movement.

This paper documents our initial efforts on studying the mobility of the orbital fat during eye movements. Inspired by previous work in this regard [AV02], we applied a deformable registration technique to derive 3-D optical flow fields from two series of MRI datasets of two healthy subjects in different directions of gaze. The resultant 3-D deformation vector fields were used to visualise and measure the motion of orbital eye-fat during eye movement.

Fields resulting from optical flow or deformable registration algorithms are in actual fact displacement fields. However, in keeping with the registration nomenclature we will also refer to them as deformation fields.

Our contribution lies in the fact that this study comes to relevant conclusions about three-dimensional orbital fat mobility during eye motion. These conclusions confirm previous work performed on two-dimensional data and lead to interesting new questions. In addition, we demonstrate that using interactive advection volumes is an effective way of investigating orbital fat mobility based on 3-D optical flow vector fields derived from MRI.

The rest of this paper is structured as follows: section 2 contains a brief summary of work related to this study. Section 3 explains the methods and tools we used for our study. In section 4 we show the results of our analysis and in section 5 we summarise our findings and indicate avenues for future work.

2 Related Work

There are numerous examples of estimating 3-D motion given a set of volumes. For example, in [dLvL02], block matching is used to derive motion vectors. We have chosen the Demons deformable registration algorithm [Thi96] due to its straight-forward implementation and the fact that it is a proven choice for the non-rigid registration of MRI datasets. However, we do plan to test more optical flow approaches in the future.

In [AV02], the 2-D Lucas and Kanade optical flow algorithm [LK81] is extended to three dimensions and applied to various test datasets as well as an MRI dataset of an eye in different directions of gaze. For reasons mentioned previously, we selected the Demons algorithm. In addition, we make use of advection to quantify the fat motion. Another important difference is that [AV02] focuses more on the validation of the actual technique rather than on the investigation of actual fat mobility. It does come to the conclusion that the orbital fat "deforms like a liquid and less like a solid". It also describes how the fat fills the area behind the moving optic nerve from above and below during horizontal motion. This is described as "this tissue [the fat], thus, fills the vacuum left by the nerve, as behind a spoon moving through syrup".

In a previous two-dimensional study based on MRI data, it was determined that the orbital fat surrounding the optic nerve moves horizontally only 54% as much as the optic nerve itself during horizontal eye motion [SHM+06].

3 Methods

3.1 Acquisition

MRI volume datasets were acquired of the eyes of two healthy subjects in respectively seven and fourteen different directions of gaze.

For the first subject, T1-weighted scans were acquired on a 1.5 Tesla General Electric MRI scanner. We made use of a mobile transceiver surface coil for higher resolution. This resulted in seven $512 \times 512 \times 84$ MRI datasets with resolution $0.273 \times 0.273 \times 0.5 mm$. For the deformation analysis, four of the datasets were used: the central direction of gaze, and three directions of gaze to the left.

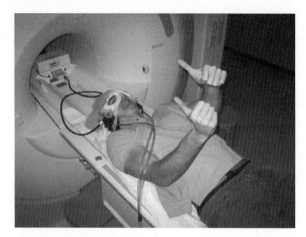

Fig. 2. The acquisition setup. The subject's head has been fixed to the scanning table. The mobile transceiver is visible over the subject's right eye.

For the second subject, T1-weighted scans were acquired on a 3 Tesla General Electric MRI scanner. This resulted in fourteen $512 \times 512 \times 128$ MRI datasets with resolution $0.312 \times 0.58 \times 0.4 mm$. For the deformation analysis, thirteen of these datasets were used: the central direction of gaze, six directions to the left and six to the right.

Figure 2 shows the acquisition setup. The inside of the MRI tube was marked with fixation points at the desired directions of gaze for the subject to focus on during the acquisition.

3.2 Software Tools

The Delft Visualisation and Image processing Development Environment, or DeVIDE, is a software environment for the rapid prototyping and application of visualisation and image processing techniques [Bot04]. It makes use of the VTK [SML99] and ITK [ISNC03] software toolkits. Its primary user interface is a graphical canvas where boxes, representing functional modules or algorithms, can be placed. These boxes can be connected together to form function networks. This is similar to other packages such as OpenDX [AT95] and AVS [UFK[+]89]. What distinguishes DeVIDE is its focus on medical visualisation and image processing, and the extensive interaction facilities made possible by the use of Python [vR01], a very high level dynamically typed language, in the program main loop. All processing and visualisation described in this paper was performed using the DeVIDE software.

3.3 Pre-processing

During image acquisition, the subject's head was fixed to the MRI patient table with surgical tape in order to minimise head motion during the relatively long total scan duration. Acquisition takes approximately one minute per direction of gaze, but significant time is spent on setup actions between directions of gaze. In spite of the head fixation, slight subject head motion did occur. In order to eliminate this rigid head motion in the acquired data, corresponding sets of six bony landmarks each were defined in all datasets. The datasets representing the central direction of gaze was chosen as the reference. Rigid transformations, optimal in a least squares sense, were derived to map the three other landmark sets onto the reference set [Hor87]. These transformations were used to resample the data volumes with cubic interpolation, thus eliminating most of the rigid motion. All registrations were visually inspected and improved if not satisfactory by re-adjusting the selected landmarks.

As explained in section 3.1, the inside of the MRI tube was marked with fixation points at the desired directions of gaze. However, the desired directions of gaze and the final actual directions of gaze were obviously not equal, as the subjects could focus with either one of their eyes. In order to determine the actual directions of gaze, we started by segmenting the lenses in all datasets and determining the centroids of these segmentations. Segmentation was based on a thresholding and 3-D region growing starting from a user selected marker in the lens. Subsequently a sphere was fitted to the eye in order to find its centre. Figure 3 illustrates this procedure. The vector between the centre of the lens and the centre of the eye determines the direction of gaze. For each of the subjects, one of the datasets was chosen as the reference, or centre direction of gaze.

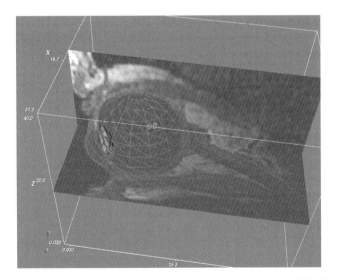

Fig. 3. Procedure for determining the actual directions of gaze. The lens is automatically segmented and its centroid is determined. The actual direction of gaze is determined by this centroid and the centre of the sphere that has been fitted to the eye.

For the first subject we determined the three directions to the left of the centre direction to be respectively 33°, 24° and 14°. For the second subject, the actual directions of gaze were 37.5°, 30.1°, 26.0°, 18.8°, 12.4°, 4.6° to the left of the centre direction, and 7.0°, 12.8°, 18.9°, 25.5°, 32.3° and 38.6° to the right.

3.4 Deformable Registration

The Demons deformable registration algorithm [Thi96] was used to determine the 3-D vector datasets describing the orbital fat deformation from the central direction of gaze through all directions of gaze to its left and to its right. In the case of the first subject, deformation fields were determined from 0° to 14°, from 14° to 24° and from 24° to 33°. For the second subject, fields were determined starting from the central direction to all directions to the left and to the right. The Demons algorithm was chosen due to its straight-forward implementation, and the fact that it is often used for this kind of deformable registration problem. We have also implemented the 3-D version of the Lukas and Kanade optical flow algorithm and as part of our future work plan to compare it with the Demons approach.

Because the Demons algorithm is based on the assumption that corresponding points in the source and target datasets have the same intensity, the intensity values of each pair of datasets were normalised by using a histogram matching implementation from ITK [ISNC03].

3.5 Visualisation and Measurement

The resulting vector fields can be visualised with traditional flow visualisation techniques such as glyphs or streamlines, but these techniques all suffer from the problems plaguing most three-dimensional flow visualisation techniques: occlusion, lack of directional cues, lack of depth cues and visual complexity. To compound matters, we are dealing with time-varying data. In spite of all this, existing techniques are a great help in localising regions of interest that can be examined in more depth with other techniques.

In our case, we were interested in the fat deformation in specific areas. We chose to apply user-guided advection volumes. Small sub-volumes are placed in regions of interest. Each sub-volume is defined by a containing polygonal surface. Points are placed within the interior of this sub-volume at a user-definable density. Each of these points, as well as the vertices defining the containing surface, can then be displaced by the interpolated deformation vectors at their various positions, for that time-step. Figure 4 shows this process for a single time-step and a single spherical volume. The deformed volume, shown on the top right, is used as the initial volume for the next vector field. In this way we can keep on deforming the volume as many times as we have vector datasets.

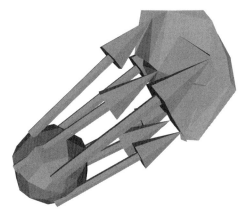

Fig. 4. The sub-volume advection, shown for a single spherical region of interest and for a single time-step. The sphere on the left is the original as selected by the user. The object on the right has been deformed by the current vector field. The vector field associated with the next time step will subsequently be applied to the vertices of the deformed region of interest.

The DeVIDE software allows one to place any number of these volumes in an MRI dataset. As soon as a volume is placed, it is advected through all loaded vector fields and the initial volume as well as all deformed volumes for that point are visualised. An already placed volume can also be moved. This allows for the interactive exploration of a time-varying deformation vector field. Figure 5 shows an example of such an interactive visualisation with four initial volumes advected over 12 vector datasets.

Another important reason to select specific regions of interest is the fact that the deformation field has the highest quality on textured structures embedded in the orbital fat. Similar to the previous 2-D study [SHM+06], we experimented by placing advection volumes on vascular structures, preferably on bifurcations.

Results can be visually studied. In addition, similar to the actual direction of gaze determination discussed in section 3.3, the relative direction of a specific advected volume can be determined with regards to the centre of the eye.

4 Results

The two subjects were studied using different methods. The first subject was part of a pilot study, and we qualitatively inspected specific anatomical regions with advection volumes. In the second subject, we quantitatively tracked a number of fat regions chosen specifically on vascular structures in the orbital fat, also using advection volumes.

228   C. P. Botha et al.

Fig. 5. A visualisation with four advection volumes over 12 time-varying vector fields. The volumes have been chosen in the same plane as the optic nerve, on venous landmarks in order to verify the findings of a previous 2-D study. In each of the four cases, the green sphere in the middle is placed in the centre direction of gaze, and is advected both to the left and to the right.

In the first subject, three regions were selected for closer qualitative inspection:

1. The region between the medial rectus and the eye. As the eye turns anti-clockwise, this muscle "rolls up" onto the eye.
2. Around the optic nerve right behind the eye.
3. The region at the apex where the eye-muscles and optic nerve meet, about 25mm behind the eye.

In all of these regions a number of spherical volumes were placed and advected with the three deformation vector fields. In the first case, i.e. between the medial rectus and the eye, the resolution of the datasets and of the resultant vector fields was too low to make any kind of judgement.

In the second case, seven small spherical volumes were placed directly behind the eye: six surrounding the optic nerve at regular angles and one in the optic nerve itself. As the left eye turns anti-clockwise, the optic nerve moves medially, i.e. in the direction of the nose. The fat surrounding the optic nerve moved in the same direction and primarily transversally. What was significant, is that the fat moved only half as much as the optic nerve.

In the third case, advected volumes indicated that the fat moved primarily in the same direction as the muscles. Once again, the fat moved half as much as the muscles.

In the second subject data, we selected a number of advection volumes specifically on vascular features in the orbital fat. In the first case, four markers were chosen in the same axial plane as the optic nerve in order to confirm the findings of [SHM+06]. In that 2-D analysis, a single MRI slice containing the optic nerve was selected for the analysis and markers were manually tracked for all acquired datasets.

In our case, the relative direction of a specific advection volume in any vector field can be determined similarly to the way in which the directions of gaze were determined. Figure 6 shows the relative rotations for each of the four selected volumes over all vector fields. For each advection volume, the linear regression with derived coefficients a and b is shown as well. Our findings, although based on optical flow advection, concur with their manually tracked results. In our case, fat rotation is between 0.35 times and 0.07 times as much as eye rotation for markers with a distance between 3.3mm and 14.0mm behind the eye. In [SHM+06], fat rotation at 4mm was 0.36 times as much and at 14.5mm 0.07 times as much as the eye rotation.

We also selected four markers in a plane completely above the optic nerve. Although not as pronounced, there is a clear relation between the distance

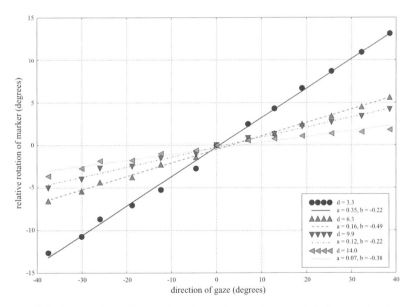

Fig. 6. Relative angles of four advection volumes in an axial slice containing the optic nerve over relative direction of gaze. d refers to the distance behind the eye for that specific advection volume. Also shown is the linear regression derived from this data with coefficients a and b.

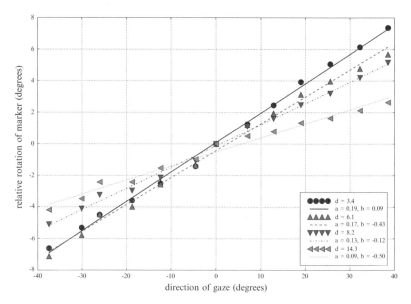

Fig. 7. Relative angles of four advection volumes in an axial slice above the optic nerve over relative direction of gaze. d refers to the distance behind the eye for that specific advection volume. Also shown is the linear regression derived from this data with coefficients a and b.

behind the eye and the ratio of fat rotation to eye rotation. Figure 7 shows the measurements and linear regression for the markers in this plane.

In the third case, we chose a number of vascular markers in random positions around the optic nerve. The relation between distance and rotation ratio is still apparent although far less pronounced. See Figure 8 for the measurements and linear regression results.

5 Conclusions and Future Work

In this paper we have documented an initial study of 3-D orbital fat dynamics based on multiple MRI datasets acquired of a two healthy subjects' eyes during different directions of gaze. Time-varying three-dimensional vector fields were generated by applying the Demons deformable registration technique to pairs of MRI datasets of sequential directions of gaze. These vector fields were visualised with the DeVIDE software system and analysed by making use of advection volumes.

In the first subject it was qualitatively determined that directly behind the eye and at the apex where the muscles and the optic nerve meet, fat moves 50% less than respectively the optic nerve and the muscles embedded in it. In other words, orbital fat moves partly along with these moving structures, but it partly deforms around them as well.

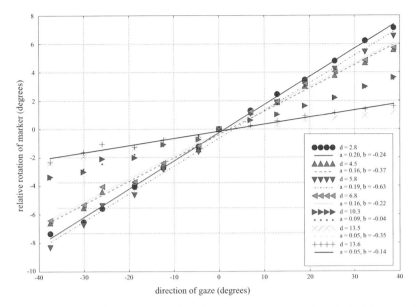

Fig. 8. Relative angles of seven advection volumes randomly selected in vascular features around the optic nerve over relative direction of gaze. d refers to the distance behind the eye for that specific advection volume. Also shown is the linear regression derived from this data with coefficients a and b.

In the second subject, the rotation angles of specific vascular markers were tracked over all thirteen vector fields. For markers in the same plane as the optic nerve, our findings correlated well with the findings of a previous 2-D study of the same data where markers were manually tracked from frame to frame. For markers in a plane above the optic nerve, there was still an apparent inverse relation between the distance from the eye and the ratio between the deformation of the fat and the rotation of the eye itself. For vascular markers randomly chosen all around the optic nerve, the relation was weaker. This is to be expected, as markers further above and below the optic nerve will be less affected by the motion of the optic nerve itself through the fat.

In all cases, motion of the vascular structures, calculated according to the Demons optical flow, was a fraction of the eye rotation. This implies that the optic nerve moves through the fat. In other words, orbital fat has to deform less, which would probably require less energy.

With the Demons algorithm, we could only reliably track textured features, such as the vascular structures, in the fat. We plan to implement more robust 3-D optical flow techniques in order to be able to track a large part of the orbital fat reliably. Subsequently, we will measure advection for a dense sampling of complete orbital fat regions of interest in order to see if our findings still hold.

The interactive advection volumes constitute an effective method to visualise and quantify local fat deformation. However, a more global visualisation technique would be useful to help understand the complex orbital fat deformation fields. One interesting idea is the development of a realistic fluid flow simulation and visualisation that uses the derived vector fields as basis, so that the fat deformation can be studied in pseudo real-time. We will also continue our investigation of alternative techniques for the visualisation of local deformation.

The fixation of the subject's head during scanning has to be improved. The rigid motion can be eliminated as explained in this paper, but it is desirable to minimise head motion during the acquisition phase. For the residual motion that still might occur, it is important to localise easily identifiable rigid landmarks during acquisition. We are currently investigating techniques to improve head fixation and landmark localisation.

Importantly, the approach documented in this paper is based on the acquisition of a series of static scenes. During the acquisition of a particular direction of gaze, the eye is in a state of equilibrium. Due to this, the dynamic behaviour of orbital fat during eye movements, e.g. the viscous or inertial effects, is not captured. In spite of its limitations, our approach still yields valuable information about the 3-D deformation of orbital fat, especially since it is currently not possible to acquire real-time 3-D MRI data of the eye in motion. In future, we will make use of 2-D tagged MRI to study the dynamic behaviour in more detail and integrate these effects into our 3-D model.

References

[AT95] Greg Abram and Lloyd Treinish. An extended data-flow architecture for data analysis and visualization. In *Proceedings of IEEE Visualization '95*, page 263. IEEE Computer Society, 1995.

[AV02] Michael D. Abràmoff and Max A. Viergever. Computation and visualization of three-dimensional soft tissue motion in the orbit. *IEEE Transactions on Medical Imaging*, 21(4):296–304, 2002.

[Bot04] Charl P. Botha. DeVIDE: The Delft Visualisation and Image processing Development Environment. Technical report, Delft Technical University, 2004.

[Dem02] Joseph L. Demer. The orbital pulley system: a revolution in concepts of orbital anatomy. *Annals of the New York Academy of Sciences*, 956:17–33, 2002.

[dLvL02] Wim de Leeuw and Robert van Liere. Bm3d: motion estimation in time dependent volume data. In *VIS '02: Proceedings of the conference on Visualization '02*, pages 427–434, Washington, DC, USA, 2002. IEEE Computer Society.

[Hor87] Berthold K.P. Horn. Closed-form solution of absolute orientation using unit quaternions. *Journal of the Optical Society of America A*, 4:629–642, 1987.

[ISNC03]   Luis Ibanez, Will Schroeder, Lydia Ng, and Joshua Cates. *The ITK Software Guide*. Kitware Inc., 2003.
[LK81]   B. Lucas and T. Kanade. An iterative image registration technique with an application to stereo vision. In *Proc. DARPA Image Understanding Workshop*, pages 121–130, 1981.
[Mil89]   J.M. Miller. Functional anatomy of normal human rectus muscles. *Visual Research*, 29:223–240, 1989.
[Sch05]   Sander Schutte. Orbital mechanics and improvement of strabismus surgery. Master's thesis, Delft University of Technology, 2005.
[SHM+06]   Ivo Schoemaker, Pepijn P W Hoefnagel, Tom J Mastenbroek, Cornelis F Kolff, Sander Schutte, Frans C T van der Helm, Stephen J Picken, Anton F C Gerritsen, Piotr A Wielopolski, Henk Spekreijse, and Huibert J Simonsz. Elasticity, viscosity, and deformation of orbital fat. *Invest Ophthalmol Vis Sci*, 47(11):4819–4826, Nov 2006.
[SML99]   Will Schroeder, Ken Martin, and Bill Lorensen. *The Visualization Toolkit*. Prentice Hall PTR, 2nd edition, 1999.
[SvdBK+03]   S.Schutte, S.P.W. van den Bedem, F.van Keulen, F.C. T. van der Helm, and H.J. Simonsz. First application of finite-element (fe) modeling to investigate orbital mechanics. In *Proceedings of the Association for Research in Vision and Ophthalmology (ARVO) Annual Meeting*, 2003.
[Thi96]   J.-P. Thirion. Non-rigid matching using demons. In *Proceedings of IEEE Computer Vision and Pattern Recognition (CVPR)*, pages 245–251, 1996.
[UFK+89]   C. Upson, T Faulhaber, D. Kamins, D. Laidlaw, D. Schleigel, J. Vroom, R. Gurwitz, and A. van Dam. The Application Visualization System: A Computational Environment for Scientific Visualization. *IEEE Computer Graphics and Applications*, pages 30–42, July 1989.
[vR01]   Guido van Rossum. *Python Reference Manual*. Python Software Foundation, April, 2001.

Part V

Visualizing Molecular Structures

Visual Analysis of Biomolecular Surfaces

Vijay Natarajan[1], Patrice Koehl[2], Yusu Wang[3], and Bernd Hamann[4]

[1] Department of Computer Science and Automation
  Supercomputer Education and Research Centre
  Indian Institute of Science, Bangalore, India
  `vijayn@csa.iisc.ernet.in`
[2] Department of Computer Science
  Genome Center
  University of California, Davis, California, USA
  `koehl@cs.ucdavis.edu`
[3] Department of Computer Science and Engineering
  The Ohio State University, Columbus, Ohio, USA
  `yusu@cse.ohio-state.edu`
[4] Institute for Data Analysis and Visualization
  Department of Computer Science
  University of California, Davis, California, USA
  `hamann@cs.ucdavis.edu`

Summary. Surface models of biomolecules have become crucially important for the study and understanding of interaction between biomolecules and their environment. We argue for the need for a detailed understanding of biomolecular surfaces by describing several applications in computational and structural biology. We review methods used to model, represent, characterize, and visualize biomolecular surfaces focusing on the role that geometry and topology play in identifying features on the surface. These methods enable the development of efficient computational and visualization tools for studying the function of biomolecules.

1 Introduction

The molecular basis of life rests on the activity of biological macro-molecules, including nucleic acids (DNA and RNA), carbohydrates, lipids and proteins. Although each plays an essential role in life, nucleic acids and proteins are central as support of the genetic information and products of this information, respectively. A perhaps surprising finding that crystallized over the last decades is that geometric reasoning plays a major role in our attempt to understand the activities of these molecules. We address this connection between biology and geometry, focusing on hard sphere models of biomolecules. In particular, we focus on the representations of biomolecular surfaces, and their applications in computational biology.

1.1 Significance of Shape

Molecular structure or shape and chemical reactivity are highly correlated as the latter depends on the positions of the nuclei and electrons within the molecule. Indeed, chemists have long used three-dimensional plastic and metal models to understand the many subtle effects of structure on reactivity and have invested in experimentally determining the structure of important molecules. The same applies to biochemistry where structural genomics projects are based on the premise that the structure of biomolecules implies their function. This premise rests on a number of specific and quantifiable correlations:

- enzymes fold into unique structures and the three-dimensional arrangement of their side-chains determines their catalytic activity;
- there is theoretical evidence that the mechanisms underlying protein complex formation depend mainly on the shapes of the biomolecules involved [LWO04];
- the folding rate of many small proteins correlates with a gross topological parameter that quantifies the difference between distance in space and along the main-chain [AB99, ME99, KWPB98, AMKB02];
- there is evidence that the geometry of a protein plays a role in defining its tolerance to mutation [KL02].

We note that structural biologists often refer to the 'topology' of a biomolecule when they mean the 'geometry' or 'shape' of the same. A common concrete model representing this shape is a union of balls, in which each ball corresponds to an atom. Properties of the biomolecule are then expressed in terms of properties of the union. For example, the potential active sites are detected as cavities [LEW98, EFL98, LEF+98b] and the interaction with the environment is quantified through the surface area and/or volume of the union of balls [EM86, OONS87, LEF+98a]. In what follows, we discuss in detail the geometric properties of the surface of union of balls, their visualization, and their relation to the physical properties of the biomolecules they represent.

1.2 Biomolecules

Biomolecules are usually polymers of smaller subunits, whose atomic structures are known from standard chemistry. While physics and chemistry have provided significant insight into the structure of the atoms and their arrangements in small chemical structures, the focus now is set on understanding the structure and function of biomolecules, mainly nucleic acids and proteins. Our presentation of these molecules follow the general dogma in biology that states that the genetic information contained in DNA is first transcribed to RNA molecules which are then translated into proteins.

DNA

The Deoxyribo Nucleic Acid is a long polymer built from four different building blocks, the nucleotides. The sequence in which the nucleotides are arranged contains the entire information required to describe cells and their functions. Despite this essential role in cellular functions, DNA molecules adopt surprisingly simple structures. Each nucleotide contains two parts, a backbone consisting of a deoxyribose and a phosphate, and an aromatic base, of which there are four types: adenine (A), thymine (T), guanine (G) and cytosine (C). The nucleotides are capable of being linked together to form a long chain, called a *strand*. Cells contain strands of DNA in pairs that are exact mirrors of each other. When correctly aligned, A can pair with T, G can pair with C, and the two strands form a double helix [WC53]. The geometry of this helix is surprisingly uniform, with only small, albeit important, structural differences between regions of different sequences. The order in which the nucleotides appear in one DNA strand defines its sequence. Some stretches of the sequence contain information that can be translated first into an RNA molecule and then into a protein. These stretches are called *genes*; the ensemble of all genes of an organism constitutes its *genome* or *genetic information*. The DNA strands can stretch for millions of nucleotides. The size of the strands vary greatly between organisms and do not necessarily reflect differences in the complexity of the organisms. For example, the wheat genome contains approximately $1.6 \cdot 10^{10}$ bases, which is close to five times the size of the human genome. For a complete list of the genomes, see `http://wit.integratedgenomics.com/GOLD/` [BEK01].

RNA

Ribo Nucleic Acid molecules are very similar to DNA, being formed as sequences of four types of nucleotides, namely A, G, C, and uracil (U), which is a derivative of thymine. The sugar in the nucleotides of RNA is a ribose, which includes an extra oxygen compared to deoxyribose. The presence of this bulky extra oxygen prevents the formation of long and stable double helices. The single-stranded RNA can adopt a large variety of conformations, which remain difficult to predict based on its sequence. RNA molecules mainly serve as templates that are used to synthesize the active molecules, namely the proteins. The information needed to synthesize the RNA is read from the genes coded by the DNA. Interestingly, RNA is considered an essential molecule in the early steps of the origin of life. More information on the RNA world can be found in [GA93].

Proteins

While all biomolecules play an important part in life, there is something special about proteins, which are the products of the information contained in the genes. They are the active elements of life whose chemical activities regulate all cellular activities. As a consequence, studies of their sequence and structure occupy a central role in biology. Proteins are heteropolymer chains of amino acids, often referred to as *residues*. There are twenty types of amino acids, which share a common *backbone* and are distinguished by their chemically diverse *side-chains*, which range in size from a single hydrogen atom to large aromatic rings and can be charged or include only non-polar saturated hydrocarbons. The order in which amino acids appear defines the *primary sequence* of the protein. In its native environment, the polypeptide chain adopts a unique three-dimensional shape, referred to as the *tertiary* or *native structure* of the protein. In this structure, non-polar amino acids have a tendency to re-group and form the core of the proteins, while polar amino acids remain accessible to the solvent. The backbones are connected in sequence forming the protein *main-chain*, which frequently adopts canonical local shapes or *secondary structures*, such as α-helices and β-strands. From the seminal work of Anfinsen [Anf73], we know that the sequence fully determines the three-dimensional structure of the protein, which itself defines its function. While the key to the decoding of the information contained in genes was found more than fifty years ago (the genetic code), we have not yet found the rules that relate a protein sequence to its structure [KL99, BS01]. Our knowledge of protein structure therefore comes from years of experimental studies, either using X-ray crystallography or NMR spectroscopy. The first protein structures to be solved were those of hemoglobin and myoglobin [KDS+60, PRC+60]. Currently, there are more than 37,000 protein structures in the database of biomolecular structures [BKW+77, BWF+00]; see http://www.rcsb.org. More information on protein structures can be found in protein biochemistry textbooks, such as those of Branden and Tooze [BT91], and Creighton [Cre84].

2 Visualizing Biomolecular Surfaces

The need for visualizing biomolecules is based on the early understanding that their shape determines their function. Early crystallographers who studied proteins and nucleic acids could not rely—as it is common nowadays—on computers and computer graphics programs for representation and analysis. They had developed a large array of finely crafted physical models that allowed them to have a feeling for these molecules. These models, usually made out of painted wood, plastic, rubber and/or metal were designed to highlight different properties of the molecule under study. In the *space-filling models*, such as those of Corey-Pauling-Koltun (CPK) [CP53, Kol65], atoms are represented as spheres, whose radii are the atoms' van der Waals radii. They provide a volumetric representation of the biomolecules, and are useful to detect cavities and pockets that are potential active sites. In the *skeletal models*, chemical bonds are represented by rods, whose junctions define the position of the atoms. These models were used for example by Kendrew and colleagues [KDS+60]

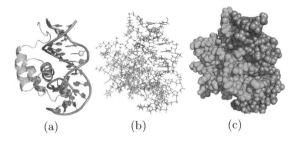

(a) (b) (c)

Fig. 1. Visualizing protein-DNA complexes. Homeodomains are small proteins that bind to DNA and regulate gene expression. Here we show the complex of the antennapedia homeodomain of *drosophila melanogaster* (fruit fly) and its DNA binding site [BQO+93], using three different types of visualization. The structure of this complex was determined by X-ray crystallography [BQO+93]; the coordinates are taken from the PDB file 1AHD. The protein is shown in green, and the DNA fragments in orange. (a) *Cartoon*. This representation provides a high level view of the local organization of the protein in secondary structures, shown as idealized helices. This view highlights the position of the binding site where the DNA sits. (b) *Skeletal model*. This representation uses lines to represent bonds; atom are located at their endpoints where the lines meet. It emphasizes the chemical nature of both molecules. (c) *Space-filling diagram*. Atoms are represented as balls centered at the atoms, with radii equal to the van der Waals radii of the atoms. This representation shows the tight binding between the protein and the ligand, that was not obvious from the other diagrams. Each of the representations is complementary to the others, and usually the biochemist uses all three of them when studying a protein, alone or, as illustrated here, in interaction with a ligand. All panels were drawn using Pymol (http://www.pymol.org)

in their studies of myoglobin. They are useful to the chemists by highlighting the chemical reactivity of the biomolecules and, consequently, their potential activity. With the introduction of computer graphics to structural biology, the principles of these models have been translated into software such that molecules could be visualized on the computer screen. Figure 1 shows examples of computer visualization of a protein-DNA interaction, including space-filling and skeletal representations.

Among all geometric elements of a biomolecule, its surface is probably the most important as it defines its interface, *i.e.*, its region of potential interactions with its environment. Here we limit ourselves to the definition of surface within the classical representation of biomolecules as union of balls. While other models are possible (such as a atom-based Gaussian descriptions [GP95], the hard-sphere model remains the most popular. Given the atom (ball) locations, the biomolecular surface can be defined in various ways. The first definition stated that the surface was simply the boundary of the union of balls. This surface, called the *van der Waals surface*, is easily computable but not continuous [Con96]. The *solvent accessible surface* is the collection of

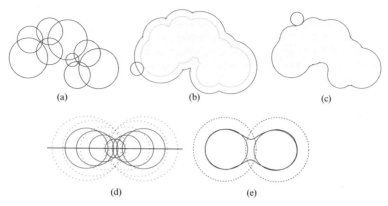

Fig. 2. (a) Each atom is modeled as a hard ball. (b) The van der Waals surface is the boundary of the union and the solvent accessible is the boundary of the union of expanded balls. Each atom is expanded by a value equal to the radius of the probe sphere. (c) The Connolly surface is traced by the front of the probe sphere. (d) The skin surface is defined as the envelope of an infinite set of spheres. Two atoms (outermost, dotted) define a family of spheres (dotted) obtained as the convex hull. Shrinking each sphere in the family results in yet another family of spheres (bold). (e) The skin surface, defined as the envelope of the shrunken family of spheres, is smooth everywhere. The radius of the two atoms is typically enlarged beforehand so that, upon shrinking, we get spheres whose radius equal the atom radius.

points traced by the center of a probe sphere as it rolls on the van der Waals surface [LR71]. The accessible surface is equivalently defined as the boundary of the union of balls whose radii are expanded by the probe sphere radius. This surface is not smooth at curves where the expanded spheres meet. The *solvent excluded surface*, also called the *Connolly surface*, is defined as the surface traced by the front of a probe sphere [Ric77]. This surface is continuous in most cases but can have cusp points and self-intersections. Figure 2 illustrates the definition of the above mentioned surfaces. Several algorithms have been proposed to construct analytic representations of the solvent excluded surface [Con83, Ric84, VJ93] and for triangulating the surface [Con85, AE96].

The *skin surface* is the envelope of families of an infinite number of evolving spheres [CDES01]. It satisfies many desirable mathematical properties. For example, it is smooth everywhere and, although defined using an infinite number of spheres, it can be described by a finite number of quadric surface regions. Efficient algorithms have been developed recently to triangulate the skin surface [CS04, CS05]. Besides being efficient in terms of computation time, these algorithms also generate watertight skin surface meshes (*i.e.*, without cracks) containing good quality triangles (*i.e.*, without small angles). The good quality of triangles leads to better visualizations of the surface. Watertight models facilitate the application of volumetric analysis methods on the surface. For example, the volume occupied by the molecule, volume of voids, and their derivatives can be computed robustly given a crack-free biomolecular surface.

Multiresolution models of the mesh enables interactive visualization of the surface. Several methods have been developed to create level-of-detail representations of surface meshes [Hop96, Gar99].

3 Significance of Biomolecular Surfaces

The activity of a biomolecule is encoded in its shape. Of all geometric properties of a molecule, its surface play an essential role as it delineates the region covered by the protein and therefore defines its region of interactions. Characterizing biomolecular surface therefore play an essential role for analyzing and predicting biomolecular complexes, as well as for modeling the energetics of formation of such complexes. As the surface of a molecule also defines its interface with the solvent it bathes in, the former is crucial for understanding solvation.

3.1 Solvent Models

The apparition of computers, and the rapid increase of their power has given hope that theoretical methods can play a significant role in biochemistry. Computer simulations are expected to predict molecular properties that are inaccessible to experimental probes, as well as how these properties are affected by a change in the composition of a molecular system. This has lead to a new branch in biology that works closely with structural biology and biochemistry, namely computational biology. Not surprisingly, an early and still essential focus of this new field is biomolecular dynamics [CK00, KM02]. Soluble biomolecules adopt their stable conformation in water, and are unfolded in the gas phase. It is therefore essential to account for water in any modeling experiment. Molecular dynamics simulation that include a large number of solvent molecules are the state of the art in this field, but they are inefficient as most of the computing time is spent on updating the position of the water molecule. It should be noted that it is not always possible to account for the interaction with the solvent explicitly. For example, energy minimization of a system including both a protein and water molecules would not account for the entropy of water, which would behave like ice with respect to the protein. An alternative is to develop an approach in which the effect of the solvent is taken into account implicitly. In such implicit solvent models, the effects of water on a biomolecule is included in an effective solvation potential, $W = W_{elec} + W_{np}$, in which the first term accounts for the molecule-solvent electrostatics polarization, and the second term for the molecule-solvent van der Waals interactions and for the formation of a cavity in the solvent.

3.2 Electrostatics in Implicit Solvent Models

Implicit solvent models reduce the solute solvent interactions to their mean-field characterization, which are expressed as a function of the solute degrees of freedom alone. They represent the solvent as a dielectric continuum that mimics the solvent-solute interactions. Many techniques have been developed to compute electrostatics energy in the context of dielectric continuum, including techniques that modify the dielectric constants in Coulomb's law, generalized Born models, and methods based on Poisson-Boltzmann equation (for a recent review, see [Koe06]. A common element of all these techniques is that they need a good definition of the interface between the protein core and the dielectric continuum, *i.e.*, a good definition of the surface of the protein.

3.3 Non Polar Effects of Solvent

W_{np}, the non-polar effect of water on the biomolecule is sometimes referred to as the *hydrophobic effect*. Biomolecules contain both hydrophilic and hydrophobic parts. In their folded states, the hydrophilic parts are usually at the surface where they can interact with water, and the hydrophobic parts are buried in the interior where they form a core (an "oil drop with a polar coat" [Kau59]). In order to quantify this hydrophobic effect, Lee and Richards introduced the concept of the solvent-accessible surface [LR71]. They computed the accessible areas of each atom in both the folded and extended state of a protein, and found that the decrease in accessible area between the two states is greater for hydrophobic than for hydrophilic atoms. These ideas were further refined by Eisenberg and McLachlan [EM86], who introduced the concept of a solvation free energy, computed as a weighted sum of the accessible areas A_i of all atoms i of the biomolecule:

$$W_{np} = \sum_i \alpha_i A_i,$$

where α_i is the atomic solvation parameter. It is not clear however which surface area should be used to compute the solvation energy [WT90, TSPA92, SB94]. There is also some evidence that for small solute the hydrophobic term W_{np} is not proportional to the surface area [SB94], but rather to the solvent excluded volume of the molecule [LCW99]. A volume-dependent solvation term was originally introduced by Gibson and Scheraga [GS67] as the hydration shell model. Within this debate on the exact form of the solvation energy, there is however a consensus that it depends on the geometry of the biomolecule under study. Inclusion of W_{np} in a molecular simulation therefore requires the calculation of accurate surface areas and volumes. If the simulations rely on minimization, or integrate the equations of motion, the derivatives of the solvation energy are also needed. It should be noted that calculation of the second derivatives are also of interest to study the normal modes of a biomolecule in a continuum solvent.

4 Feature-based Analysis and Visualization

To improve our understanding of a biomolecule and its functions, it is highly desirable to visualize various of its properties over its surface. Such visualization tools can for example help to emphasize important structural motifs and to reveal meaningful relations between the physiochemical information and the shape of a molecule.

One *general* framework for such visualization is as follows: the input molecular surface is considered as a *domain* $\mathbb{M}$, and one or more scalar functions $f_1, \ldots, f_k$ are defined over it, where each $f_i : \mathbb{M} \to \mathbb{R}$ is called a *descriptor function*. Such functions describe various properties that may be important, be it geometric, physiochemical, or any other type. We then visualize these descriptor functions over $\mathbb{M}$. Two key components involved here are (i) how to design meaningful descriptor functions and (ii) how to best visualize them. Below we briefly describe approaches in these two categories, focusing on topological methods[1].

4.1 Descriptor Functions

Descriptors capturing physiochemical properties, such as the electrostatic potential and the local lipophilicity, are relatively easier to develop. Below we focus on molecular *shape* descriptors. In particular, most current molecular shape descriptors aim at capturing *protrusions and cavities* of the input structure, given that proteins function by interacting (binding) with other molecules, and there is a rough "lock-and-key" principle behind such binding [Fis94] (see Figure 3 (a) for a 2D illustration).

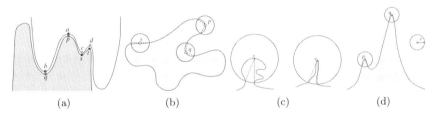

Fig. 3. (a) The shape of two molecules complement each other at the interface. (b) In the 2D case, the Connolly function value is proportional to the angle spanned by the two intersection points between the circle and the curve. (c) p and q have the same Connolly function value, but the details within the neighborhood are different. (d) Using a fixed neighborhood radius is unable to distinguish the two features located at p and q.

---

[1] We note that both these components are widely studied in many other fields such as computer graphics, vision, and pattern recognition. We focus only on methods from the field of molecular biology.

Curvature-based Descriptors

The most natural choice to describe protrusions and cavities may be curvatures. A large family of molecular shape descriptors are based on curvatures. One of the most widely used one is the Connolly function [Con86a, Con86b]. For any point $x \in \mathbb{M}$, consider the ball $B_r(x)$ centered at x with radius r, and let $S_r(x) = \partial B_r(x)$ be the boundary of $B_r(x)$, and S_I the portion of $S_r(x)$ contained inside the surface. The *Connolly function* $f_r : \mathbb{M} \to \mathbb{R}$ is defined as (see Figure 3 (b) for a 2D illustration):

$$f_r(x) = \frac{\text{Area}(S_I)}{r^2}.$$

Roughly speaking, the Connolly function can be considered as an analog of the mean curvature within a fixed size neighborhood of each point [CCL03]. A large function value at $x \in \mathbb{M}$ means that the surface is concave around x, while a small one means that it is convex.

The Connolly function ignores the exact details of the surface contained in $B_r(x)$. Hence it is insensitive to the two features pictured in Figure 3 (c). The *atomic density (AD)* function [MKE01], $f_a : \mathbb{M} \to \mathbb{R}$, improves the Connolly function by taking a sequence of, say k, neighborhood balls around a point x with increasing radii, computing (roughly) the Connolly function with respect to each radius, and obtaining the final function value at x based on these k Connolly function values.

The concept of curvatures for 2-manifolds is more complicated than that of 1-manifolds — there are two principal curvatures at a given point. Several approaches have been proposed to combine these two curvatures into a single value to describe local surface features [CBJ00, DO93a, DO93b, EKB02, HB94].

More Global Descriptors

The functions above are good at identifying points located at local protrusions and cavities. However, they all depend on a pre-fixed value r (the neighborhood size) — if r is small, then they may identify noise as important features; while if r is too large, then they may overlook interesting features. Furthermore, it is desirable that the function value can indicate the size (importance) of the feature that a point captures. However, none of the functions described above can measure the size of features directly (see Figure 3 (d)).

Since binding sites usually happen within cavities, it is natural to measure how *deep* a point $x \in \mathbb{M}$ is inside a cavity directly, independent of some neighborhood size. For example, a natural way to define such measures is as follows [LA05]. Given a molecular surface $\mathbb{M}$, let $CH(\mathbb{M})$ be the convex hull of $\mathbb{M}$, and $\text{Cover}(\mathbb{M}) = CH(\mathbb{M}) \setminus \mathbb{M}$ intuitively covers the cavities of $\mathbb{M}$. One can define $f_c : \mathbb{M} \to \mathbb{R}$ as $d(x, \text{Cover}(\mathbb{M}))$ if $x \notin CH(\mathbb{M})$, and 0 otherwise; where $d(x, X) = \min_{y \in X} d(x, y)$ is the closest Euclidean distance from x to the set X.

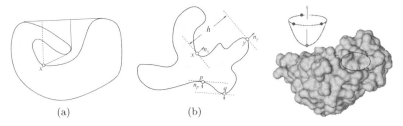

Fig. 4. 2D illustrations of (a) $f_c(x)$ and the shortest collision-free distance, (b) elevation function (where $\text{Elev}(x) = h$). (c) A local maximum of elevation function is associated with three more points to indicate the shape of a cave it captures.

This measure however ignores details inside cavities. A better measure is probably by using the shortest *collision-free* distance from a point x to the convex cover (Figure 4 (a)), which intuitively measures how difficult it is for a solvent molecule to access a specific point on the molecular surface. However, the computation of collision-free shortest path is expensive, and we are not aware of any result applying it to molecular surfaces yet. Furthermore, such concavity measures is asymmetric in measuring convexity. The *elevation function* $\text{Elev} : \mathbb{M} \to \mathbb{R}$, developed in [AEHW04, WAB+05] is independent of any pre-fixed parameter, and can measure both convexity and convexity in a meaningful manner.

Specifically, each point $x \in \mathbb{M}$ is paired with a canonical pairing partner y that shares the same normal direction n_x with x, and the function value $\text{Elev}(x)$ is equal to the height difference between x and y in direction n_x. See Figure 4 (b) for a 2D illustration. The identification of the canonical pairing partners is based on a topological framework, called the *persistence algorithm*, developed by Edelsbrunner *et al.* [ELZ02]. Roughly speaking, the persistence algorithm provides a meaningful way to uniquely pair up critical points of a given function f, each pair specifies some feature with respect to f, and its *persistence* value indicates the size of this feature by measuring how long it can persist as one changes the function value locally.

The elevation function has several nice properties that make it appealing for various applications. In some sense, it finds a different and appropriate r for every point x on the surface and the function value $\text{Elev}(x)$ roughly measures the depth of the cave (or protrusion) captured by x in its normal direction (Figure 4 (b)). More interestingly, each extreme point of the elevation function is associated with pairing partners that helps the user to visualize the feature located at this point (Figure 4 (c)).

The computation of the elevation function is unfortunately costly. The function also has some discontinuity that may be undesirable when segmenting the input surface based on it. In general, there is no universally good descriptor functions. Designing meaningful and easy-to-compute molecular shape descriptors for visualization and structure characterization purposes is still in great need.

4.2 Visualizing Scalar Functions

To enhance the visualization of a given scalar function, one natural approach is to highlight features, such as the critical points of the input function. Another widely used technique is to segment the input surface into meaningful regions, which we focus on below.

Region-growing

Given a function $f : \mathbb{M} \to \mathbb{R}$, one family of surface segmentation methods is based on the region-growing idea [CBJ00, DNW02, EKB02]. In particular, certain *seeds* are first selected, and neighboring points are gradually clustered into the regions around these seeds. Merging and splitting operations are performed to compute regions of appropriate sizes, and/or to obtain a hierarchical representation of segmentations. For example, it is common to use the critical points of the input function as seeds. When two regions meet each other, criteria such as the size of each region and/or the difference between the function values of points from two neighboring regions decide whether to merge these two regions or not.

The advantage of such methods is that the criteria used to merge regions is not limited to the input function — such as using the size of current segment, or even combing multiple descriptor functions into the criteria. Thus they are able to create more versatile types of segments. On the other hand, the decision of merging/splitting is usually locally made and ad hoc, thus may not be optimal globally. Furthermore, various parameters control the output, such as the order of processing different regions, and it is not trivial to identify the best strategy for choosing these parameters.

Topological Methods

A second family of segmentation algorithms is based on Morse theory [BEHP04, EHZ03, NWB+06]. Such topological frameworks usually produce a hierarchical representation of segmentation easily, and are also more general — a segmentation is induced for any given function $f : \mathbb{M} \to \mathbb{R}$, with usually no parameter other than the one to specify the resolution of the segmentation.

Notation

Given a smooth 2-manifold $\mathbb{M} \subseteq \mathbb{R}^3$ and a scalar function $f : \mathbb{M} \to \mathbb{R}$, a point on $\mathbb{M}$ is *critical* if the gradient of f at this point is zero. f is a *Morse function* if none of its critical points are degenerate, that is, the Hessian matrix is non-singular for all critical points, and no two critical points have the same function value. For a Morse function defined on a 2-manifold, there are three types of critical points: minima, saddle points, and maxima. In molecular biological applications, a molecular surface is typically represented as a (triangular) mesh K, and the input scalar function $f : K \to \mathbb{R}$ over K is piecewise-linear (PL): f is given at the vertices and linearly interpolated within edges and triangles of K. The type of a critical point p of such a PL function can be determined by inspecting the *star* of p, which consists of all triangles and edges containing p [Ban70].

Morse complex and Morse-Smale complex

An integral line of f is a maximal path on the surface $\mathbb{M}$ whose tangent vectors agree with the gradient of f at every point of the path. Integral lines have a natural origin and destination at critical points where the gradient equals zero. Grouping the integral lines based on their origin and destination results in a segmentation of the surface. The Morse-Smale (MS) complex [EHZ03] is a topological data structure that stores this segmentation (see Figure 5). A characteristic property of this segmentation is that every cell of the MS complex is monotonic (i.e, it does not contain any critical point in its interior).

Alternatively, grouping integral lines based exclusively on their origin or destination results in yet another segmentation called the Morse complex (see Figure 6). The MS complex can also be obtained as an overlay of the two Morse complexes for the set of maxima and minima, respectively.

Peak-valley decomposition

The Morse complex for the set of all maxima results in a segmentation where peaks (regions around maxima) are separated (Figure 6 (a)). However, these segments extend all the way to adjoining valleys (regions around minima). The MS complex, on the other hand, does more refinement than necessary. In many cases, it is desirable to segment input surface into peaks and valleys. To this end, Natarajan *et al.* proposed an extension of the MS

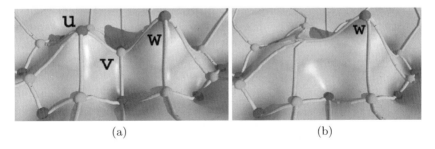

Fig. 5. (a) A simple height function with two maxima surrounded by multiple local minima and its Morse-Smale complex. (b) Smoother height functions are created by canceling pairs of critical points. Canceling a saddle-maximum pair removes a topological feature. The function is modified locally by rerouting integral lines to the remaining maximum.

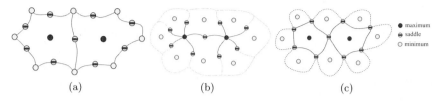

Fig. 6. Morse complex computed for the set of all (a) maxima and (b) minima. (c) The Morse-Smale complex is obtained as an overlay of these two Morse complexes. Paths connecting saddles within each cell of the MS complex segment the surface into peaks and valleys.

complex to compute such segmentations [NWB+06]. In particular, each cell in the MS complex is a *quad* containing one maximum, one minimum, and two saddles on its boundary (Figure 5 (a)). Connecting the two saddles will bisects this quad. Bisecting all quads that contain a specific maximum u, we get all regions that constitute the peak containing u. Similarly, we can obtain valleys containing a minimum v. In other words, these saddle-saddle paths describe the boundary between peaks and their adjoining valleys, and the resulting segmentation is called *peak-valley decomposition* (see Figure 6 (c)). Various criteria have been explored in [NWB+06] for the construction of such saddle-saddle paths.

Hierarchical segmentation

A major advantage of using the MS complex as a starting point for segmentation is that we can segment the surface at multiple levels of detail. A smoother Morse function can be generated from f by repeated cancellation of pairs of critical points. Each cancellation makes the function smoother by removing a topological feature. This operation can be implemented by a local modification of the gradient vector field when the two critical points are connected by a common arc in the MS complex [BEHP04]. Figure 5 (b) shows how the MS complex in Figure 5 (a) is modified after canceling a saddle-maximum pair.

The order of critical point pairs is guided by the notion of *persistence* [ELZ02], which quantifies the importance of the associated topological feature. The peak-valley decomposition can be computed at multiple levels of detail [NWB+06] by first canceling critical point pairs in the MS complex and then constructing saddle-saddle paths within the simplified complex. See Figure 7 for an example where we visualize the peak-valley segmentation of a protein molecule (chain D from the protein complex Barnase-Barstar with pdb-id 1BRS) based on the atomic density function at different levels of details. There are on-going work to use such segmentation to help to characterize protein binding sites.

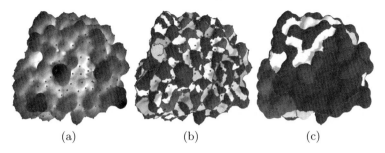

Fig. 7. (a) The atomic density function computed for chain D of the Barnase-Barstar complex. Darker regions correspond to protrusions and lighter regions to cavities. (b) Peak-valley segmentation of the surface. (c) Coarse segmentation obtained by performing a series of critical pair cancellations.

5 Conclusions

We have provided an overview of selected methods based on well-established concepts from differential geometry, computational geometry and computational topology to characterize complex biomolecular surfaces. We have discussed how such concepts are relevant in the context of studying the complex interaction behaviors between biomolecules and their surroundings. Mathematically sound approaches to the analysis of intricate biochemical processes have become increasingly important to make progress in the study of protein folding and protein-protein interactions. Some of the most exciting and challenging questions that remain to be answered include dynamic biomolecular behavior, where surface analysis techniques like the ones discussed by us here need to be generalized substantially to support effective visualization and analysis of rapidly changing shapes.

Acknowledgments

Work done by Vijay Natarajan and Bernd Hamann was supported in part by the National Science Foundation under contracts ACI 9624034 (CAREER Award), a large Information Technology Research (ITR) grant, and the Lawrence Livermore National Laboratory under sub-contract agreement B551391. Vijay Natarajan was also supported by a faculty startup grant from the Indian Institute of Science. Patrice Koehl acknowledges support from the National Science Foundation under contract CCF-0625744. Yusu Wang was supported by DOE under grant DE-FG02-06ER25735 (Early Career Award). We thank members of the Visualization and Computer Graphics Research Group at the Institute for Data Analysis and Visualization (IDAV) at the University of California, Davis.

References

[AB99] E. Alm and D. Baker. Prediction of protein-folding mechanisms from free energy landscapes derived from native structures. *Proc. Natl. Acad. Sci. (USA)*, 96:11305–11310, 1999.

[AE96] N. Akkiraju and H. Edelsbrunner. Triangulating the surface of a molecule. *Discrete Applied Mathematics*, 71:5–22, 1996.

[AEHW04] P. K. Agarwal, H. Edelsbrunner, J. Harer, and Y. Wang. Extreme elevation on a 2-manifold. In *Proc. 20th Ann. Sympos. Comput. Geom.*, pages 357–365, 2004.

[AMKB02] E. Alm, A. V. Morozov, T. Kortemme, and D. Baker. Simple physical models connect theory and experiments in protein folding kinetics. *J. Mol. Biol.*, 322:463–476, 2002.

[Anf73] C. B. Anfinsen. Principles that govern protein folding. *Science*, 181:223–230, 1973.

[Ban70] T. F. Banchoff. Critical points and curvature for embedded polyhedral surfaces. *American Mathematical Monthly*, 77(5):475–485, 1970.

[BEHP04] P. T. Bremer, H. Edelsbrunner, B. Hamann, and V. Pascucci. A topological hierarchy for functions on triangulated surfaces. *IEEE Transactions on Visualization and Computer Graphics*, 10(4):385–396, 2004.

[BEK01] A. Bernal, U. Ear, and N. Kyrpides. Genomes online database (GOLD): a monitor of genome projects world-wide. *Nucl. Acids. Res.*, 29:126–127, 2001.

[BKW$^+$77] F. C. Bernstein, T. F. Koetzle, G. William, D. J. Meyer, M. D. Brice, J. R. Rodgers, et al. The protein databank: a computer-based archival file for macromolecular structures. *J. Mol. Biol.*, 112:535–542, 1977.

[BQO$^+$93] M. Billeter, Y.Q. Qian, G. Otting, M. Muller, W. Gehring, and K. Wuthrich. Determination of the nuclear magnetic resonance solution structure of an antennapedia homeodomain-dna complex. *J. Mol. Biol.*, 234:1084–1093, 1993.

[BS01] D. Baker and A. Sali. Protein structure prediction and structural genomics. *Science*, 294:93–96, 2001.

[BT91] C. Branden and J. Tooze. *Introduction to Protein Structure*. Garland Publishing, New York, NY, 1991.

[BWF$^+$00] H. M. Berman, J. Westbrook, Z. Feng, G. Gilliland, T. N. Bhat, H. Weissig, et al. The Protein Data Bank. *Nucl. Acids. Res.*, 28:235–242, 2000.

[CBJ00] D. A. Cosgrove, D. M. Bayada, and A. J. Johnson. A novel method of aligning molecules by local surface shape similarity. *J. Comput-Aided Mol Des*, 14:573–591, 2000.

[CCL03] F. Cazals, F. Chazal, and T. Lewiner. Molecular shape analysis based upon the Morse-Smale complex and the Connolly function. In *Proc. 19th Annu. ACM Sympos. Comput. Geom.*, 2003.

[CDES01] H.-L Cheng, T. K. Dey, H. Edelsbrunner, and J. Sullivan. Dynamic skin triangulation. *Discrete Comput. Geom.*, 25:525–568, 2001.

[CK00] T. E. Cheatham and P. A. Kollman. Molecular dynamics simulation of nucleic acids. *Ann. Rev. Phys. Chem.*, 51:435–471, 2000.

[Con83] M. L. Connolly. Analytical molecular surface calculation. *J. Appl. Cryst.*, 16:548–558, 1983.

[Con85]   M. L. Connolly. Molecular surface triangulation. *J. Appl. Cryst.*, 18: 499–505, 1985.
[Con86a] M. L. Connolly. Measurement of protein surface shape by solid angles. *J. Mol. Graphics*, 4:3 – 6, 1986.
[Con86b] M. L. Connolly. Shape complementarity at the hemo-globin albl subunit interface. *Biopolymers*, 25:1229–1247, 1986.
[Con96]   M. L. Connolly. Molecular surface: A review. *Network Science*, 1996.
[CP53]   R. B. Corey and L. Pauling. Molecular models of amino acids, peptides and proteins. *Rev. Sci. Instr.*, 24:621–627, 1953.
[Cre84]   T. E. Creighton. *Proteins. Structures and Molecular Principles.* Freeman, New York, NY, 1984.
[CS04]   H. L. Cheng and X. Shi. Guaranteed quality triangulation of molecular skin surfaces. In *Proc. IEEE Visualization*, pages 481–488, 2004.
[CS05]   H. L. Cheng and X. Shi. Quality mesh generation for molecular skin surfaces using restricted union of balls. In *Proc. IEEE Visualization*, pages 399–405, 2005.
[DNW02]   D. Duhovny, R. Nussinov, and H. J. Wolfson. Efficient unbound docking of rigid molecules. In *WABI '02: Proceedings of the Second International Workshop on Algorithms in Bioinformatics*, pages 185–200, 2002.
[DO93a]   B. S. Duncan and A. J. Olson. Approximation and characterization of molecular surfaces. *Biopolymers*, 33:219–229, 1993.
[DO93b]   B. S. Duncan and A. J. Olson. Shape analysis of molecular surfaces. *Biopolymers*, 33:231–238, 1993.
[EFL98]   H. Edelsbrunner, M. A. Facello, and J. Liang. On the definition and construction of pockets in macromolecules. *Discrete Appl. Math.*, 88:83–102, 1998.
[EHZ03]   H. Edelsbrunner, J. Harer, and A. Zomorodian. Hierarchical Morse-Smale complexes for piecewise linear 2-manifolds. *Discrete and Computational Geometry*, 30(1):87–107, 2003.
[EKB02]   T. E. Exner, M. Keil, and J. Brickmann. Pattern recognition strategies for molecular surfaces. I. Pattern generation using fuzzy set theory. *J. Comput. Chem.*, 23:1176–1187, 2002.
[ELZ02]   H. Edelsbrunner, D. Letscher, and A. Zomorodian. Topological persistence and simplification. *Discrete and Computational Geometry*, 28(4):511–533, 2002.
[EM86]   D. Eisenberg and A. D. McLachlan. Solvation energy in protein folding and binding. *Nature (London)*, 319:199–203, 1986.
[Fis94]   E. Fischer. Einfluss der configuration auf die wirkung derenzyme. *Ber. Dtsch. Chem. Ges.*, 27:2985–2993, 1894.
[GA93]   R. F. Gesteland and J. A. Atkins. *The RNA World: the nature of modern RNA suggests a prebiotic RNA world.* Cold Spring Harbor Laboratory Press, Plainview, NY, 1993.
[Gar99]   M. Garland. Multiresolution modeling: survey and future opportunities. In *Eurographics State of the Art Report*, 1999.
[GP95]   J.A. Grant and B.T. Pickup. A Gaussian description of molecular shape. *J. Phys. Chem.*, 99:3503–3510, 1995.
[GS67]   K. D. Gibson and H. A. Scheraga. Minimization of polypeptide energy. I. Preliminary structures of bovine pancreatic ribonuclease s-peptide. *Proc. Natl. Acad. Sci. (USA)*, 58:420–427, 1967.

[HB94] W. Heiden and J. Brickmann. Segmentation of protein surfaces using fuzzy logic. *J. Mol. Graphics.*, 12:106–115, 1994.

[Hop96] H. Hoppe. Progressive meshes. In *ACM SIGGRAPH*, pages 99–108, 1996.

[Kau59] W. Kauzmann. Some factors in the interpretation of protein denaturation. *Adv. Protein Chem.*, 14:1–63, 1959.

[KDS+60] J. Kendrew, R. Dickerson, B. Strandberg, R. Hart, D. Davies, and D. Philips. Structure of myoglobin: a three dimensional Fourier synthesis at 2 angstrom resolution. *Nature (London)*, 185:422–427, 1960.

[KL99] P. Koehl and M. Levitt. A brighter future for protein structure prediction. *Nature Struct. Biol.*, 6:108–111, 1999.

[KL02] P. Koehl and M. Levitt. Protein topology and stability defines the space of allowed sequences. *Proc. Natl. Acad. Sci. (USA)*, 99:1280–1285, 2002.

[KM02] M. Karplus and J. A. McCammon. Molecular dynamics simulations of biomolecules. *Nature Struct. Biol.*, 9:646–652, 2002.

[Koe06] P. Koehl. Electrostatics calculations: latest methodological advances. *Curr. Opin. Struct. Biol.*, 16:142–151, 2006.

[Kol65] W. L. Koltun. Precision space-filling atomic models. *Biopolymers*, 3:665–679, 1965.

[KWPB98] K. T. Simons K. W. Plaxco and D. Baker. Contact order, transition state placement and the refolding rates of single domain proteins. *J. Mol. Biol.*, 277:985–994, 1998.

[LA05] J. Lien and N. M. Amato. Approximate convex decomposition of polyhedra. Technical report, Technial Report TR05-001, Texas A&M University, 2005.

[LCW99] K. Lum, D. Chandler, and J. D. Weeks. Hydrophobicity at small and large length scales. *J. Phys. Chem. B.*, 103:4570–4577, 1999.

[LEF+98a] J. Liang, H. Edelsbrunner, P. Fu, P. V. Sudhakar, and S. Subramaniam. Analytical shape computation of macromolecules. I. Molecular area and volume through alpha shape. *Proteins: Struct. Func. Genet.*, 33:1–17, 1998.

[LEF+98b] J. Liang, H. Edelsbrunner, P. Fu, P. V. Sudhakar, and S. Subramaniam. Analytical shape computation of macromolecules. II. Inaccessible cavities in proteins. *Proteins: Struct. Func. Genet.*, 33:18–29, 1998.

[LEW98] J. Liang, H. Edelsbrunner, and C. Woodward. Anatomy of protein pockets and cavities: measurement of binding site geometry and implications for ligand design. *Prot. Sci.*, 7:1884–1897, 1998.

[LR71] B. Lee and F. M. Richards. Interpretation of protein structures: estimation of static accessibility. *J. Mol. Biol.*, 55:379–400, 1971.

[LWO04] Y. Levy, P. G. Wolynes, and J. N. Onuchic. Protein topology determines binding mechanism. *Proc. Natl. Acad. Sci. (USA)*, 101:511–516, 2004.

[ME99] V. Muñoz and W. A. Eaton. A simple model for calculating the kinetics of protein folding from three-dimensional structures. *Proc. Natl. Acad. Sci. (USA)*, 96:11311–11316, 1999.

[MKE01] J. C. Mitchell, R. Kerr, and L. F. Ten Eyck. Rapid atomic density measures for molecular shape characterization. *J. Mol. Graph. Model.*, 19:324–329, 2001.

[NWB+06] V. Natarajan, Y. Wang, P. Bremer, V. Pascucci, and B. Hamann. Segmenting molecular surfaces. *Computer Aided Geometric Design*, 23:495–509, 2006.

[OONS87] T. Ooi, M. Oobatake, G. Nemethy, and H. A. Scheraga. Accessible surface-areas as a measure of the thermodynamic parameters of hydration of peptides. *Proc. Natl. Acad. Sci. (USA)*, 84:3086–3090, 1987.

[PRC+60] M. Perutz, M. Rossmann, A. Cullis, G. Muirhead, G. Will, and A. North. Structure of hemoglobin: a three-dimensional Fourier synthesis at 5.5 angstrom resolution, obtained by X-ray analysis. *Nature (London)*, 185:416–422, 1960.

[Ric77] F. M. Richards. Areas, volumes, packing and protein structure. *Ann. Rev. Biophys. Bioeng.*, 6:151–176, 1977.

[Ric84] T. J. Richmond. Solvent accessible surface area and excluded volume in proteins. *J. Molecular Biology*, 178:63–89, 1984.

[SB94] T. Simonson and A. T. Brünger. Solvation free-energies estimated from macroscopic continuum theory: an accuracy assessment. *J. Phys. Chem.*, 98:4683–4694, 1994.

[TSPA92] I. Tunon, E. Silla, and J. L. Pascual-Ahuir. Molecular-surface area and hydrophobic effect. *Protein Eng.*, 5:715–716, 1992.

[VJ93] A. Varshney and F. P. Brooks Jr. Fast analytical computation of richard's smooth molecular surface. In *Proc. IEEE Visualization*, pages 300–307, 1993.

[WAB+05] Y. Wang, P. Agarwal, P. Brown, H. Edelsbrunner, and J. Rudulph. Fast geometric algorithm for rigid protein docking. In *Proc. 10th. Pacific Symposium on Biocomputing (PSB)*, pages 64–75, 2005.

[WC53] J. D. Watson and F. H. C. Crick. A structure for Deoxyribose Nucleic Acid. *Nature (London)*, 171:737–738, 1953.

[WT90] R. H. Wood and P. T. Thompson. Differences between pair and bulk hydrophobic interactions. *Proc. Natl. Acad. Sci. (USA)*, 87:946–949, 1990.

BioBrowser – Visualization of and Access to Macro-Molecular Structures

Lars Offen[1] and Dieter Fellner[2,1]

[1] Institute of Computer Graphics and Knowledge Visualization,
TU Graz, Austria
l.offen@cgv.tugraz.at
[2] GRIS, TU Darmstadt, Germany and
Fraunhofer Institute of Computer Graphics,
Darmstadt Germany
d.fellner@igd.fraunhofer.de

Summary. Based on the results of an interdisciplinary research project the paper addresses the embedding of knowledge about the function of different parts/structures of a macro molecule (protein, DNA, RNA) directly into the 3D model of this molecule. Thereby the 3D visualization becomes an important user interface component when accessing domain-specific knowledge – similar to a web browser enabling its users to access various kinds of information.

In the prototype implementation – named *BioBrowser* – various information related to bio-research is managed by a database using a fine-grain access control. This also supports restricting the access to parts of the material based on the user privileges. The database is supplied by a SOAP web service so that it is possible (after identifying yourself by a login procedure of course) to query, to change, or to add some information remotely by just using the 3D model of the molecule. All these actions are performed on sub structures of the molecules. These can be selected either by an easy query language or by just picking them in th 3D model with the mouse.

1 Introduction

In the field of structural biology, especially when talking about drug design, the three dimensional model of an protein becomes more and more important. The reason for this purpose is based on the fact that the three dimensional appearance is very closely linked to the folding or tertiary structure of a protein. This foldings is vital for the function of a protein, because it defines the so called active sites. These are the parts which can interact with other molecules or vice versa. The knowledge about these active sites and their function is fundamental when creating a new drug. Therefore this knowledge must be obtained and stored in a way, that other researchers can access it in an easy manner.

The prediction of the folding just by the sequence of amino acids – the so called primary structure, which can be obtained by sequencing [SM04] – can greatly simplify the process of drug design. Therefore many attempts in this direction exist. They are based on energy minimization [LSSP04] or finding similar structures with known folding [ZS05]. But such a prediction is difficult to compute and does not result in the correct folding in all cases. A more reliable result may be found when a good initial configuration of the folding is known. For creating such configurations some tools for protein manipulation are available [KHM+03]. The process can be simplified even more when the secondary structure is known. A prediction of this local structure is noticeable easier and there is a lot of work done in that area [MBJ00, CEB+02, BjMM+05]. But the main source for structure determinated proteins are the NMR spectroscopy and x-ray diffraction. The models of almost all of these protein can be found in the RCBS protein data bank [SLJ+98]. Additional information to these proteins are stored in many other databases spread over the net. These are for example the UniProt database [ABW+04], which was created by combining the Swiss-Prot- [BBF+05], TrEMBL- [BBA+03] and PIR- [HHSW04] databases, or the GenomeNet (http://www.genome.jp).

There are two main classes of programs for accessing the available information. On the one hand there are many tools for visualizing the three dimensional model of a protein. On the other hand there are tools for the database access.

Programs belonging to the first class are for example RasMol [SMW95, Ber00], an elderly program and therefore the visualization quality does not meet todays standard, but it is still under maintenance and very wide spread. The open source project JMol [JMo] uses the java-applet mechanism to ensure platform independence, but this results in a slower visualization. PyMol [DeL02] is a python based tool, which is capable of editing proteins as well, and gOpenMol [Laa92, BLL97] is manly an graphical interface for a set of programs from OpenMol. There are many other programs like Chimera [PGH+04], Cn3D [Hog97], or FPV [CWWS03] to name just a few of them.

All of them either exhibit a significant drop in rendering performance when handling very large proteins or have limited visualization styles for, e.g., ribbon structures [CB86, Ric81], molecular surfaces [Con83, LR71] or even spacefill.

All these programs have the lack of some annotation possibilities in common. For example they cannot create references to a database like the ones mentioned above or include some additional informations, which are not yet ready for publication.

The second class of tools include the web interfaces of the different databases to access them and tools like BLAST to query the databases for similarity in the primary structure. But to our knowledge there is no program combining these two aspects in such a way, that the three dimensional model of the protein is used as an interactive access tool for the databases. Speaking in the terms of the digital library: The 3d-model of the protein becomes the central document in the daily work.

This integration of the available information into the model was one main goal of the interdisciplinary *BioBrowser* project.

2 BioBrowser

The first challenge in this project was to provide a core application, which provides a plugin interface such that almost all necessary functionality could be implemented as a plugin [HOF05]. Afterward the main visualization styles, which are used in the community of structural biologists, have to be realized in such a way, that they provide a quality as high as possible combined with an interactive frame rate. This is essential when using the 3D model as the interactive access tool for the databases. Thereby the three dimensional visualization of the molecule is established as the comprehensive document type, when considering it from the point of the digital library.

Figures 1(a)–1(e) show the common visualization styles as screen shots from the *BioBrowser*. Figure 1(f) combines the different styles in one single view, by using a kind of semantic lens. This lens allows the user to peer through the surface of a molecule. Inside the hole the remaining styles can be combined by their distance from the point of interest. It is also possible to assign different visualization styles to different sub-selections of the whole molecule. The handling of such sub-selections is described later.

2.1 The Visualization Part of the BioBrowser

To gain the quality shown in the figures very different approaches from computer graphics were involved. Some of them use functionality which is only available on recent graphics hardware, such as fragment shaders. Fallback solutions are available if the program runs on older hardware. They usually involve a minor loss in display quality, but are crucial to establish this new approach of dealing with annotations to molecular structures.

Ball and Stick, Sticks, Spacefill

All these styles use different combinations of spheres and cylinders of different size. To reach the quality shown in Figs. 1(a),1(b), and 1(c) extensive use of the vertex and fragment shader functionality of modern graphics boards is made. Therefore the sphere and the cylinder are parameterized, e.g. the sphere by position and radius (a 4d vector) and the cylinder by a position, an axis, a radius and a length (two 4d vectors). These simple data is then transfered to the graphics board, where a ray casting algorithm, implemented as a fragment shader, calculates the individual intersection points and normals on a per pixel level [KE04]. As fallback solutions billboarding, depth sprites, and multi resolution meshes are provided [HOF05] (Fig. 2).

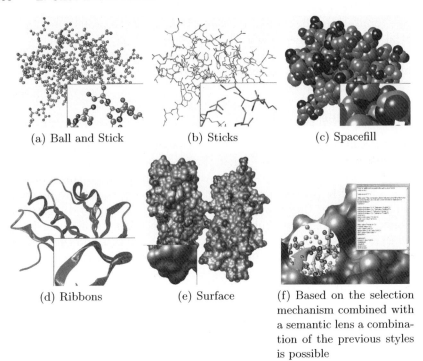

Fig. 1. The images show the common visualization styles as screen shots from the *BioBrowser* rendering the molecule 1I7A. In the lower corner of each style a close up view is shown.

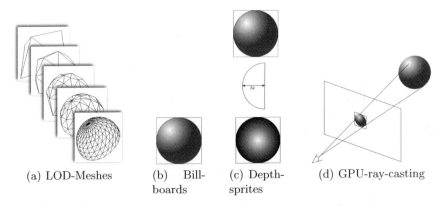

Fig. 2. The visualization uses different approaches. From left to right these are: Level-of-detail meshes, billboarding, depth sprites, and ray-casting on the GPU. These methods differ in their hardware requirements (increasing) and the needed geometric data (decreasing).

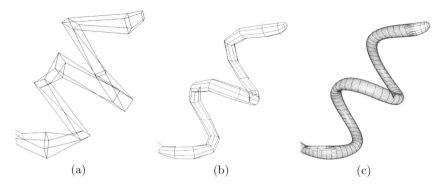

Fig. 3. The ribbon structures are created from a coarse base mesh (left). It can be refined in an adaptive way from a coarse (middle) to a fine version (right).

Ribbons

The ribbon structure shown in Fig. 1(d) is realized as a combined BRep (CBRep) [Hav02]. The CBRep is a combination of polygonal boundary representations and Catmull and Clark subdivision surfaces by introducing a sharpness flag for each edge. Using this representation it is sufficient to calculate a coarse base mesh, which is refined in a view dependent way (Fig. 3). The vertex positions of the base mesh are calculated such that the subdivided version of the mesh passes exactly through the positions of the supporting atoms – the C_α atoms [HOF04]. The refinement uses a scoring function for the estimation of the subdivision level. It is evolved for each patch independently and includes the projected size and the affiliation to the silhouette of the patch as well as the average frame rate. Thereby a nearly constant frame rate is achieved. When a fixed viewing position is used, the frame rate is ignored such that the highest possible resolution is reached – even on slower computers.

Molecular Surfaces

For a fast calculation and visualization of the molecular surfaces an algorithm based on the reduced surfaces [SO96] is combined with an adaptive tessellation of the resulting surface, which is based on second order Bézier patches. The reduced surface can be calculated by rolling a probe sphere of constant radius – normally 1.4 Å– across the atoms of the molecule (Fig. 4(a)). Thereby a polytope is created (Fig. 4(b). The vertices of this polytope are the center positions of the atoms, which the probe sphere touches in this process without intersecting another atom. An edge between two vertices is constructed, if the probe sphere touches both of them at the same time. A face is created for each fixed position of the probe sphere. A fixed position is reached, when the probe

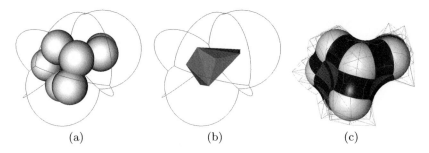

Fig. 4. The reduced surface (middle) of a molecule (left) is calculated by rolling a probe around the molecule. The arcs in the image show the path of this probe sphere. The reduced surface is converted into one base mesh consisting of triangular and quadrangular Bézier patches (right). It is refined to interpolate the molecular surface, consisting of atomic caps (light gray), concave spherical triangles (gray), and toroidial patches (dark gray).

sphere touches three or more atom spheres. Afterward the reduced surface is converted into one base mesh consisting of triangular and quadrangular Bézier patches (Fig. 4(c)). The vertices and faces of the reduced surface are replaced by triangular patches resulting in the atomic caps and the concave spherical triangles. Each edge results in a quadrangular patch, which interpolates the toroidal parts of the molecular surface. A paper describing this in detail is in preparation. For a more detailed overview have a look at [HOF05].

2.2 Selections

The core system saves the molecule in a tree structure – the molecule tree (Fig. 5 left). It also provides classes for a so called selection tree (Fig. 5 right). This is a tree with the same layout as the molecule tree, which nodes are linked to the corresponding nodes in the molecule tree. Each node of the selection tree encodes whether the assigned node in the molecule tree is selected or not. An additional flag is set when all children are selected as well. The selection tree provides fast functions for:

- Merging two selections
- Subtracting one selection from another
- Intersecting two selections
- Inverting one selection
- Iterating through a selection

The core system differentiates between two different classes of selections. One class has a rather global character, whereas the other is more local. The global selection is managed by the core system itself and is responsible for all kinds of manipulation, e.g. changing color, creating annotations, and so on. The local

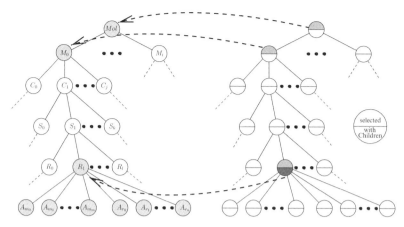

Fig. 5. A selection tree (on the left) is a tree with the same layout as the molecule tree, which encodes the molecule. The different levels of this tree are the molecule (Mol) itself, the different models (M), the chains (C), the secondary structures (S), the residues (R), and the atoms (A). The selection tree is linked to the molecule tree and states whether a node is selected (upper half of a node). To reduce the complexity it is also possible to select all children of a node as well (lower half of a node). In the molecule tree the selected nodes are shaded in light gray.

selections are closely related to the rendering and managed by the plugins itself. They define which sub structures should be rendered by the respective plugin. These different types are freely convertible into each other, such that the user can create a global selection and advice the frame work to render it as ball-and-sticks.

So every plugin has its own selection tree, which is used for displaying purpose, and the core system holds a selection tree for manipulation issues. The latter one can be manipulated by the user by simple picking operations in the 3d model or by using a query language with a simple xml-syntax. The picking operation provide functionality for setting, adding, or removing substructures to the current selection. For example when an atom is picked by the mouse any parent structure of this atom could be selected for the operation, e.g. the amino acid the atom belongs to. Such an interactive selection process is shown in Fig. 6.

For more complex selections, like selecting all oxygen atoms in a chain, a simple query language is included. The syntax of this language is xml-based and stated in Table 1.

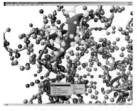  

(a) One secondary structure is already selected. Another is added by right-clicking on the molecule and select add from the context-menu

(b) The result of the selection

(c) From this selection one residue is removed. Again by right-clicking and selecting remove from the context menu

Fig. 6. Interactive selection process based on mouse input. The light gray atoms represent the current selection.

```
SELECTION: [<list>ITEM*</list>]|ITEM ITEM:       <item [MOL] [MODEL]
[CHAIN] [SSTRUCT] [RES] [ATOM]
                [CHILDREN]/>
MOL:        molecule-id=ID MODEL:      moleculemodel-id=ID CHAIN:
chain-id=ID SSTRUCT:    [secs-id=ID] [secs-type=TYPE] RES:
[res-id=ID] [res-bh=BHID] [res-name=RESNAME] ATOM:
[atom-id=ID] [atom-element=ELEMENT] [atom-name=ATOMNAME]
            [atom-atomicnumber=NUMBER] [atom-function=FUNC]
CHILDREN:   withChildren="yes"|"no"
ID:         STRING|!STRING                % id string
TYPE:       TYPENAME|!TYPENAME TYPENAME:
"helix"|"sheet"|"turn"|"none" BHID:      BROOKHAVEN|!BROOKHAVEN
BROOKHAVEN:"ALA"|"ARG"|"ASP"|"ASN"|"CYS"|"GLU"|"GLN"|"GLY"|
           "HIS"|"ILE"|"LEU"|"LYS"|"MET"|"PHE"|"PRO"|"SER"|
           "THR"|"TRP"|"TYR"|"VAL"
RESNAME:    STRING|!STRING                % name of amino acid
ELEMENT:    ELEM|!ELEM ELEM:      "C"|"N"|"O"|...
ATOMNAME:   STRING|!STRING                % pdb-columns 13-16
FUNC:       FUNCTION|!FUNCTION NUMBER:    INT|!INT FUNCTION:
"M"|"E"|"S"|"B"|"3"
```

Table 1. The syntax of the query-language. All elements can be accessed by their internal ids. Amino acids can also be selected by their names or the standard abbreviations used by the brookhaven database. The atoms can be accessed by their elements or their function, whereas the following abbreviations are used: main chain (M), end atom (E), side chain (S), and branches (B,3). The "!" is interpreted as a not equal.

For example to select all atoms but oxygen the statement would be:

```
<item atom-element="!O"/>
```

In this statement the "!" is interpreted as "not". To select the backbone of a chain without the nitrogen atoms use:

```
<item atom-function="M" atom-element="!N" chain-id="A"/>
```

The selection coded into the selection tree in Fig. 5 is:

```
<list>
<item molecule-id="Mol" withChildren="no"/>
<item molecule-id="Mol" molecule-model-id="M0"
      withChildren="no"/>
<item molecule-id="Mol" molecule-model-id="M0" chain-id="C1"
      secs-id="S1" res-id="R1" withChildren="yes"/>
</list>
```

3 Annotations

An Annotation to a protein is some extra information associated with any substructure of this protein or – as an annotation to an annotation – to an existing annotation. Nowadays these annotations are for example part of the "unstructured" part of the pdb-files:

```
REMARK   1 REFERENCE 1
REMARK   1  AUTH   L.W.GUDDAT,J.C.BARDWELL,T.ZANDER,J.L.MARTIN
REMARK   1  TITL   THE UNCHARGED SURFACE FEATURES SURROUNDING THE
REMARK   1  TITL 2 ACTIVE SITE OF ESCHERICHIA COLI DSBA ARE CONSERVED
REMARK   1  TITL 3 AND ARE IMPLICATED IN PEPTIDE BINDING
REMARK   1  REF    PROTEIN SCI.                  V.   6  1148 1997
REMARK   1  REFN   ASTM PRCIEI  US ISSN 0961-8368                  0795
```

This annotation are difficult to parse for a computer and are normally interpreted as free text, resulting in imprecise search results. Therefore a more structured way like the xml-structure, as shown in the following example, should be preferred to save annotations:

```
<references>
 <reference id="1">
  <title value="The Uncharged Surface Features Surrounding the
                Active Site of Escherichia Coli DSBA are Conserved
                and are Implicated in Peptide Binding"/>
  <authors>
   <author firstname="L.W." lastname="Guddat"/>
   <author firstname="J.C." lastname="Bardwell"/>
   <author firstname="T." lastname="Zander"/>
```

```
        <author firstname="J.L." lastname="Martin"/>
      </authors>
      <astm_code value="PRCIEI"/>
      <ccdc_code value="0795"/>
      <journal>
        <title value="Protein Sci."/>
        <volume value="6"/>
        <year value="1997"/>
        <pages first="1148" last="1148"/>
        <issn value="0961-8368"/>
        <publisher>
          <address></address>
          <country value="US"/>
        </publisher>
      </journal>
    </reference>
</references>
```

The advantages of this approach are the easier parsing, and the more efficient way of searching since it could be specified, if the searching term is the name of an author or part of the title. The *BioBrowser* frame work uses a database layout (see section 3.1) which can handle both types of annotations – free text as well as xml.

In the *BioBrowser* framework making annotations is as easy as making the selection to annotate. By selecting "Annotate..." from the context menu the annotation dialog (Fig. 7(a)) is shown.

The annotation dialog supports the user when annotating a structure. It serves as a kind of "annotation wizard". It is split into four main parts. On the top the name of the annotation and the related selection can be modified. The right part of the dialog provides an interface for controlling the access management. The appropriate rights can be assigned to the different users and groups by marking them in some check boxes. On the left side of the dialog the annotation window can be found. In this window the annotation is entered. Internally the entered annotation is split up into an unstructured part and the part, which is structured by xml-tags. The known tags are displayed in the middle of the dialog. Hovering over them brings up a tool tip showing the assigned attributes as well as the common sub tags. New tags can be created by just using them in an annotation. They will be inserted into the database automatically.

Existing annotations are visualized as a kind of needles sticked into the protein (Fig. 7(b). When such a needle is selected for an annotation this annotation will be annotated. Thereby a complete tree of annotations is possible.

3.1 Database

All the annotations are saved into a large database. The layout of this database is shown in Fig. 8. It can be split into five main Parts:

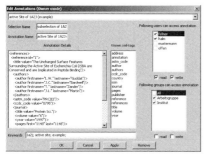 

(a) Selecting "Annotate..." from the context menu, brings up the annotation dialog. It serves as an "annotation wizard" and supports the user when annotating the structures. The dialog is explained in a more detailed way in the main text.

(b) Annotations of annotations can be made by right-clicking on an existing annotation. This will be shown by a needle on the needle (dark arrow)

Fig. 7. Annotations can be created using the annotation dialog.

- User management (very dark gray block)
- XML Mapping (dark gray block on the left)
- Selection Mapping (gray block on the right)
- Annotations (light gray block at the top)
- Access management (gray block in the bottom left corner)

User Management

The user management is inspired by the file access management of the unix file system. It differentiates between groups, e.g. all, institution, working group, and the users. Each group respectively user has some identifying data assigned to, such as a name or a description and each user can be assigned to as many groups as needed. For each user some more rights can be specified on this level: the right to change the user or group data and the administrative access to the database. For identifying an user by an login mechanism an encrypted password for each user is stored as well.

XML Part

The xml-part maps the tree structure of a xml-file into the flat structure of a relational database. This is done by saving the parent id and the root id of the tree into each node (xmlentry). The entry also saves the used token and has a mapping to the different attributes this node has. The token is not saved directly, but uses only a link to another table collecting all tokens. Thereby the complete xml-file can be reconstructed due to SQL-Queries.

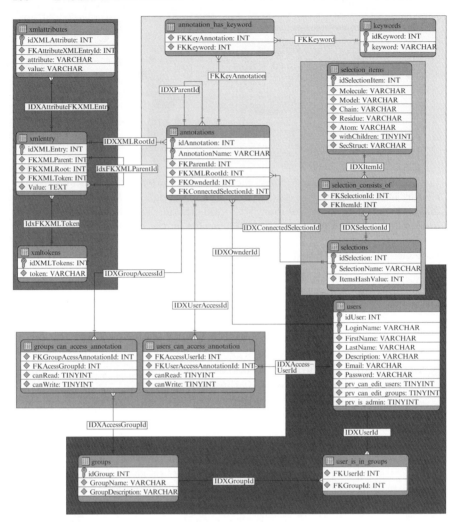

Fig. 8. Database layout.

Selection Mapping

A selection is saved into the database by splitting up its selection tree into the different selected items as shown at the end of Sect. 2.2. Each such item is inserted into the database using a hash value for faster access. The hash value of an item is computed from the ids specifying the item. A complete selection is normally a collection of more than one selected item. This $1 : n$-mapping is stored in the database, as well as a hash value, which is computed from the individual hash values of each item.

Thereby a query for a specific selection S can be done very fast. The query consists of two different steps:

1. Split up the selection S into the different items I_i.
2. Compute the hash values H_i for each I_i.
3. Compute the hash value H of S from the H_i.
4. Query the database for all Selections S_i having the same hash value H.
5. Split up each Selection S_i into the different items and check them against the items I_i.

Depending on the used hash functions the database query will result in only one selection, which have to be approved by matching its selected items against the I_i.

Annotations and Access Management

Through the access management of the database a fine access control can be assigned to every annotation. The owner or creator of an annotation can specify who can read or change the annotation by assigning the desired rights to the users and/or groups. Afterward only the users, which are allowed by the owner, have the possibility to access the annotation. Thereby they are restricted to the rights specified by the owner.

An annotation consists of a link to the root node of the assigned xml-structure. It can additionally contain a plain text version as a note. If the annotation is annotating another annotation the id of it will be added as parent id. Thereby a tree of annotations is created. To reduce the complexity of common queries the keywords assigned to an annotation are not included into the xml-structure, but saved in a seperate table.

3.2 Database Access

To access the database structure described in the last section the framework supports two different approaches. On the one hand it has an ODBC interface, which can connect to local databases. On the other hand a web service based on the SOAP protocol [W3C] – including the needed server – has been developed. These two approaches are explained in more detail below. But first the main scenarios for accessing the database are shortly presented:

Logging into the database This is the first step, which must be done when using the database. Here the given user and password are checked against the data in the database. If this check is positive the user has access to the database.

Administrative access If the current user has the administrator flag set, all SQL instructions are forwarded to the database. This is only necessary for low level access and there should be at most one user with this flag set.

Changing the user management data When the user has the right to change the user data the framework will allow the user to add new groups or users respectively change the old ones.

Querying all annotations for a given selection This setting is mainly used when inserting the known annotations into the 3d model of the protein. When the visualization mode of the annotations is activated the database is queried every some frames for the known annotations of the current molecule. Needless to say the access rights of the current user are considered. The resulting annotations are placed into the 3D model according to their real selection. The constant refresh of the visualization ensures that changes or additional comments from remote users will be incorporated with only a few frames delay.

Updating annotations Using this the user can add new annotations to any selection or to any old annotation. He can also change the content of an old annotation if he has write access to this annotation. The data that can be changed of course includes the access rights of these annotations. So for example an annotation could be opened for the public, when the results are published elsewhere. When a user opens an annotation for editing it is marked in the database, such that other users have a read-only access to it until the editing procedure is finished.

Searching for annotations having a given property This query results in a list of annotations having the given property, which can be a keyword, a phrase in the plain text section, or a phrase in some specific section of the xml structure of the annotation. A query for an annotation using the xml-syntax from Sec. 3 can be formulated as follows:

```
select FKXMLRoot from xmlentry, xmltokens, xmlattributes where
       xmltokens.token='Author' &&
       xmlentry.FKXMLToken=xmltokens.idXMLTokens &&
       xmlattributes.FKAttributeXMLEntryId=xmlentry.idXMLEntry &&
       xmlattributes.attribute='Lastname'&&
       xmlattributes.value='Guddat'
```

It returns all annotations having an author with the given last name.

SOAP Web Service

To share the database with remote users a so called SOAP web service, based on a xml protocol, has been implemented. Thereby a client sends xml formatted queries to a server. This server interprets the queries, processes them and generates the answer, which is send back to the client.

The protocol for the allowed queries is defined in the web service description language (wsdl) and can be retrieved from the server. So the access to the database is not restricted to the *BioBrowser* framework but every client implementing this protocol could access the database. Since the server interprets every query, it can check whether the current user is allowed do execute this query or not. When the user has the required rights, the answer is send back to the querying application, otherwise an appropriate error message is sent.

ODBC Interface

When using the ODBC protocol, which the operating system must provide, the framework is responsible for the access control. Whenever a query to the database is submitted, the result is checked against the rights of the user. If the user currently logged into the system hasn't the right to read the result is discarded. Otherwise it is presented to the user. If he has the right to change it any changes will be submitted to the database. When the user has only the rights to read all changes are discarded.

The problem with the direct ODBC protocol is, that the framework could be circumvent through this protocol. This means that every user connected to the database using this protocol can sent arbitrary low level SQL queries. Therefor the use of this interface is really restricted to the access to local databases or databases protected from unrestricted access by a firewall. For remote databases the web service should be used, since then all SQL instructions are parsed by the server and discarded when the current user does not have the appropriate rights.

4 Conclusion/Summary

This paper demonstrate how to integrate the knowledge about molecular structures into the three dimensional model of the structure. This is exemplified with the prototype implementation of the *BioBrowser*. For an usable integration two different problems had to be solved:

First of all the visualization of the three dimensional model must be fast enough and of high quality so that the three dimensional model of the molecule can be introduced as the main document for the daily research work. This part has been overcome by consequent use of the available hardware and modern computer graphics algorithms. The second problem is connected to the integration of the available knowledge into this model. Therefore the *BioBrowser* uses an annotation mechanism, which connects the knowledge about parts of the molecule directly to the corresponding parts of the model. This annotation are saved into a database which can be queried locally by an ODBC interface or remotely by a SOAP web service. To restrict the accessibility of certain annotations, a fine-grain access control is included into the database.

References

[ABW+04] R. Apweiler, A. Bairoch, C.H. Wu, W.C. Barker, B. Boeckmann, S. Ferro, E. Gasteiger, H. Huang, R. Lopez, M. Magrane, M.J. Martin, D.A. Natale, C. O'Donovan, N. Redaschi, and L.S. Yeh. UniProt: the Universal Protein knowledgebase. *Nucleic Acids Res.*, 32:115–119, 2004. http://www.uniprot.org.

[BBA+03]   B. Boeckmann, A. Bairoch, R. Apweiler, M.-C. Blatter, A. Estreicher, E. Gasteiger, M.J. Martin, K. Michoud, C. O'Donovan, I. Phan, S. Pilbout, and M. Schneider. The SWISS-PROT protein knowledgebase and its supplement TrEMBL in 2003. *Nucleic Acids Res.*, 31:365–370, 2003. http://www.expasy.ch/cgi-bin/sprot-search-ful.

[BBF+05]   B. Boeckmann, M.-C. Blatter, L. Famiglietti, U. Hinz, L. Lane, B. Roechert, and A. Bairoch. Protein variety and functional diversity: Swiss-Prot annotation in its biological context. *Comptes Rendus Biologies*, 328:882–899, 2005.

[Ber00]   H. J. Bernstein. Recent changes to RasMol, recombining the variants. *Trends in Biochemical Sciences*, 25:453–455, 2000. http://www.rasmol.org/.

[BjMM+05] K. Bryson, L. j. McGuffin, R. L. Marsdenl, J. J. Ward, J. S. Sodhi, and D. T. Jones. Protein structure prediction servers at University College London. *Nucleic Acids Research*, 33:W36–W38, 2005.

[BLL97]   D.L. Bergmann, L. Laaksonen, and A. Laaksonen. Visualization of solvation structures in liquid mixtures. *J Mol Graph Model*, 15:301–306, 1997.

[CB86]   M. Carson and C.E. Bugg. Algorithm for Ribbon Models of Proteins. *J.Mol.Graphics*, pages 121–122, 1986.

[CEB+02]   S. N. Crivelli, E. Eskow, B. Bader, V. Lamberti, R. Byrd, R. Schnabel, and T. Head-Gordon. A physical approach to protein strcuture prediction. *Biophysical Journal*, 82:36–49, 2002.

[Con83]   M.L. Connolly. Solvent-accessible surfaces of proteins and nucleic acid. *Science*, 221:709–713, 1983.

[CWWS03]   Tolga Can, Yujun Wang, Yuan-Fang Wang, and Jianwen Su. FPV: fast protein visualization using Java 3D^{TM}. In *Proceedings of the 2003 ACM symposium on Applied computing*, pages 88–95. ACM Press, 2003. http://www.ceng.metu.edu.tr/~tcan/fpv/.

[DeL02]   W. L. DeLano. *The PyMOL Molecular Graphics System*. DeLano Scientific, San Carlos, CA, USA, 2002. http://www.pymol.org.

[Hav02]   S. Havemann. Interactive Rendering of Catmull/Clark Surfaces with Crease Edges. *The Visual Computer*, 18:286–298, 2002.

[HHSW04]   H. Huang, Z.Z. Hu, B.E. Suzek, and C.H. Wu. The PIR integrated protein databases and data retrieval system. *Data Science*, 3:163–174, 2004.

[HOF04]   Andreas Halm, Lars Offen, and Dieter Fellner. Visualization of Complex Molecular Ribbon Structures at Interactive Rates. In *Proceedings of the Information Visualisation, Eighth International Conference on (IV'04)*, pages 737–744. IEEE Computer Society, 2004.

[HOF05]   Andreas Halm, Lars Offen, and Dieter Fellner. BioBrowser: A Framework for Fast Protein Visualization. In Ken Brodlie, David Duke, and Ken Joy, editors, *Eurographics / IEEE VGTC Symposium on Visualization*, pages 287–294, Leeds, United Kingdom, 2005. Eurographics Association.

[Hog97]   C. W.V. Hogue. Cn3D: a new generation of three-dimensional molecular structure viewer. *Trends in Biochemical Sciences*, 22:314–316, 1997. ftp://ftp.ncbi.nih.gov/cn3d/.

[JMo]   Jmol. http://www.jmol.org.
[KE04]   T. Klein and T. Ertl. Illustrating Magnetic Field Lines using a Discrete Particle Model. In *Proceedings of the Workshop on Vision, Modelling, and Visualization 2004 (VMV '04)*, pages 387–394, 2004.
[KHM+03]   Oliver Kreylos, Bernd Hamann, Nelson L. Max, Silvia N. Crivelli, and E. Wes Bethel. Interactive Protein Manipulation. In *Proceedings of the 14th IEEE Visualization Conference 2003*, 2003.
[Laa92]   L. Laaksonen. A graphics program for the analysis and display of molecular dynamics trajectories. *J Mol Graph*, 10:33–34, 1992.
[LR71]   B. Lee and F. M. Richards. The interpretation of protein structures: Estimation of static accessibility. *J. Mol. Biol.*, 55:379–400, 1971.
[LSSP04]   S. M. Larson, C. D. Snow, M. R. Shirts, and V. S. Pande. *Computational Genomics: Theory and Application*, chapter Folding@Home and Genome@Home: Using distributed computing to tackle previously intractable problems in computationla biology. Horizon Press, 2004.
[MBJ00]   L. J. McGuffin, K. Bryson, and D. T. Jones. The PSIPRED protein structure prediction server. *Bioinformatics Applications Note*, 16:404–405, 2000.
[PGH+04]   E. F. Pettersen, T. D. Goddard, C. C. Huang, G. S. Couch, D. M. Greenblatt, E. C. Meng, and T. E. Ferrin. UCSF Chimera - A visualization system for exploratory research and analysis. *Journal of Computational Chemistry*, 25:1605–1612, 2004. http://www.cgl.ucsf.edu/chimera/.
[Ric81]   J. S. Richardson. The anatomy and taxonomy of protein structure. *Adv. Protein Chem.*, pages 167–339, 1981.
[SLJ+98]   J. L. Sussman, D. Lin, J. Jiang, N. O. Manning, J. Prilusky, O. Ritter, and E.E. Abola. Protein Data Bank (PDB): database of three-dimensional structural information of biological macromolecules. *Acta Crystallogr.*, D 54:1078–1084, 1998. http://www.pdb.org.
[SM04]   H. Steen and M. Mann. The abc's (and xyz's) of peptide sequencing. *Nature Reviews Molecular Cell Biology*, 5:699–711, 2004.
[SMW95]   R. Sayle and E. J. Milner-White. RasMol: Biomolecular graphics for all. *Trends Biochem. Sci.*, 20:374, 1995. http://www.rasmol.org/.
[SO96]   Michel F. Sanner and Arthur J. Olson. Reduced Surface: an Efficient Way to Compute Molecular Surfaces. *Biopolymers*, 38:305–320, 1996.
[W3C]   W3C. *SOAP Version 1.2 Part 0: Primer*.
[ZS05]   Y. Zhang and J. Skolnick. The protein structure prediction problem could be solved using the current pdb library. *Proc. Natl. Acad. Sci*, 102:1029–1034, 2005.

Visualization of Barrier Tree Sequences Revisited

Christian Heine[1], Gerik Scheuermann[1], Christoph Flamm[2], Ivo L. Hofacker[2], and Peter F. Stadler[3]

[1] Image and Signal Processing Group, Department of Computer Science, University of Leipzig, {heine,scheuermann}@informatik.uni-leipzig.de
[2] Department of Theoretical Chemistry and Structural Biology, University of Vienna, {xtof,ivo}@tbi.univie.ac.at
[3] Bioinformatics Group, Department of Computer Science, University of Leipzig, studla@bioinf.uni-leipzig.de

Summary. The increasing complexity of models for prediction of the native spatial structure of RNA molecules requires visualization methods that help to analyze and understand the models and their predictions. This paper improves the visualization method for sequences of barrier trees previously published by the authors. The barrier trees of these sequences are rough topological simplifications of changing folding landscapes – energy landscapes in which kinetic folding takes place. The folding landscapes themselves are generated for RNA molecules where the number of nucleotides increases. Successive landscapes are thus correlated and so are the corresponding barrier trees. The landscape sequence is visualized by an animation of a barrier tree that changes with time.

The animation is created by an adaption of the foresight layout with tolerance algorithm for dynamic graph layout problems. Since it is very general, the main ideas for the adaption are presented: construction and layout of a supergraph, and how to build the final animation from its layout. Our previous suggestions for heuristics lead to visually unpleasing results for some datasets and, generally, suffered from a poor usage of available screen space. We will present some new heuristics that improve the readability of the final animation.

1 Introduction

1.1 Biological Background

Ribonucleic acid (RNA) is a linear biopolymer, i.e. a chain of covalently connected units (nucleotides) of which there are four types: adenine (A), guanine (G), cytosine (C), and uracil (U). RNA molecules play an important role in many biological contexts, e.g. protein synthesis. The biological function of an RNA molecule is determined predominantly by its spatial structure which in turn is determined by the sequence of nucleotides. When an RNA molecule is produced in the cell, it folds back to form double helical regions consisting of paired nucleotides. The list of helices or (equivalently) of base pairs is known as the *secondary structure* of the RNA molecule. Since helices stabilize the structure while the intervening unpaired loops are destabilizing, each secondary structure can be assigned a free energy equivalent to the energy released when the molecule folds. To a large extent, the secondary structure already determines the function of RNA.

Various methods have been proposed to explain and predict the structures of RNA molecules. Typically, one considers the structure with the lowest free energy, i.e. the one for which the folding process that starts from the completely unfolded state releases the maximum amount of energy. This structure is the most stable one, and according to the laws of statistical mechanics, the one that is most frequently attained in thermodynamic equilibrium. The folding process itself can, however, take a long time so that the equilibrium state that will be reached after an infinite waiting time may not be biologically relevant. Instead, the folding process may pause in metastable structures from which it is hard to escape due to high energy barriers. The folding process of an RNA molecule can be modeled as a Markov process whose states are the individual secondary structures [CHS96]. Transitions are allowed only between "neighboring configurations", i.e. those that differ by only one base pair [FFHS00], and transition rates are proportional to $\exp(\Delta E/RT)$, where ΔE is the difference in energy, T is the ambient temperature, and R is a constant. In practice, however, the transition matrix is much too large to solve the resulting master equation directly.

A refined model transforms the configuration space into a large graph, whose vertices are secondary structures and whose edges connect neighboring structures. The neighbor graph along with the energy specific to each configuration can be imagined as a discrete energy landscape. A folding or refolding process can then be described by a path in the graph or a walk in the energy landscape. For each such path there exists one structure of maximal free energy, the *maximum* of the path. The *barrier* between two configurations is the smallest maximum of all paths between the two configurations. If a structure refolds, it has to overcome at least this energy barrier. These barriers partition the graph into "basins" that are centered around local energy minima (secondary structures of which all neighbors are less stable). An approximate model is now obtained by considering the basins as effective states of the RNA molecules. Transition rates between basins can be derived from the more detailed model under the assumption that the folding process is nearly equilibrated locally within each basin [WSSF+04].

The relevant information can now be stored in the so-called *barrier tree* T of the landscape. The leaves of T correspond to the local minima of the energy landscape together with their basins of attraction, while inner vertices represent the barriers (also called saddle-points) between the basins. Figure 1 shows an example of a barrier tree for a very simple landscape. This example is just for illustrative purposes; we consider mainly landscapes where individual

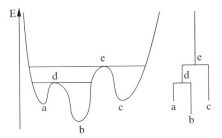

Fig. 1. A very simple landscape and barrier tree. In contrast to normal trees, each vertex of a barrier is drawn at a height that reflects the free energy of the folding configuration it represents. To determine the energy barrier between two local minima, one has to find the barrier tree vertex that has both leaves representing the local minima as descendants and the greatest topological distance to the root of the tree.

points do have a high and varying number of neighbors, making the landscape a high dimensional object. Barrier trees are constructed by successively "flooding" the basins of the landscape. A barrier is found at the point where the lakes of two basins would join. These two joined basins are considered to be one when the "flooding" is continued. See Flamm et al. [FHSW02] for a detailed description.

In reality, however, RNA molecules are not "born" as a whole. Rather, they are "transcribed" nucleotide by nucleotide from their DNA template, so that the molecule is still growing while it already starts to fold [MM04]. The structures that are formed are thus dependent upon the relative rates of folding and transcription. Similar effects are observed when an RNA molecule travels through a narrow pore, where it must unfold on one side and refolding on the other [GBH04]. Again the kinetics of folding is coupled to the speed with which the molecule is pulled through the pore. Instead of single static energy landscape, we thus have to deal with a situation where the energy landscape, and hence the rules of folding, changes with each step of the second dynamical process. Since the latter proceeds in small steps, it only causes moderate changes in the energy landscape. Thus, there is a natural correspondence between a local energy minimum x before and a (unique) local minimum x' after a step of the second dynamics: Structure x is modified to some structure x^* i.e. by appending a single unpaired nucleotide. Then x^* relaxes to the local minimum x' to whose basin it belongs to. Note that multiple local minima can map to the same local minimum in the next step, and that local minima might arise that are not mapped from any local minimum of the previous step.

From the biophysical point of view, the problem is thus to understand the dynamics of folding combined with another process such as transcription or pore traversal. As in the static case, this can be done by approximating the folding energy landscape at each step by its barrier tree. The second dynamics is then represented by transitions between corresponding local minima. While the folding process in the static case is relatively easily interpreted as a movement on the barrier tree, we now have to consider a movement on a series of barrier trees whose vertices are connected in a specific way.

In numerical simulations, one observes, that for some RNAs the fraction of folding trajectories that reach the ground state of a certain fully grown chain depends in a non-trivial way on the relative speed of transcription. Both for very slow and very fast transcription the molecule reaches the ground state quickly, while in an intermediate regime most of the trajectories become trapped in a metastable, very different, secondary structure. In order to understand this phenomenon it is necessary to compare the trajectories in the barrier tree series and to pinpoint the step(s) in which escape from local minima occurs at the same time scales as chain elongation. The same type of questions naturally arise in other settings where the folding energy changes, i.e., whenever the temperature or salt concentration changes.

1.2 Visualization Problem

Without an appropriate visualization tool it is virtually impossible to find the time-steps and transition at which time-scale difference have a drastic effect, as there is little or no *a priori* coherence between the layouts of the individual barrier trees in a series. It is thus very tedious to actually follow a trajectory through a series and to determine the likely transitions. The mapping of local minima, however provides information that, as we shall see, can be utilized to enhance the coherence of adjacent trees in a series.

The barrier trees thus share common information that should be presented accordingly, i.e. it should not attract more attention than the parts that differ. Instead of visualizing a sequence of barrier trees that have some redundancy, one can also say that there is just one barrier tree that changes with time in a way that the barrier trees of the sequence are snapshots of the dynamic tree at certain points of time. In this work, we will thus view this problem as a dynamic graph drawing problem. As an abstraction, we define the problem as follows: *Given a sequence of barrier trees and leaf mappings, where leaves of one tree are mapped on leaves of the following tree, determine the layout of all trees such that in a presentation the mental map is retained.*

1.3 Algorithm Outline

To solve the visualization problem, our algorithm is split in several parts, which we will decribe in sections 3 to 5. Given the barrier trees $T_i = (V_i, E_i, e_i)$, $(e_i : V_i \to R$ is a function that gives the energy of each vertex) and the leaf mappings $f_i : V_{i-1} \to V_i$ between them, we first find equivalent vertices. These vertices are then arranged in an order that minimizes an objective function which is mainly determined by the number of visible edge crossings for the whole sequence at presentation time. We use simulated annealing to determine this order. Given this order, we directly derive the layout of the single trees that make up the barrier tree sequence and present them in an animation with transitions that help to communicate the changes.

2 Related Work

Drawing a graph is the process of transforming topological properties of the graph to geometric objects in a graphical representation. This process is mostly determined by the generation of a layout for that graph, that *places* vertices in a vector space and *routes* edges to connect the vertices. The layout of a graph has properties that can be measured with certain cost functions, e.g., area of the layout, number of edge crossings, distribution of vertices and edges, congruency of isomorphic structures, etc. To make visually pleasing drawings, esthetic criteria have been defined. Such criteria often demand maximizing or minimizing one of the cost functions. As not all esthetic criteria can be obeyed simultaneously, a layout algorithm generally makes a trade-off between them. The field of static graph layout creation has been intensively studied in the past decades. There exist good overviews for this topic ([dBETT94, HMM00, Tam99]).

The first attempts toward dynamic graph drawing were very specific. Moen [Moe90] presents an algorithm that shows a part of an ordered tree. Although the tree itself stays the same, the selected subset may change through replacement of subtrees by leaves and vice versa. Cohen et al. [CdBT[+]92] gives detailed algorithms and data structures for a number of dynamic graph classes. These allow visualizing popular data structures, e.g. AVL-Trees, and adjusting the layout of a graph, if it is being edited or browsed. Both approaches share a common motivation: they reduce the computation time of the layout by reusing information about the previous layout. This has the side effect of making the layout of the changed graph similar to the unchanged, but accumulation of many elementary changes can result in an esthetically unpleasing drawing.

North [Nor95] measures the quality of an algorithm to make good dynamic drawings based on *incremental* or *dynamic stability*, i.e. the property of an algorithm to produce very similar layouts for graphs that differ only slightly. He applies his concepts to the drawing of dynamic directed acyclic graphs. Misue et al. [MELS95] introduce the concept of *mental distance*. It formally describes the difference of two layouts and can be used to measure the perceived stability of a dynamic graph layout. They define the esthetic criterion "preserving the mental map" for any dynamic graph drawing problem, and refine it to three models. In the *orthogonal ordering* the left-to-right, and up-down order of vertices stays the same. *Proximity relations* are preserved, if the relative distances of vertices and edges do not change. The *topology* is preserved, if vertices and groups of vertices of one region stay in that region.

The *mental distance* of two layouts is the number of times or the amount by which a rule is broken. Frishman and Tal [FT04] present an algorithm that draws dynamic clustered general graphs using an incremental force directed method. Their algorithm generally preserves the mental map by reusing the earlier layout, but improves the layout slightly, if a static graph drawing esthetic criteria is not met any more. They recently generalized their method to unclustered graphs ([FT07]).

If the layout process cannot be formulated to minimize the mental distance between successive layouts, a local transition or morphing of the layouts has to take place. Friedrich and Eades [FE02] describe a method to make sure that the transition preserves the mental map. To do that, an affine transformation that registers both layouts is determined and performed. Using a force-directed approach, vertices are moved to their final positions while avoiding occlusions and other visual artifacts linear interpolation would bring forth. Fortunately, our algorithm produces layouts that are stable enough not to require these forms of transition.

Erten et al. [EHK+03] describe a method to layout general dynamic graphs using a force-directed method. Vertices of the evolving graph that are equivalent are connected by virtual springs that contract in the force-directed method. As a result, vertices referring to the same instance at different times are positioned closely together. This ensures a good stability of the dynamic layout. We do not use this general approach, because we feel that the final animation should at least resemble the look and feel of barrier trees.

Diehl and Görg [DG02] propose a general scheme to layout dynamic graphs when all graphs of the sequence are known prior to layout creation. This scheme is independent of the class of the graphs and the layout algorithm used. Their *Foresight Layout with Tolerance* algorithm makes a trade-off between static and dynamic graph drawing esthetic criteria based on a tolerance parameter. In a first phase a *supergraph* is constructed that contains all graphs of the sequence as subgraphs. Then the layout of this (static) supergraph is determined and used as a blueprint for the layout of the subgraphs. The layout of the subgraphs can be further improved with respect to static graph drawing esthetic criteria, but its mental distance may not differ by more than the tolerance parameter from the blueprint layout. Presentation of the sequence can be done using morphing geometry information between the single subgraphs. Görg et al. [GBPD04] further improve the scheme with the notion of the *importance* of a vertex or edge. This importance is a measure for the number of times a vertex or edge is present in the graph sequence and is used to improve the visual quality of the layouts.

A similar idea is presented by Gaertler and Wagner [GW05]. Instead of an animation, a $2\frac{1}{2}$D visualization, i.e. a 3D view of a stack of static 2D layouts–each showing the graph at a certain point of time–is generated. Brandes et al. [BDS03] also use $2\frac{1}{2}$D visualization to show a set of similar metabolic pathways. They create the layouts of the acyclic directed graphs representing the pathways using a layout of an union of all graphs, and also determine the optimal ordering of layouts. Both approaches share the notion of the supergraph, local adjustments like in the *Foresight Layout with Tolerance* algorithm are not performed. Dwyer and Schreiber [DS04] also use $2\frac{1}{2}$D to visualize a set of similar phylogenetic trees. Phylogenetic trees are very similar in structure to barrier trees. In contrast to the other two approaches instead of a supergraph only a *minimal leaf ordering* is determined. This neglects the identification of equivalent inner vertices, which becomes necessary, if transitions are to be shown between keyframes. It also requires each inner vertex to have exactly two children, a property which barrier trees do not have in general.

In this work we adapted the *Foresight Layout with Tolerance* algorithm. Since it is very general, we optimized each of the phases to fit our dynamic barrier tree application.

The layouts of the subgraphs that is generated from the supergraph layout can also be used in a $2\frac{1}{2}$D visualization. However, we found this to be inappropriate, because the barrier tree sequences under consideration were highly dynamic. In our datasets we observed that almost any tree at time t has nearly nothing in common with the tree at time $t + 5$. A $2\frac{1}{2}$D visualization would therefore exhibit much visual clutter. Also, the energy of a vertex, and thus its vertical position, can change between subgraphs. In a $2\frac{1}{2}$D visualization one would have to indicate such events with edges between slices, we found it more natural to indicate that in an animation with a movement of the vertex. In general, we think that the animation of transitions between subgraph layouts can be efficiently used to communicate the changes the barrier tree topology to the user.

3 Constructing the Equivalence Classes

The first step in the *Foresight Layout with Tolerance Algorithm* is to construct a supergraph. In the general case, this would be the union of all graphs of the sequence. Unfortunately, this works well only if the supergraph contains all the information of the subgraphs afterward. But barrier trees have additional information per vertex that is used for layout, i.e. their energy. Since a vertex of the supergraph may represent multiple vertices of the tree sequence and each of these vertices may have a different energy, a supergraph vertex may not have a single energy value. We found no useful solution to incorporate this information into the supergraph, so in earlier work [HSF+06] we simply ignored this information and constructed the supergraph nonetheless. While it can be shown that this may lead to suboptimal results, especially during edge crossing minimization, we found that hardly ever a problem for the datasets we considered, and used preprocessing to minimize the errors.

When we considered larger datasets, we observed that using too much preprocessing on them deleted much information, but construction and layout of the supergraph for the unprocessed data gave esthetically unpleasing results. For this work, we decided to take an alternative approach and do not use all of the output of the supergraph construction but put all barrier tree information directly into the layout process. The identification of equivalent barriers, however, is still required and to that end the supergraph algorithm in [HSF+06] can be used. For brevity we will not repeat the rather lengthy algorithm here, instead we refer to its original publication. From the output of the algorithm we ignore the structure, i.e., we ignore the edges of the supergraph G. The vertices of G are our equivalence classes and the function k, which maps from each barrier tree vertex to a supergraph vertex, becomes a function which maps each barrier tree vertex to its equivalence class.

4 Layout

4.1 Supergraph Layout

We use the barrier trees directly as an input for the supergraph layout. We try to find an order σ of the equivalence classes such that the sum of all edge crossings in all trees is minimized, if the barrier tree vertices were drawn using this order as the horizontal order.

$$\sigma = \underset{O}{\operatorname{argmin}} \sum_{i=1}^{N} (\alpha \cdot crossings(T_i, O_{V_i}) + \beta \cdot localorder(T_i, O_{V_i}))$$

where

- $T_i = (V_i, E_i, e_i)$ is the i-th tree in the sequence,
- $G = (V, E)$ is the supergraph of the tree sequence according to [HSF+06],
- V is the set of equivalence classes,
- $k : \bigcup_{i=1}^{N} V_i \to V$ maps each tree vertex to its equivalence class,
- $O \subset V \times V$ is an ordering relation,
- O_{V_i} is that ordering relation restricted to V_i and satisfies $(u, v) \in O_{V_i}$, if and only if $(k(u), k(v)) \in O$ for all $u, v \in V_i$,
- $crossings\,(T_i, O_{V_i})$ denotes the number of edge crossings if the tree T_i was drawn with the horizontal order of the vertices given by O_{V_i},
- $localorder\,(T_i, O_{V_i})$ names approximately the number of times a parent vertex is not drawn between its children, and
- α, β constants, which we set to 1 and 5 respectively.

At first we minimzed the above function only considering minimizing the number of edge crossings and used simulated annealing [KGV83] to that end. We were surprised that it is possible to draw the simplest sequence (ATT) with a total of 27 edge crossings for the whole sequence. We were quickly disappointed by the images themselves, as it was apparent that we neglected to encourage the father of two vertices to be drawn between them. Because of that, the use of our orthogonal drawing style resulted in hardly readable images. So we added the second term to our objective function to avoid this particular effect. It accumulates the difference of the number of vertices that are drawn to the left of their parent and the number of vertices that are drawn to the right of their parent for each vertex. If each vertex is always between its two successors, the contribution of this term to the objective function is always zero. We experimentally determined $\alpha = 1$ and $\beta = 5$ to give good final layouts. It roughly means that we rather allow 5 edge crossings than one parent that is not between its children.

There are multiple possibilities to implement the simulated annealing strategy for this particular objective function. We tested several of them, and found the following to behave the best. We start with a random order of equivalence classes and iteratively improve this order. At each iteration, we pick a random equivalence class and insert it at a random position between two other equivalence classes. Then we re-evaluate the objective function for all trees and compare it to the old value. If we improved, we keep the new order, otherwise we only keep it with the probability

$$p = \frac{1}{1 + exp(\Delta C\, T_t^{-1})} \qquad T_t = \frac{n_t - t}{t}$$

with ΔC being the cost increase and T_t being a temperature which decreases linearly with each iteration t. We stop the process after a fixed number of iterations n_t.

Instead of recalculating the total number of edge crossings, we just calculate ΔC by considering only adjacent and incident edges on all vertices v with $h(v)$ being the equivalence class currently moved. We can do this similarly for the *localorder* term of the objective function. This greatly decreases the time per iteration and makes the process very fast.

In our previous work we computed the layout of the supergraph using the *dot* algorithm by Gansner et al. [GKNV93]. In this algorithm most of the time is spend minimizing edge crossings in a repeated heuristic two layer edge crossing minimization which had a time complexity of $O(N^4)$, where N is the maximum number of vertices on one layer. Although the algorithm seldom runs in that order for real world examples, it takes a very long time to find the minimal number of edge crossings for our barrier tree sequences. Not only because we observed that there was at least one layer where one eighth of all supergraph vertices resided in, but also because the swapping of vertices often did not change the number of edge crossings directly, but a few iterations later might have allowed improvements.

Our new method has much faster iterations because the number of operations per iteration is in the order of

$$O\left(\sum_{i \in \{1,\ldots,N\}} deg(T_i)|E_i|\right) = O\left(\sum_{i \in \{1,\ldots,N\}} deg(T_i)|V_i|\right)$$

where $deg(T)$ is the degree of T, i.e. the maximum number of incident and adjacent edges on any vertex v of T. So one iteration roughly scales linearly with the total number of vertices of the whole sequence, as the degree of our trees is 2 or 3 in almost all cases, i.e. a very small constant. So one iteration lies in $O(N)$, but we require many more iterations to achieve the same quality improvement of one iteration of the *dot* algorithm.

4.2 Tree Layout

Coordinate assignment of tree vertices is done for each tree separately, respecting the ordering relation generated in the layout phase. This constraint preserves the mental map, specifically the *orthogonal ordering*. Initially the horizontal position of a tree vertex v is directly gained from the number of equivalence classes smaller than $h(v)$ with respect to the global ordering relation O. The vertical position of v directly reflects its energy.

After the vertices have been positioned, edges must be routed. For simplicity each tree edges consist of just one horizontal and one vertical line segment that directly connect the two adjacent vertices. In general, it is not always possible to draw the trees without edge crossings. We sacrificed this property for the preservation of the mental map. Drawing the edges as orthogonal line segments conforms to the style, barrier trees are usually drawn. We also found that a straight line drawing does not necessarily reduce the number of edge crossings and additionally makes tracing the edges harder than in an orthogonal drawing.

Positioning each tree separately allows us to locally improve the layout of the subgraphs. This corresponds to the third phase of the *Foresight Layout with Tolerance* algorithm. It is trivially possible to generate the horizontal position of a Tree vertex v from the number of vertices of the same tree that are smaller than v with respect to O_{V_i}. This would make a better visual impression, if the keyframes were studied by themselves, but it destroys a lot of the mental map, so we do not use this possibility for our animations.

5 Animation

After the layout for each tree has been generated, the single trees could be presented using the generated layout. In practice, there can be quite a number of changes between consecutive trees. Vertices and edges may appear or disappear, and whole subtrees can change the energy of their vertices. We created methods to make the transition smooth and to indicate the type of change. Vertices that experience a change of energy are moved accordingly in the drawing area using linear interpolation of the coordinates. Barriers that appear or disappear are presented using *blending*. Edges are modified based on the changes of their adjacent vertices. Subtrees that are created or merged "grow" out of or into the vertices, where they are created or merged into, again using linear interpolation of their coordinates.

Usually the huge number of changes would require each change to be visualized separately. In our proof-of-concept implementation, all changes are shown simultaneously using the following scheme: Each transition is given a time interval $[t_i, t_i + \Delta t)$. Vertices that change their energy are moved during $[t_i + \frac{3}{8}\Delta t, t_i + \frac{7}{8}\Delta t)$. Subtrees that grow into a vertex because of merging are scaled during $[t_i + \frac{2}{8}\Delta t, t_i + \frac{5}{8}\Delta t)$, subtrees that grow out of a vertex, do so during $[t_i + \frac{5}{8}\Delta t, t_i + \frac{8}{8}\Delta t)$. Fading out of barriers is done during $[t_i + \frac{2}{8}\Delta t, t_i + \frac{6}{8}\Delta t)$ and fading in takes place during $[t_i + \frac{4}{8}\Delta t, t_i + \frac{8}{8}\Delta t)$. The remaining interval $[t_i, t_i + \frac{2}{8}\Delta t)$ is used for a static presentation of tree T_i. The segments overlap intentionally. In the dataset we observed we found that using non-overlapping sections resulted in large parts of the tree simply disappear and appear and destroy the mental map of the user.

6 Highlighting

One common question for a domain expert that analyzes the barrier tree sequence is "which of two given structures is the winner", i.e., which one is more probable to be found in nature. It is typically found when the folding process starts in one part of the energy landscape and, later on, a new part of the energy landscape is created which is separated by a very high barrier from the rest. Regardless of how optimal the local minima of the new part of the landscape are, it is unlikely that the molecule will fold into one of them, because the barrier is too high and the probability that it will be overcome is very low on the timescale for folding reactions.

Using a simple technique, some elements of the animation can be highlighted to emphasize such observations. We split the last tree at the root and look for the leaf of lowest energy in the left subtree and the leaf of lowest energy in the right subtree. The first one is marked blue and the later one red. When the animation is shown, predecessors of these two leaves will be drawn with the appropriate color. The predecessors are given by the leaf mapping, i.e., they are local minima that will refold into the two final configurations during the process. We also found that drawing the path from the root to the actual leaf with the highlight color is visually more attracting that just coloring the leaf and its one adjacent edge.

7 Results

We evaluated our improved algorithm on three datasets. The ATT dataset consists of 20 barrier trees, with at most 25 leaves per tree and a total of 894 vertices in all trees. It represents a small RNA molecule, with sequence length growing from 40 to 74 nucleotides with varying step size. Figure 2 shows the keyframes for this dataset. The LEPTO dataset consists of 47 barrier trees, with a maximum of 50 leaves per tree and a total of 3727 vertices in all trees.

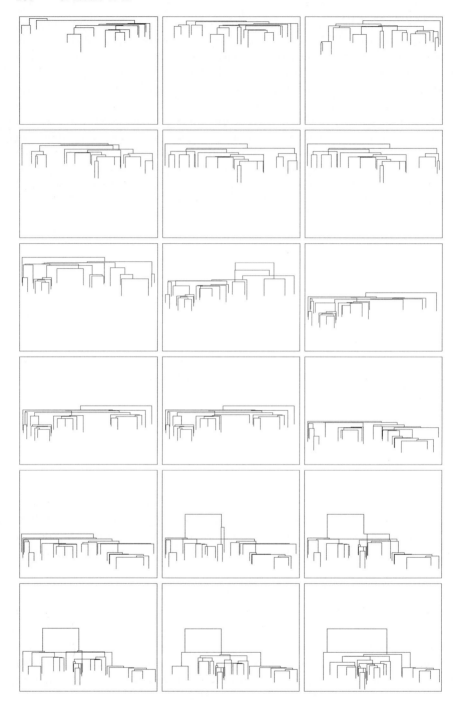

Fig. 2. The first 18 subgraph layouts of the ATT sequence.

The sequence length of the molecule increases from 10 to 56 nucleotides. The largest example, the HOK dataset, consists of 65 trees with a maximum of 100 leaves and a total of 8635 vertices. The sequence length grows from 10 to 74 nucleotides. The inner vertices of all trees of these datasets satisfy $odeg(v) = 2$, i.e., all inner vertices have exactly two children. All datasets present rather short RNA molecules.

We found our new supergraph layout, which effectively only determines an optimal vertex order for the supergraph vertices, to perform very well. We feared that most of the random permutations do not improve the tree layout at all or only slowly. Indeed, we require a greater number of iterations until the images produced were readable. Whereas the old algorithm used 1000 layer sweep iterations the new algorithm uses at least 100.000 iterations to generate readable animations. But because each iteration requires much less time, the new implementation is still faster. On an AMD Opteron with 2.0GHz, the old algorithm with 1000 iterations required approximately 37, 3462, and 16790 seconds for the ATT, LEPTO, and HOK dataset with 381, 1531, and 3793 equivalence classes respectively. The new method using 1.000.000 iterations required approximately 27, 79, and 182 seconds respectively.

The visual output of the two radically different methods is not directly comparable. There is one thing directly observable for all datasets. The new algorithm distributes vertices more evenly across the drawing area. But this is rather a nice side effect of the method and was not originally intended. While the ATT sequence does not improve much visually if compared to the old method, it benefits from the reduced computation time. The visual quality of the LEPTO and HOK sequences improved greatly, because now all edge crossings are accounted for.

8 Conclusion and Future Work

We have shown that it is possible to generate readable layouts for sequences of barrier trees using the *Foresight Layout with Tolerance* algorithm. We also showed, that construction of a supergraph may sometimes lead to suboptimal results if the supergraph does not use all information from the graphs it is constructed for. Our naïve implementation of a combination of supergraph construction and layout clearly outperformed the version where these two were separate, both quality and runtime-complexity wise. The number of iterations needed for a given dataset seems to scale with its size, but we are yet uncertain exactly how. We are currently looking for methods that automatically determine the optimal number of iterations.

The layout of the single trees may be combined with additional information. The simulation of the folding process during the growing of the molecule under various temperatures and growing rates results in distribution functions for local minima. Because the animation of the barrier trees preserves the orthogonal ordering, annotating the barrier tree leaves with the density

of the corresponding structure configurations preserves the mental map for
the annotations. The change in the densities can be additionally indicated by
a flow of liquid along the tree edges. Methods that combine tree layout and
additional information are currently investigated.

As a transition from one tree to the next consists of many elementary
operations, instead of showing them simultaneously, it might be better to
break the leaf mappings in elementary operations and show them in sequence.

The constructed supergraph is a static visualization of the whole sequence,
and presentation forms other than an animation, may be investigated. One
idea is synthesizing a 2D landscape from all barrier trees, where the folding
process is visualized as a walk.

9 Acknowledgements

We would like to thank the anonymous reviewers for their in-depth comments
on our work. This work was supported in part by the EMBIO project in FP-6
(http://www-embio.ch.cam.ac.uk/).

References

[BDS03] Ulrik Brandes, Tim Dwyer, and Falk Schreiber. Visualizing related metabolic pathways in two and a half dimensions. In Liotta [Lio04], pages 111–122.

[CdBT+92] Robert F. Cohen, Giuseppe di Battista, Roberto Tamassia, Ioannis G. Tollis, and Paola Bertolazzi. A framework for dynamic graph drawing. In *Symposium on Computational Geometry*, pages 261–270, 1992.

[CHS96] Jan Cupal, Ivo L. Hofacker, and Peter F. Stadler. Dynamic programming algorithm for the density of states of RNA secondary structures. In R. Hofstädt, T. Lengauer, M. Löffler, and D. Schomburg, editors, *Computer Science and Biology 96 (Proceedings of the German Conference on Bioinformatics)*, pages 184–186. Universität Leipzig, 1996.

[dBETT94] G. di Battista, P. Eades, R. Tamassia, and I. G. Tollis. Algorithms for drawing graphs: An annotated bibliography. *Computational Geometry: Theory and Applications*, 4(5):235–282, 1994.

[DBL04] *10th IEEE Symposium on Information Visualization (InfoVis 2004), 10-12 October 2004, Austin, TX, USA*. IEEE Computer Society, 2004.

[DG02] Stephan Diehl and Carsten Görg. Graphs, they are changing. In Stephen G. Kobourov and Michael T. Goodrich, editors, *Graph Drawing*, volume 2528 of *Lecture Notes in Computer Science*, pages 23–30. Springer, 2002.

[DS04] Tim Dwyer and Falk Schreiber. Optimal leaf ordering for two and a half dimensional phylogenetic tree visualisation. In Neville Churcher and Clare Churcher, editors, *InVis.au*, volume 35 of *CRPIT*, pages 109–115. Australian Computer Society, 2004.

[EHK+03] Cesim Erten, Philip J. Harding, Stephen G. Kobourov, Kevin Wampler, and Gary V. Yee. Graphael: Graph animations with evolving layouts. In Liotta [Lio04], pages 98–110.
[FE02] Carsten Friedrich and Peter Eades. Graph drawing in motion. *J. Graph Algorithms Appl.*, 6(3):353–370, 2002.
[FFHS00] Christoph Flamm, Walter Fontana, Ivo L. Hofacker, and Peter Schuster. RNA folding at elementary step resolution. *RNA*, 6:325–338, 2000.
[FHSW02] Christoph Flamm, Ivo L. Hofacker, Peter F. Stadler, and Michael T. Wolfinger. Barrier trees of degenerate landscapes. *Z. Phys. Chem.*, 216:1–19, 2002.
[FT04] Yaniv Frishman and Ayellet Tal. Dynamic drawing of clustered graphs. In *INFOVIS* [DBL04], pages 191–198.
[FT07] Yaniv Frishman and Ayellet Tal. Online dynamic graph drawing. In *EUROVIS* [DBL04], pages 191–198.
[GBH04] Ulrich Gerland, Ralf Bundschuh, and Terence Hwa. Translocation of structured polynucleotides through nanopores. *Phys. Biology*, 1(1-2):19–26, 2004.
[GBPD04] Carsten Görg, Peter Birke, Mathias Pohl, and Stephan Diehl. Dynamic graph drawing of sequences of orthogonal and hierarchical graphs. In *Graph Drawing*, pages 228–238, 2004.
[GKNV93] Emden R. Gansner, Eleftherios Koutsofios, Stephen C. North, and Kiem-Phong Vo. A technique for drawing directed graphs. *IEEE Trans. Software Eng.*, 19(3):214–230, 1993.
[GW05] Marco Gaertler and Dorothea Wagner. A hybrid model for drawing dynamic and evolving graphs. In Patrick Healy and Nikola S. Nikolov, editors, *Graph Drawing*, volume 3843 of *Lecture Notes in Computer Science*, pages 189–200. Springer, 2005.
[HMM00] Ivan Herman, Guy Melançon, and M. Scott Marshall. Graph visualization and navigation in information visualization: A survey. *IEEE Transactions on Visualization and Computer Graphics*, 06(1):24–43, 2000.
[HSF+06] Christian Heine, Gerik Scheuermann, Christoph Flamm, Ivo L. Hofacker, and Peter F. Stadler. Visualization of barrier tree sequences. *IEEE Transactions on Visualization and Computer Graphics*, 12(5):781–788, 2006.
[KGV83] S. Kirkpatrick, C. D. Gelatt, and M. P. Vecchi. Optimization by simulated annealing. *Science, Number 4598, 13 May 1983*, 220, 4598:671–680, 1983.
[Lio04] Giuseppe Liotta, editor. *Graph Drawing, 11th International Symposium, GD 2003, Perugia, Italy, September 21-24, 2003, Revised Papers*, volume 2912 of *Lecture Notes in Computer Science*. Springer, 2004.
[MELS95] K. Misue, P. Eades, W. Lai, and K. Sugiyama. Layout adjustment and the mental map. *J. Visual Languages and Computing*, 6(2):183–210, 1995.
[MM04] Irmtraud M. Meyer and Istvan Miklos. Co-transcriptional folding is encoded within RNA genes. *BMC Molecular Biology*, 5(10), 2004.
[Moe90] Sven Moen. Drawing dynamic trees. *IEEE Software*, 7(4):21–28, July 1990.
[Nor95] Stephen C. North. Incremental layout in dynadag. In *Graph Drawing*, pages 409–418, 1995.

[Tam99]   Roberto Tamassia. Advances in the theory and practice of graph drawing. *Theoretical Computer Science*, 217(2):235–254, 1999.

[WSSF+04]   Michael T. Wolfinger, W. Andreas Svrcek-Seiler, Christoph Flamm, Ivo L. Hofacker, and Peter F. Stadler. Exact folding dynamics of RNA secondary structures. *J. Phys. A: Math. Gen.*, 37:4731–4741, 2004.

Part VI

Visualizing Gene Expression Data

Interactive Visualization of Gene Regulatory Networks with Associated Gene Expression Time Series Data

Michel A. Westenberg[1], Sacha A. F. T. van Hijum[2], Andrzej T. Lulko[2], Oscar P. Kuipers[2], and Jos B. T. M. Roerdink[1]

[1] Institute for Mathematics and Computing Science, University of Groningen, P.O. Box 800, 9700 AV Groningen, The Netherlands m.a.westenberg@rug.nl, j.b.t.m.roerdink@rug.nl
[2] Department of Genetics, Groningen Biomolecular Sciences and Biotechnology Institute, University of Groningen, P.O. Box 14, 9750 AA Haren, The Netherlands s.a.f.t.van.hijum@rug.nl, a.t.lulko@rug.nl, o.p.kuipers@rug.nl

Summary. We present GENeVis, an application to visualize gene expression time series data in a gene regulatory network context. This is a network of regulator proteins that regulate the expression of their respective target genes. The networks are represented as graphs, in which the nodes represent genes, and the edges represent interactions between a gene and its targets. GENeVis adds features that are currently lacking in existing tools, such as mapping of expression value and corresponding p-value (or other statistic) to a single visual attribute, multiple time point visualization, and visual comparison of multiple time series in one view. Various interaction mechanisms, such as panning, zooming, regulator and target highlighting, data selection, and tooltips support data analysis and exploration. Subnetworks can be studied in detail in a separate view that shows the network context, expression data plots, and tables containing the raw expression data. We present a case study, in which gene expression time series data acquired in-house are analyzed by a biological expert using GENeVis. The case study shows that the application fills the gap between present biological interpretation of time series experiments, performed on a gene-by-gene basis, and analysis of global classes of genes whose expression is regulated by regulator proteins.

1 Introduction

The unraveling of interactions between components of living cells is an important aspect of systems biology. The interaction networks are very complex, since interactions take place not only at genomic, proteomic, and metabolomic levels, but also between these levels. We are establishing a software framework that is able to visualize such networks, and which offers interactive exploration to a researcher [BBO+06]. As part of this effort, we have developed an application for visualization of gene regulatory networks.

Gene regulatory networks can be represented by graphs, in which nodes represent genes, and edges represent interactions between a gene product (a regulator protein) and its target genes. The nodes have several attributes, such as position on the chromosome, a Gene Ontology classification [The00], and in our case, they also have gene expression attributes for multiple time points acquired during distinctive phases of growth together with p-values indicating statistical significance or other statistical data. Gene expression is measured in terms of the amount of messenger RNA (mRNA) produced after transcription of the gene. A number of tools have been proposed that visualize gene networks and overlay gene expression data on the network [BST03, SMO+03, HMWD04, HMW+05, BSRG06]. These tools overlay the expression value of one time point on a node, often as the node color, and do not always map the associated statistical data to a visual representation or one that is easy to interpret. However, proper analysis and interpretation is not possible without statistical confidence information. A further problem is that none of the existing tools allows a researcher to overlay multiple time points and associated statistical confidence on nodes. However, simultaneous visualization of multiple time points would make discovery of trends and outliers much easier. Similarly, it is also not possible to compare multiple time series with each other in a single view.

In this paper, we present GENeVis (*G*ene *E*xpression and *N*etwork *Vis*ualization), an application that allows a researcher to simultaneously visualize gene regulatory networks and gene expression time series data. Our application extends on concepts introduced in previous work in gene regulatory network visualization, and it adds features that are currently lacking in existing tools, such as mapping of expression value and corresponding p-value (or other statistic) to a single visual attribute, multiple time point visualization, and visual comparison of multiple time series in one view. We have used GENeVis to analyze time series data of the bacterium *Bacillus subtilis*, acquired in-house [LBKK07], in its regulatory network context, acquired from DBTBS (DataBase of Transcriptional Regulation in *Bacillus subtilis*) [MNON04].

The organization of this paper is as follows. We briefly discuss previous work in Section 2. We then describe the design of our application in detail (Section 3), and present the case study (Section 4). Conclusions are drawn in Section 5.

2 Previous Work

There exist a large number of tools that allow visualization of general graphs, see Herman et al. [HMM00] for an overview. In the bioinformatics field, a number of tools have emerged more or less independently due to specific requirements from the biological community (see [BBO+06] for an overview).

Osprey [BST03] was one of the first biological interaction network visualization tools. Genes are colored by their biological process as defined by the Gene Ontology [The00]. Osprey cannot overlay time series expression data on the interaction network. Cytoscape [SMO+03] is a popular data analysis tool, which does support visualization of gene expression attributes. Statistical attributes associated with the expression data can be mapped to visual styles of nodes, such as color, shape, size, border width, and border color. A main shortcoming of Cytoscape is that, for time series, it can show only the expression value of a single time point. Though extensible through a plug-in mechanism, the design of Cytoscape makes it hard to incorporate alternative node visualization methods.

VisANT [HMWD04, HMW+05] is a tool for biological network analysis and visualization. It focusses strongly on the analysis of network of various types, such as protein-protein interaction networks, gene transcription networks, metabolic pathways, and interconnections between these. Network analysis is performed by calculating topological statistics and features, or querying a server-side database for functional information. The strength of VisANT is the integration of multiple network data sources for a large number of species, which is a hard problem due to naming convention issues between data sets. For analyzing and visualizing gene expression data, VisANT is not suitable, since it provides no support for loading such data.

BiologicalNetworks [BSRG06] is a tool with a strong focus on data integration and analysis. It supports visualization of time series gene expression data in matrix form and by plots, but only allows the user to overlay one time point at a time on an interaction network.

Recently, Saraiya et al. [SLN05] performed a user performance study for various graph and time series visualizations. This study is of particular interest, since the test case reflects tasks that are performed commonly in bioinformatics pathway analysis. Metabolic pathways are also represented as graphs, forming a kind of flow chart of the chemical reactions and genes involved in, for instance, some biological process. The data for the test case consisted of time series gene expression data for 10 time points and a 50-node directed graph. Their source of data remains unclear. The study involved four visualization approaches: (i) single attribute (showing one time point at a time) and single view (show only the graph), (ii) single attribute and multiple views (show the graph and a parallel coordinate linked view), (iii) multiple attribute (show all time points simultaneously) and single view, and (iv) multiple attribute and multiple views. Statistical data associated with the expression data were not included in the experiment. It was found that overlaying a single attribute at a time works well for analyzing graphs at particular time points, and for search tasks that require topological information. Showing multiple attributes simultaneously reduces user performance for such tasks. On the other hand, multiple attribute visualizations result in better performance in outlier search tasks, and also in node comparison tasks between two time points. The conclusion was that visualization design should be task specific.

Despite the existence of tools for gene expression analysis in a regulatory network context, we believe and demonstrate in our case study that biologists would benefit from a richer and more interactive visualization environment to analyze their time series data. Existing tools overlay only expression values of one time point at a time on a node, and usually have a poor visual mapping of the statistical properties of the data. Furthermore, none of the tools supports comparison of multiple time series, i.e., in which each gene is associated with multiple time points obtained from multiple time series. GENeVis provides a solution to these issues.

3 Visualization Design

We will now present the design of GENeVis, and describe our choices regarding graph layout, expression mapping, and possibilities for interaction and data exploration.

3.1 Graph Layout

The layout of the network is computed by a force-directed algorithm, in which the edges act as springs, and the nodes repel each other [BETT99]. This layout algorithm produces satisfactory layouts for the type of networks in our application, since the grouping of nodes corresponds quite well with the biological concept of a regulon (a collection of genes under regulation by the same protein). The nodes (genes) are drawn as boxes and the edges (interactions) as lines between the nodes. A regulator protein can inhibit or activate its target, which is represented graphically at the target end of an edge by a bar or an arrow, respectively. It can also be the case that the interaction type is unknown, in which case the edge is not decorated. An example is shown in Fig. 1, in which a part of the regulatory network of *Bacillus subtilis* is drawn. The gene boxes are annotated with the gene name, and we can see, for example, that *gerE* inhibits *cotA* and *spoIVCB*, activates *cotB*, *cotC* and a number of other genes, and that it also interacts with *cotD* and *sigK* in an unknown way.

3.2 Gene Expression Mapping

Exploration of gene regulatory networks is often based on time series gene expression data, where the growth of an organism or tissue type is followed in time. For bacteria, one could take measurements during early, middle, late exponential, and stationary growth phases, resulting in a time series containing four points. This amount is small enough to allow visualization of all time

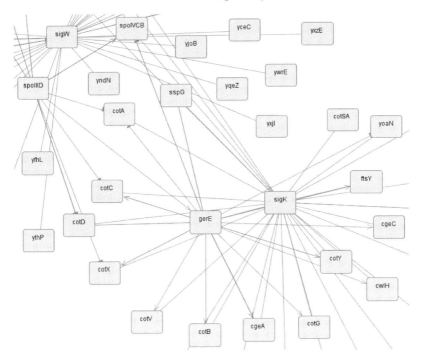

Fig. 1. Part of the regulatory network of *B. subtilis*. Gene boxes are annotated with their respective gene names. Graph edges represent gene interactions, where bars and arrows at the target ends represent inhibition and activation, respectively. Undecorated edges are used when the type of interaction is not known.

points simultaneously, and we map each time point to a colored expression box drawn inside the gene box. As a time series experiment is very time consuming and also expensive, a maximum number of 50 time points would be a realistic limit. Our approach can be used also for these larger time series. A larger number of time points will usually correspond to higher resolution in time, which can be mapped in a visually intuitive way by reducing the width of the expression boxes.

Gene expression values can either be absolute levels of expression or ratios between a test condition and a reference condition. To each expression value, a statistical value is associated, which expresses the reliability of the measurement. Commonly, the coefficient of variation is used in the case of expression levels and a p-value (indicated by p in the remainder of this paper) is used in the case of expression ratios. The reliability value is used to scale the height of an expression box: the more reliable, the higher the box. Expression levels are mapped to colors that range from white to black via yellow and red. Expression ratios are mapped to colors that range from green to red via black. The use of these colormaps is standard practice in the bioinformatics field. We divide the expression data range into a number of quantiles, and assign

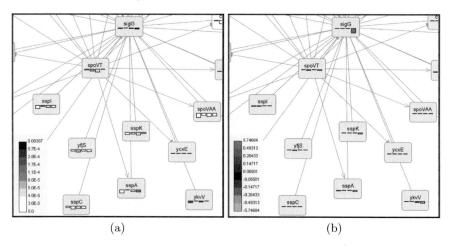

Fig. 2. Visualization of expression levels (a) and expression ratios (b) for four time points overlaid on a part of the gene regulatory network of *B. subtilis*.

each quantile a color from the colormap. By inspecting the expression value range corresponding to each quantile, a user can obtain some insight in the statistical distribution of the data. We selected a colored rectangular glyph as a graphical representation of the expression data, since color and size are perceptually easy to separate and interpret independently [War04].

Figure 2(a) and Fig. 2(b) show a visualization of expression levels and expression ratios, respectively. In order to demonstrate the visual mapping, an arbitrary part of the regulatory network of *B. subtilis* is shown. More details about the gene expression data can be found in Section 4. At all times, the user can refer to the color legend for the statistical distribution of the data. Note that the expression values are mapped in a nonlinear way to colors through the use of quantiles. Figure 2(a), for instance, shows that, at time point three, *spoVT* has a low expression (yellow color), and that this is measured reliably. The expression box would be square for a high confidence, and it reduces to a line for a very low confidence. In Fig. 2(b), we can see that *sigG* is strongly up-regulated (its expression level is high in comparison with the expression level of the reference) in time point four, and that this is the only significant change in this part of the network.

3.3 Interaction

The user can interact with the visualization by simple mouse operations. Panning and zooming are performed by dragging the mouse with the left or right button down, respectively. A right mouse button click in the background causes an automatic pan and zoom, such that the entire network fits within the display bounds. A small overview display containing a view of the whole network and a semitransparent rectangle corresponding to the area visible in the main display helps the user to navigate through the network, see Fig. 6. The rectangle in the overview display supports user interaction, and it can be dragged to pan the display.

3.4 Exploration

Even though a force-directed layout algorithm produces acceptable layouts, it is sometimes difficult to understand the network structure in dense areas with highly interconnected nodes. Therefore, we have implemented a mechanism that highlights a gene and its direct targets when the mouse hovers over the gene. Highlighting increases the line widths of the gene boxes and the edges, and it colors the edges according to the type of regulation. Activation maps to the color green, inhibition to the color red, and unknown interactions map to a shade of grey. Figure 3 shows an example when the user hovers the mouse over the gene *gerE*.

Tooltips are used to display additional information about a specific gene. This information includes the gene locus (the position of the gene on the chromosome), the gene name and possibly synonyms, gene function, and a

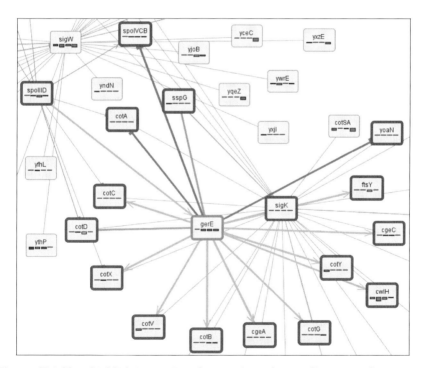

Fig. 3. Neighbor highlighting assists the user in understanding network structure. The interaction type is mapped to a color: red for inhibition, green for activation, and grey for other cases.

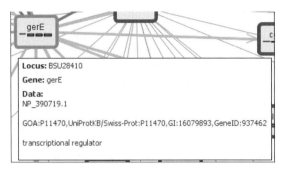

Fig. 4. Tooltips display additional information about a gene.

list of gene identifiers for other databases (e.g., Gene Ontology Annotation, Universal Protein Resource). The tooltips appear also by hovering the mouse over a gene box (i.e., in combination with the highlighting effect described previously) see Fig. 4 for an example.

GENeVis also supports keyword search to aid the user in finding a specific gene. When the gene exists in the network, the display automatically pans, such that the found gene is centered. As an additional visual cue, the gene box is enlarged slightly.

Our application also supports single time point analysis. The user can choose a time point, and construct a filter that selects genes for which the measurements are statistically significant, and for which the expression levels or ratios fall within a certain range. The filter parameters are specified by sliders and radio buttons (see Fig. 6 top left). Interaction with the sliders and buttons provides immediate visual feedback: the background of each selected gene box is filled with a color corresponding to its expression value; the background of a gene box that is not selected is set to a standard color that is not in the color map. In this way, a user can quickly spot particular behavior at specific time points, and answer questions such as "which genes are up-regulated during early growth?".

To study the interaction between a specific gene and its direct targets, it is possible to select the corresponding subnetwork by right-clicking a node. This action opens a new window that contains the subnetwork, plots of the expression values, and a table containing the raw gene expression data. The table lists expression values and corresponding statistical data. Both plot and table show only expression profiles than contain at least one time point for which the significance falls within the significance range limits. This range is controlled by the significance slider that is also used for single time point analysis, as explained above. This view allows a biologist to consider the expression data qualitatively in the network visualization, but also quantitatively by inspecting the raw data or the corresponding plots. To assist the user in maintaining a mental map of the complete network, the layout of the subnetwork is not changed. An example is shown in Fig. 5, which contains the gene *ccpA* and its targets. The expression plot shows the ratio measures at all time points of only significant data ($p < 0.00001$). The gene box backgrounds are colored according to time point 1.

For a large regulator, the plot can become cluttered, even when only significant data are shown. Therefore, the user can also add or remove the expression profile of a gene by left-clicking its gene box in the network visualization. The profile and corresponding raw data will then be added to (or removed from) the plot and the table below the plot, respectively.

3.5 Implementation

Our application was built with use of the Prefuse library [HCL05], which is an open source Java toolkit for interactive information visualization. The toolkit provides basic data structures for storing graphs and node and edge attributes, supports many layout algorithms, and has a flexible rendering mechanism. Expression profile plotting was implemented with JFreeChart[Gil06], a free Java chart library, distributed under the LGPL.

Prefuse provides basic functionality to calculate a layout, and to color data based on some attribute. It uses Java2D to draw the graphs in a display that supports panning and zooming. The display also provides a handle to tooltips. We have implemented extensions of the standard edge and node renderers (those perform the actual drawing on the display). The edge renderer was modified such that it decorates the edge with an arrow or a bar depending on the interaction type. The node renderer was extended such that in addition to a text label containing the gene name, it also draws the gene expression boxes. Other components, such as the overview display, were created by combining modules of basic functionality already present in Prefuse.

4 Case Study

GENeVis has been used to further explore a short time series DNA microarray dataset described by Lulko and coworkers [LBKK07]. In this study, the global mRNA levels (thus gene transcription or gene expression) at four distinct stages of growth of the bacterium *B. subtilis* strain 168 and the same strain containing a gene deletion are compared. These four growth stages were sampled to obtain a view of the changes in gene expression during growth of this bacterium. The four time points sampled ranged from (i) the early exponential phase (the onset of fast cell growth), (ii) mid-exponential phase (fast cell growth), (iii) end-exponential phase (nutrients start slightly limiting the growth), and (iv) the stationary phase of growth (no growth of cells and start of cell death). The *B. subtilis* strain with a gene deletion has its *ccpA* gene disabled, and it is therefore called a *ccpA* deletion mutant. This comparison was performed by DNA microarrays.

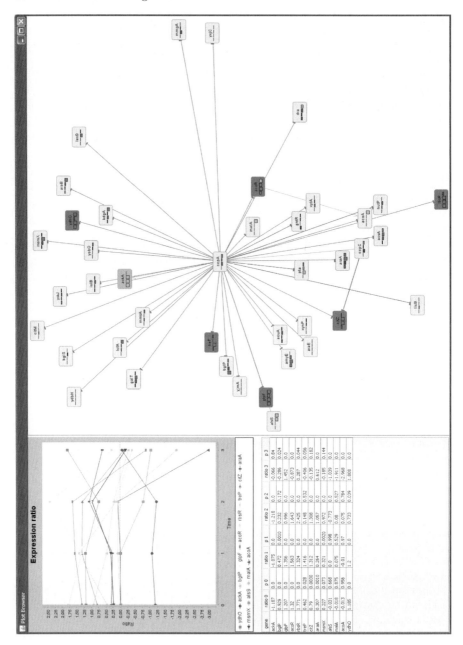

Fig. 5. Visualization of the subnetwork consisting of *ccpA* and its direct targets. The plot and table show the expression ratios for all time points of only the significant data ($p < 0.00001$). Data from time point 1 were used to color the gene box backgrounds.

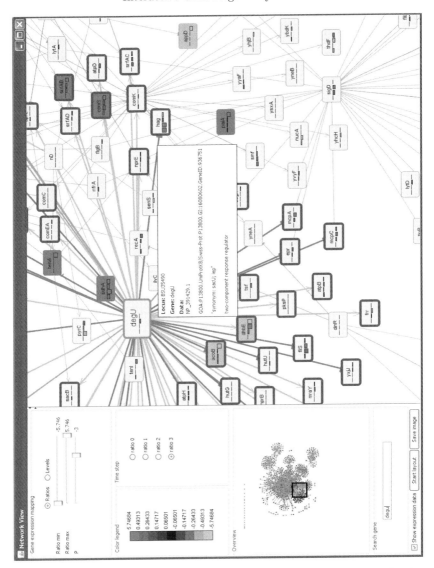

Fig. 6. Screen shot of GENeVis. The user can switch between expression level and expression ratio visualization with the radio buttons in the top left corner. The sliders together with the time point radio buttons can be used to select a subset of genes that have expression data within the ranges set by the sliders and buttons. The background of the selected genes is then colored according to their expression value. The color legend is shown to the left of the time point selection buttons. An overview display is shown to help the user in maintaining a view of the network context. The square indicates the area of the network visible in the main display, and it can be dragged to pan the view in the main display. The search box can be used to find a specific gene, and when found, the display automatically pans such that the gene found is centered in the view.

With DNA microarrays, global mRNA levels (which indicate the "activity" of genes) are determined by comparing a reference (e.g. the *B. subtilis* wild-type) to a test (e.g. the deletion mutant) condition. All genes of *B. subtilis* are present on the DNA microarray which allows monitoring the expression of these genes during the four growth-stages sampled. After quantification and normalization of the signals of the DNA microarray, the researcher is left with signals for each gene and for each condition (4 time points and 2 samples). These signals indicate the relative gene expressions. A large problem for an experimentalist is to identify relevant biological phenomena from all these measurements (in this case over 100,000 signals of over 4000 genes). Often, a ratio is used to relate the changes in expression between two conditions. This ratio is calculated by dividing the signal of the test condition (in this study the mutant) by the signal of the reference condition (in this study wild-type) for each gene and for each time point.

The CcpA protein (for which the *ccpA* gene codes)[1] is a master transcriptional regulator (a protein that drives the expression of target genes) involved in governing carbon catabolite repression in many so-called Gram-positive bacteria [SH00, WL03]. As is shown in the study of Lulko and coworkers and other studies, the inactivation of the *ccpA* gene has broad implications for a large number of key cellular processes [LBKK07, BHL+03, LCB+05, MSM+01]. From a biologists' point of view, it is crucial to have a possibility to oversee the global changes caused by any kind of interference or indirect effects of, in this case, the deletion of a regulator gene. Furthermore, after global changes have been identified, a biologist needs to delve into the behavior of specific genes as we will demonstrate below. Current research is switching from single time point analysis to monitoring the changes of gene expressions over time in time series DNA microarray experiments. As we will show, following gene expression in time allows a richer description of the direct and indirect effects of, e.g., a gene deletion. Analysis of time series data was originally performed in a gene-by-gene approach of the most differentially expressed genes (genes whose expression was most notably changed) involving literature search and mining of information available at public repositories such as PubMed (www.pubmed.org). Furthermore, the global effects of the *ccpA* deletion on the known regulators and metabolic pathways were studied by Lulko and coworkers by using FIVA [BBvH+07]. This tool presents an overview of the key cellular processes affected, but it does not allow visual or manual identification of groups of genes exhibiting correlated behavior within these processes. For instance, the software will indicate a regulon (a collection of genes under regulation by the same protein) affected, but not which members of the regulon. Therefore, after identification of affected key cellular processes, the experimentalist has to mine the data manually. This makes the

---

[1] The biological convention is that gene names are written in italics with the first letter in lower case; the corresponding protein for which the gene codes is written in roman with the first letter in upper case.

investigation of the (indirect) effects of the *ccpA* gene deletion on parts of regulons in these four time points very difficult and time consuming.

The application presented in this paper allows the projection of time series expression data derived from DNA microarrays on a gene interaction network (the network of regulator proteins which drive the expression of target genes, which in turn can also encode regulator proteins). This not only facilitates the biological interpretation of the overall direct effects of the *ccpA* gene deletion but also offers an opportunity to focus on some indirect responses caused by the disruption of this transcriptional regulator. Mining the time-series data starts with a visual exploration of the whole interaction network. By using GENeVis with an overview of the network and cycling through the four time points it immediately becomes apparent that the impact of the *ccpA* mutation dynamically develops and intensifies during growth of *B. subtilis* cells. These overviews are shown in Fig. 7 for time points 0 and 1 and in Fig. 8 for time points 3 and 4. The significant genes ($p < 0.001$) are colored according to expression ratio (calculated by the gene expression levels of the *ccpA* mutant over those of the wild-type). A red color indicates a higher expression level in the mutant strain compared to the wild-type strain, whereas a green color indicates a lower expression level. These images clearly show that (i) the number of colored genes increases strongly from time point 2 to 3, and (ii) about half of the genes in the dashed box are differentially expressed (green or red).

The regulon in the dashed box (SigB) is shown enlarged in Fig. 9. The expression of the SigB regulon in time point 3 is a prominent indication of the dynamics just described. This regulon is involved in the response to harmful environmental conditions, such as heat, osmotic, acid, or alkaline shock. In previous studies, only minor effects during the exponential phases of growth have been reported for SigB. However, the visualization of ratio-based data of the SigB regulon genes as a function of time allows a spectacular view on the reprogramming of the SigB-dependent gene expression at later growth stages (the transition from the late exponential to stationary phase of growth). From Fig. 9, three distinctive gene clusters can be identified; (i) a few genes (*csbA*, *csbX*, *yfhK*) whose expression levels in the wild-type strain, compared to the mutant strain, were lower in one of the time points during exponential growth (red-colored expression boxes inside the gene boxes for time points 0 to 2); (ii) a few genes (*dps*, *spoVG*, *yqgZ*, *yvyD*) for which there is a clear switch in the expression profile between the late exponential and the stationary phase time points (the expression box color changes from red to green for time point 3); and (iii) a larger number of genes (colored by green backgrounds) whose expression is strongly increased in the wild-type strain in the stationary phase compared to the three exponential growth phase time points during which transcripts levels remained essentially unaffected. The latter cluster is particularly interesting since it explicitly reveals that the SigB regulon is recruited stronger during the late growth stages of the wild-type strain than the *ccpA* deletion strain. Measurements of glucose concentration (a major

Fig. 7. A gene regulatory network of *B. subtilis* superimposed with DNA microarray ratio data of *B. subtilis* wild-type over its *ccpA* deletion mutant. The time series consists of four time points corresponding to different phases of growth. Top image: the early exponential growth phase (the onset of fast cell growth). Bottom image: mid-exponential growth phase (fast cell growth). Significantly expressed genes ($p < 0.001$) are colored to their expression ratio; others have a neutral background color. Continued in Fig. 8.

Fig. 8. Continued from Fig. 7. Top image: end-exponential growth phase (nutrients start slightly limiting the growth). Bottom image: stationary growth phase (no growth of cells and start of cell death). The *ccpA* gene is indicated by a solid black rectangle. The dashed rectangle indicates the SigB regulon, which is shown enlarged in Fig. 9.

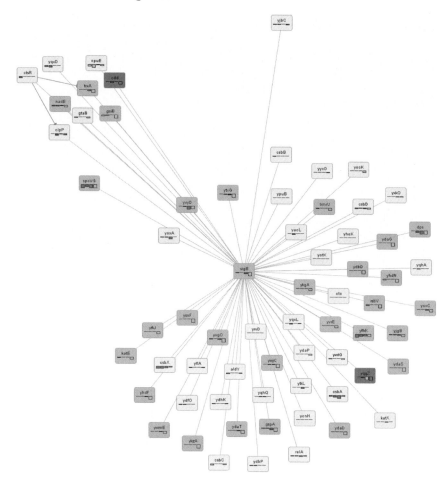

Fig. 9. The SigB regulon with gene coloring based on expression ratios ($p < 0.001$) in the stationary growth phase (time point 3). A red color indicates a higher expression level in the mutant strain compared to the wild-type strain, whereas a green color indicates a lower expression level.

source of energy for growth of B. subtilis; results not shown) demonstrated that glucose is completely consumed by the wild-type strain in the stationary phase, while it is still present in the culture of the mutant. This could explain the apparent induction of several members of this general stress/starvation regulon. This energy depletion due to lack of glucose apparently is the signal for the wild-type strain to adapt to the upcoming stress conditions (nutrient limitation). The induction of genes whose products counteract the oxidative stress (stress to bacterial cells caused by oxygen radicals) conditions (*dps*, *katE*, *nadE*, *trxA*) is just one of the examples indicative of bacterial adaptation to survive deteriorating environmental circumstances.

The analysis we have performed in this section could not have been done with the tools described in Section 2. Some of these tools, Cytoscape, for example, would be able to produce overviews of the entire network for the individual time points. The main problem is to visualize only significant expression ratios, which currently can only be done by preprocessing the data and removing the insignificant entries. This is very inconvenient, since the user may wish to modify the p-value range during this exploration phase. In the second part of the analysis, we have studied the behavior of the SigB regulon over time. This was only possible by looking at all time points simultaneously, which is only supported by GENeVis and not by any of the tools discussed in Section 2.

5 Conclusions

We have presented GENeVis, an application to visualize gene expression time series data in a gene regulatory network context. GENeVis adds features that are currently lacking in existing tools, such as mapping of expression value and corresponding p-value (or other statistic measures) to a single visual attribute, multiple time point visualization, and visual comparison of multiple time series. Various interaction mechanisms, such as panning, zooming, regulator and target highlighting, data selection, tooltips, and support for subnetwork analysis facilitate data analysis and exploration.

We have presented a case study, in which gene expression time series data acquired in-house have been analyzed by an end-user. Our case study has revealed that GENeVis clearly fills the gap in the present gene-by-gene biological interpretation of time series experiments and global regulon analysis. This goal is achieved by (i) allowing the biologist an overview of the gene regulatory network with mapped gene expressions as a function of time to quickly identify biologically relevant changes in parts of the network, and (ii) delve into detail by visual identification of the partitioning of members of regulons as a function of time. We have shown that the combination of single and multiple time point visualization and filtering based on statistical significance is very powerful and helpful for a biologist. The analysis presented in the case study could not have been performed otherwise, i.e., by existing visualization tools.

In the case study, we have used GENeVis to visualize the relatively small network of *B. subtilis* (772 nodes and 1179 edges). For networks of that size, a force-directed layout algorithm produces satisfactory layouts. However, this type of layout algorithm does not scale very well for larger networks. In future work, therefore, we plan to investigate other layout algorithms, such as the multi-level algorithm based on topological features [AMA07]. Another promising algorithm is the grid layout algorithm that takes both connection structure and biological function associated to the genes into account [LK05].

References

[AMA07] D. Archambault, T. Munzner, and D. Auber. TopoLayout: Multi-level graph layout by topological features. *IEEE Trans. Visualization and Computer Graphics*, 13(2):305–317, 2007.

[BBO+06] D. W. J. Bosman, E.-J. Blom, P. J. Ogao, O. P. Kuipers, and J. B. T. M. Roerdink. MOVE: A multi-level ontology-based visualization and exploration framework for genomic networks. *In Silico Biology*, 7:0004, 2006.

[BBvH+07] E. J. Blom, D. W. J. Bosman, S. A. F. T. van Hijum, R. Breitling, L. Tijsma, R. Silvis, J. B. T. M. Roerdink, and O. P. Kuipers. FIVA: Functional information viewer and analyzer extracting biological knowledge from transcriptome data of prokaryotes. *Bioinformatics*, page btl658, 2007.

[BETT99] G. Di Battista, P. Eades, R. Tamassia, and I. G. Tollis. *Graph Drawing: Algorithms for the Visualization of Graphs*. Prentice Hall, New Jersey, 1999.

[BHL+03] H. M. Blencke, G. Homuth, H. Ludwig, U. Mader, M. Hecker, and J. Stulke. Transcriptional profiling of gene expression in response to glucose in *Bacillus subtilis*: Regulation of the central metabolic pathways. *Metabolic Engineering*, 5(2):133–149, 2003.

[BSRG06] M. Baitaluk, M. Sedova, A. Ray, and A. Gupta. BiologicalNetworks: Visualization and analysis tool for systems biology. *Nucleic Acids Research*, 34:W466–W471, 2006. Web Server Issue.

[BST03] B.-J. Breitkreutz, C. Stark, and M. Tyers. Osprey: A network visualization system. *Genome Biology*, 4(3):R22, 2003.

[Gil06] D. Gilbert. JFreeChart. http://www.jfree.org/jfreechart, 2006.

[HCL05] J. Heer, S. K. Card, and J. A. Landay. Prefuse: a toolkit for interactive information visualization. In *CHI '05: Proc. SIGCHI conf. Human factors in computing systems*, pages 421–430, 2005.

[HMM00] I. Herman, G. Melançon, and M. S. Marshall. Graph visualization and navigation in information visualization: a survey. *IEEE Trans. Visualization and Computer Graphics*, 6(1):24–43, 2000.

[HMW+05] Z. Hu, J. Mellor, J. Wu, T. Yamada, D. Holloway, and C. DeLisi. VisANT: Data-integrating visual framework for biological networks and modules. *Nucleic Acids Research*, 33:W352–W357, 2005. Web Server Issue.

[HMWD04] Z. Hu, J. Mellor, J. Wu, and C. DeLisi. VisANT: An online visualization and analysis tool for biological interaction data. *BMC Bioinformatics*, 5:17, 2004.

[LBKK07] A. T. Lulko, G. Buist, J. Kok, and O. P. Kuipers. Transcriptome analysis of temporal regulation of carbon metabolism by CcpA in *Bacillus subtilis* reveals additional target genes. *Journal of Molecular Microbiology and Biotechnology*, 12(1–2):82–95, 2007.

[LCB+05] G. L. Lorca, Y. J. Chung, R. D. Barabote, W. Weyler, C. H. Schilling, and M. H. Saier Jr. Catabolite repression and activation in *Bacillus subtilis*: Dependency on CcpA, HPr, and HprK. *Journal of Bacteriology*, 187(22):7826–7839, 2005.

[LK05] W. Li and H. Kurata. A grid layout algorithm for automatic drawing of biochemical networks. *Bioinformatics*, 21(9):2036–2042, 2005.

[MNON04] Y. Makita, M. Nakao, N. Ogasawara, and K. Nakai. DBTBS: database of transcriptional regulation in *Bacillus subtilis* and its contribution to comparative genomics. *Nucleic Acids Research*, 32:D75–77, 2004.

[MSM+01] M. S. Moreno, B. L. Schneider, R. R. Maile, W. Weyler, and M. H. Saier Jr. Catabolite repression mediated by the CcpA protein in *Bacillus subtilis*: Novel modes of regulation by whole-genome analysis. *Molecular Biology*, 39(5):1366–1381, 2001.

[SH00] J. Stulke and W. Hillen. Regulation of carbon catabolism in bacillus species. *Annual Review of Microbiology*, 54:849–880, 2000.

[SLN05] P. Saraiya, P. Lee, and C. North. Visualization of graphs with associated timeseries data. In *Proc. IEEE Symp. Information Visualization (InfoVis'05)*, pages 225–232, 2005.

[SMO+03] P. Shannon, A. Markiel, O. Ozier, N. S. Baliga, J. T. Wang, D. Ramage, N. Amin, B. Schwikowski, and T. Ideker. Cytoscape: A software environment for integrated models of biomolecular interaction networks. *Genome Research*, 13(11):2498–2504, 2003.

[The00] The Gene Ontology Consortium. Gene ontology: Tool for the unification of biology. *Nature Genetics*, 25:25–29, 2000.

[War04] C. Ware. *Information Visualization: Perception for Design*. Morgan Kaufmann Publishers, 2nd edition, 2004.

[WL03] J. B. Warner and J. S. Lolkema. CcpA-dependent carbon catabolite repression in bacteria. *Microbiology and Molecular Biology Reviews*, 67(4):475–490, 2003.

Segmenting Gene Expression Patterns of Early-stage Drosophila Embryos

Min-Yu Huang[1,6], Oliver Rübel[1,2,3,6], Gunther H. Weber[3], Cris L. Luengo Hendriks[4,6], Mark D. Biggin[5,6], Hans Hagen[2], and Bernd Hamann[1,3,6]

[1] Institute for Data Analysis and Visualization, University of California, Davis, 1 Shields Avenue, Davis CA 95616, USA
{myhuang,oruebel,bhamann}@ucdavis.edu
[2] International Research Training Group "Visualization of Large and Unstructured Data Sets," University of Kaiserslautern, Germany
hagen@informatik.uni-kl.de
[3] Visualiztion Group, Lawrence Berkeley National Laboratory, 1 Cyclotron Road, Berkeley CA 94720, USA ghweber@lbl.gov
[4] Life Sciences Division, Lawrence Berkeley National Laboratory, 1 Cyclotron Road, Berkeley CA 94720, USA clluengo@lbl.gov
[5] Genomics Divisions, Lawrence Berkeley National Laboratory, 1 Cyclotron Road, Berkeley CA 94720, USA MDBiggin@lbl.gov
[6] Berkeley Drosophila Transcription Network Project, Lawrence Berkeley National Laboratory, 1 Cyclotron Road, Berkeley CA 94720, USA
http://bdtnp.lbl.gov/Fly-Net/

Summary. To make possible a more rigorous understanding of animal gene regulatory networks, the *Berkeley Drosophila Transcription Network Project* (BDTNP) has developed a suite of methods that support quantitative, computational analysis of three-dimensional (3D) gene expression patterns with cellular resolution in early *Drosophila* embryos.

Defining the pattern of gene expression is an essential step toward further analysis in order to derive knowledge about the characteristics of gene expression patterns and to identify and model gene inter-relationships. To address this challenging task we have developed an integrated, interactive approach toward pattern segmentation. Here, we introduce a ridge-detection-based 3D gene expression pattern segmentation algorithm. We compare this algorithm to common 2D pattern segmentation methods, such as thresholding and edged-detection-based methods, which we have adapted to 3D pattern segmentation. We show that such automatic strategies can be improved to obtain better segmentation results by user interaction and additional post-processing steps.

Key words: three-dimensional gene expression, pattern segmentation, gene expression pattern, ridge detection, edge detection, thresholding

1 Introduction

Intricate spatial and temporal patterns of gene expression are responsible for determining the shape of a developing animal embryo. Research of these patterns is typically based on visual inspection or computer-assisted analysis of two-dimensional (2D) photomicrographic images. Animal embryos comprise dynamic 3D arrays of cells, however, and thus analysis of 2D images cannot capture the full complexity of a developing embryo. To overcome this challenge, the BDTNP has developed image processing methods from 3D image data [LKF+06] [KFL+06] to extract information about gene expression from imaging data using early *Drosophila melanogaster* embryos as model organisms. Stacks of confocal images of blastoderm stage *Drosophila* embryos are converted into matrices specifying the position of individual nuclei and the amount of mRNA or protein (expression levels) of genes around each nucleus. The resulting novel datasets, termed *PointClouds*, promise to be an invaluable resource for studying animal development.

The purpose of the data generated by the BDTNP is to understand the gene regulatory network. A subset of genes, the so-called transcription factors, regulate the expression of all genes. That is, they cause a gene to be expressed or not in a particular cell. Each of these transcription factors regulates, and is regulated by, other transcription factors. These interactions form the transcription network. This network in combination with the initial conditions in the egg given by maternally transcribed factors yields the expression patterns we analyze here. Modeling the regulatory network is a common approach to gain understanding of the integrated behavior of genes and their regulatory interactions. Despite its limitations, the binary gene regulatory model, which takes the gene to be either on or off in each cell, allows examination and creation of very large systems (several thousand genes) and is mathematically the most tractable network modeling approach. However, the algorithm chosen to define the binary pattern of the genes' expression will influence the results obtained from this network model. Therefore we propose a segmentation algorithm that accurately follows pattern edges, as given by locations of rapid change in expression level, rather than a globally chosen threshold value. Additionally, binarization yields a high level of abstraction for gene expression patterns. Based on such binary gene expression patterns, dedicated user interactions for similarity-based pattern queries can be defined to allow searches for patterns that have, for example, seven stripes or for genes that are expressed in specific regions. And, concentrating only on those cells that show expression of a specific gene or gene combination, more complex analysis becomes possible.

In the field of image processing (IP), there are several approaches which can segment grey-value images into binary patterns. Many of these segmentation algorithms require data to be on a rectangular grid or on a 2D manifold. However, even in the very simple blastoderm *Drosophila* embryos, which are mostly a single layer of cells, our *PointCloud* expression data cannot be segmented by these approaches because the locations of the nuclei form irregular grids and do not necessarily form a 2D manifold, especially near the posterior pole. Furthermore, we require that our segmentation algorithms be applicable to later-stage embryos, where the cells are packed in 3D volumes. Since we cannot record expression values with an absolute metric, an algorithm which will not be affected by scaling is desired here. Intrinsic properties such as local maxima, local minima, and inflection points are scale-invariant and thus particularily suitable to be used to define the gene expression pattern.

A gene expression pattern can be defined as regions with high expression values enclosed by loci of inflection points. To define the pattern of a gene in 3D space, we have developed a segmentation method based on ridge region detection. Using a fully automatic approach does not always produce results of sufficient quality. In order to make use of biologists' knowledge, we have developed an integrated interactive approach toward pattern segmentation that significantly improves the accuracy of the final results.

Thresholding and edge detection are two common techniques used to specify gene expression patterns. We compare our method to thresholding and edge-detection-based segmentation methods, which we have extended and applied to define 3D gene expression patterns.

2 Related Work

Until last year, scientists did not have access to 3D gene expression data at cellular resolution for a whole multi-cellular organism. Kumar et al. [KJP+02] segmented gene expression patterns from 2D image data by using a simple thresholding approach. For each gene stained in the image, a specific threshold is suggested by a histogram-based algorithm. Nuclei with expression values larger than the threshold are considered to be in the pattern. This binarizes the image by defining the *pattern* and the *background*. The choice of a good or "correct" threshold has always been an open question since there are several algorithms designed for different purposes.

While many algorithms choose the threshold based on the histogram of the data, some algorithms segment data using spatial information and do not depend on histogram information. In the field of 2D image processing and computer vision, edge detection and ridge detection are two commonly used techniques that use spatial information to extract patterns from images. An edge detection algorithm, such as Canny edge detection [Can86], can locate pixels where the luminous intensity changes sharply. When such an algorithm is applied to gene expression data, it can tell us at which nuclei expression values change the most in a local neighborhood. If the expression value changes sharply, it is likely that this nucleus is at the edge of an expression domain. On the other hand, a ridge/valley detection algorithm can extract skeletons of watershed/watercourse patterns in an image [FSC04]. Because ridges usually occur along the center of elongated patterns, they can provide compact representations for these patterns, especially when the patterns in the data have soft edges.

Several ways exist to define a ridge mathematically and different algorithms are designed accordingly. Two major ridge detection categories evaluate ridges by computing principal curvatures [MvdEV96] and by constructing separatrices [Nac84]. Besides approaches mentioned above, Peng et al. [PM04] assumed that gene expression patterns in a 2D image can be described by a Gaussian mixture model (GMM), and that gene expression patterns can be segmented and extracted by decomposing Gaussian elements.

Although 2D histogram-based algorithms can be easily applied directly on 3D gene expression data with little, or no modification, algorithms which depend on spatial information, such as GMM, usually cannot be easily modified because 3D expression data are stored on irregular grids. We modified the algorithm proposed in [LLS97] to extract ridge patterns in 3D expression data.

3 Segmenting Gene Expression Patterns

Segmentation techniques, such as Canny edge detection and ridge detection, require gradients of expression values for segmentation. Hence, we have to estimate gene expression gradients before we can apply those techniques.

3.1 Expression Gradient Estimation and Canny Edge Detection

If we treat gene expression values as a function $\mathcal{E} : \mathbb{R}^3 \to \mathbb{R}$, the gradient vector for nucleus ρ, which is located at position $\boldsymbol{\rho} = (x, y, z)$, is defined as $\nabla \mathcal{E}(x, y, z) \equiv \frac{\partial \mathcal{E}}{\partial X}\hat{X} + \frac{\partial \mathcal{E}}{\partial Y}\hat{Y} + \frac{\partial \mathcal{E}}{\partial Z}\hat{Z}$ which defines the direction in which the expression value increases maximally. (Here $\hat{X}$, $\hat{Y}$, and $\hat{Z}$ represent unit vectors in X, Y, and Z direction respectively.) In a *PointCloud*, we only have spatially discrete gene expression values stored at the locations of nuclei. Assume the nucleus ρ_0 has n neighboring nuclei $\rho_i, i = 1 \ldots n$. We can choose nearst n neighbors or natural neighbors of ρ_0 (i.e. neighbors in the Voronoi diagram) to be ρ_i. Because the expression function can be linearly approximated by the first-order terms of its Taylor expansion $\mathcal{E}(\boldsymbol{\rho}) \approx \mathcal{E}(\boldsymbol{\rho_0}) + \nabla \mathcal{E}(\boldsymbol{\rho_0}) \cdot (\boldsymbol{\rho} - \boldsymbol{\rho_0}) = e + a(x - x_0) + b(y - y_0) + c(z - z_0)$, where the gradient vector is $\nabla \mathcal{E}(\boldsymbol{\rho_0}) = (a, b, c)$, after considering known values $(x_i, y_i, z_i, i = 0 \ldots n)$, we have the following $n + 1$ equations:

$$\begin{cases} \mathcal{E}(\boldsymbol{\rho_0}) = [e\ a\ b\ c]\ [1 \quad 0 \quad 0 \quad 0 \quad]^T \\ \mathcal{E}(\boldsymbol{\rho_1}) = [e\ a\ b\ c]\ [1 \quad (x_1-x_0) \quad (y_1-y_0) \quad (z_1-z_0) \ ]^T \\ \quad \vdots \\ \mathcal{E}(\boldsymbol{\rho_n}) = [e\ a\ b\ c]\ [1 \quad (x_n-x_0) \quad (y_n-y_0) \quad (z_n-z_0) \ ]^T \end{cases}, \quad (\text{-1})$$

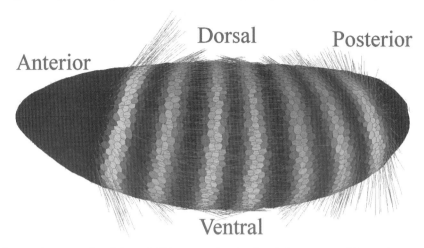

Fig. 1. An example of the mRNA expression levels of the gene *eve*(shown in green) and corresponding gradient vectors. The embryo is shown in a 3D view described in [RWK+06]. The center of mass of nuclei define a surface that is partitioned using Voronoi tessellation. A cell is represented by a polygon-like structure and it shape does not reflect the real cell size. Gradient vectors start from red ends and terminate at gray ends. Note that some vectors point into the inside of the embryo.

where unknown variables in the matrix $[e\ a\ b\ c]$ can be determined by solving a least-square fitting problem. Please note that there also exist other approaches to estimates these gradient vectors. The one we present here can be replaced by another method without influencing the segmentation algorithm too much.

Figure 1 shows the mRNA expression level of a gene called *even-skipped* (*eve*), and the corresponding gradient vectors. Once we have gradient information, the Canny edge detection algorithm can be applied to the gene expression data with little modification. Figure 2(b) shows the Canny edge detection result for *eve* mRNA expression levels. In this example, edges of the seven expression stripes are detected. However, this approach cannot directly provide the full regions of expression, and as we show later, does not always usefully segment gene expression patterns. Thus, we also adapted another segmentation method, ridge region detection (see Section 3.2), to obtain region information for gene expression patterns and to provide an alternative approach that may be more successful in segmenting gene expression patterns when edge detection fails.

3.2 Segmenting by Ridge Region Detection

Ridge detection can be done by building separatrices or by evaluating curvatures. Separatrix-based algorithms are easier to implement. However, the results are usually found not to be satisfying. An edge in the result could be sometimes both a ridge line and a valley line [RCH92]. Another problem of

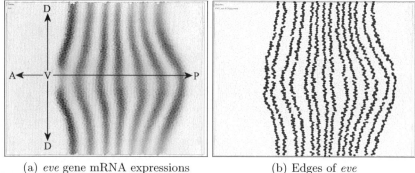

(a) *eve* gene mRNA expressions (b) Edges of *eve*

Fig. 2. *Eve* edges detected by Canny's algorithm. The embryo is shown in a cylindrical projection as described by Luengo Hendriks et al. [LKF+06] with cell-like structures to provides a better overview of the blastoderm surface. The cylindrical projection is only used for displaying the data, the actual algorithms we used in this paper are performed in 3D. The top and bottom of each panel corresponds to the dorsal side (D) of the embryo; the middle corresponds to the ventral side (V); the left side corresponds to the anterior (A); and the right side corresponds to the posterior (P). Higher expression values are shown in darker gray levels.

separatrix-based algorithms is that they are global algorithms, *i.e.* the result in a local area can be influenced by changes far away. This is not the case for curvature-based algorithms, which are local.

Traditionally, curvature-based algorithms detect ridges by *level set extrinsic curvatures* (LSEC) [MvdEV96]. Higher-order derivatives have to be evaluated to compute curvatures. López et al. [LLS97] proposed their *multi-local level set extrinsic curvature* (MLSEC) algorithm which only needs gradient vectors and is more accurate when only discrete data are available. Their basic idea is that the most obvious characteristic of a ridge pattern is that the gradient vectors near the ridge are all pointing toward it. They proved that the divergence of normalized gradient vectors is mathematically equivalent to LSEC. The first reason why we choose their algorithm is that it also works for irregular grids and thus would require less modification. The second reason is that López' algorithm can produce not only ridge lines, which are less useful in our application, but also ridge regions(regions with high values enclosed by loci of inflection points), while Separatrix-based algorithms cannot produce ridge regions.

In López' algorithm, the ridge evaluator $\tilde{\kappa}_d$ is defined as

$$\tilde{\kappa}_d = -\mathrm{div}(\hat{\omega})$$
$$= -\lim_{V \to 0} \frac{1}{V} \oint_{\partial V} \hat{\omega} \cdot d\mathbf{A}$$
$$\approx -\frac{d}{r} \sum_{k=1}^{r} \hat{\omega}_k \cdot \hat{n}_k ,$$

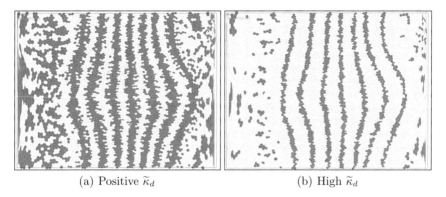

(a) Positive $\widetilde{\kappa}_d$ (b) High $\widetilde{\kappa}_d$

Fig. 3. MLSEC ridge region detection results of *eve* mRNA expression shown in cylindrical projection. False ridge regions still remain in the anterior area and the posterior area even when we use a high threshold for $\widetilde{\kappa}_d$.

where $\hat{\omega}$ is the normalized gene expression gradient vector, d is the number of dimensions, r is the number of neighboring nuclei, $\hat{\omega}_k$ is the normalized gene expression gradient vector at the k-th neighboring nucleus, and $\hat{n}_k$ is the unit normal vector on ∂V for the k-th neighboring nucleus, which can be estimated as $\boldsymbol{\rho}_k - \boldsymbol{\rho}_0$ in our application. The MLSEC ridge evaluator $\widetilde{\kappa}_d$ is bounded by $[-d, +d]$. Higher positive values of $\widetilde{\kappa}_d$ imply more ridge-like behavior and lower negative values of $\widetilde{\kappa}_d$ indicate more valley-like behavior. Figure 3 shows the MLSEC ridge detection result of *eve* mRNA gene expression.

Mathematically, edges of patterns should be composed of nuclei whose $\widetilde{\kappa}_d$ is zero, or very close to zero, since edges also represent loci of inflection points in gene expression. In other words, a ridge region is an area enclosed by edges and has higher expression values than its neighboring regions. To binarize gene expression, the ridge regions are marked "pattern", the rest is "background".

Any measured gene expression data generally contain noise which will introduce small artefactual ridges and valleys that interfere ridge region detection. As Figure 3 shows, undesirable ridge regions still remain in the anterior area and the posterior area, in which *eve* expression values are very low, even when we choose a large threshold for $\widetilde{\kappa}_d$ 3(b). Since in MLSEC $\widetilde{\kappa}_d$ is evaluated by divergence of *normalized* gradient vectors, only directional changes of gradient vectors contribute to $\widetilde{\kappa}_d$. To get rid of ridges caused by fluctuation of noise, one possibility is to also consider magnitudes of gradient vectors along with their directions, since expression fluctuation caused by noise usually only has a small divergence. Thus, we can define a new ridge evaluator κ_d as

$$\kappa_d = -\mathrm{div}(\boldsymbol{\omega})$$
$$\approx -\frac{d}{r}\sum_{k=1}^{r} \boldsymbol{\omega}_k \cdot \hat{n}_k \; ,$$

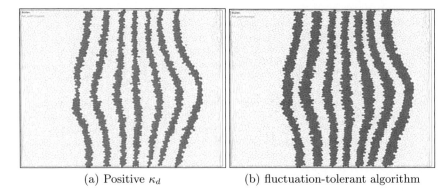

(a) Positive κ_d (b) fluctuation-tolerant algorithm

Fig. 4. Ridge regions detected in *eve* mRNA expression with our new ridge evaluator κ_d. A low positive threshold is used to remove false ridge regions caused by noise. With our fluctuation-tolerant strategy, this new ridge region evaluator generates better and more accurate results.

where, ω and ω_k are the unnormalized versions of gradient vectors. Again, higher positive κ_d values still indicate more ridge-like behavior, and lower negative κ_d values indicate more valley-like behavior. However, we note that κ_d is not bounded while $\tilde{\kappa}_d$ is. Figure 4(a) shows the resulting image of *eve* mRNA expression levels by using our new ridge evaluator κ_d and a low positive threshold obtained by Rosin's unimodal [Ros01] thresholding method, as described in Section 4.1.

The new ridge evaluator κ_d gets rid of the problem of small fluctuations in expression generated by noise; However, it also generates a new problem. Compared to Figure 2(a), ridge regions in Figure 4(a) are in general thinner than expected. This is due to the fact that the use of a threshold to remove nuclei with small κ_d removes not only false ridge regions caused by small fluctuation (noise) but also nuclei that are near, or on edges and hence have smaller κ_d. To address this problem, we propose a fluctuation-tolerant strategy to recover nuclei in real ridge regions. This strategy uses the following three rules:

(1) When a nucleus' κ_d value is larger than the positive threshold T, the nucleus is part of the ridge region.
(2) When a nucleus' κ_d value is smaller than $-T$, the nucleus is not part of the ridge region.
(3) If a nucleus' κ_d value is within the range $[-T, +T]$, the nucleus is part of the ridge region only when at least one of its neighboring nuclei fulfills rule (1); otherwise, it is not in the ridge region.

Here the positive threshold T needs to be defined by the user. In our tool, *PointCloudXplore* [RWK+06], we offer several thresholding options which are described in more detail in the next section. An example result of the fluctuation-tolerant strategy is shown in Figure 4(b).

4 Interactive Segmentation

Since no fully automatic segmentation algorithm is applicable in every situation, a good segmentation tool should provide a user interface to allow users to customize parameters used in the algorithms and to perform post-processing to improve the results.

4.1 Thresholding

Thresholding is required in all three approaches discussed above: simple thresholding, edge detection, and ridge detection. It is difficult to design a universal algorithm to pick a good threshold automatically for every case. By considering a data histogram, one can either chose a threshold manually, or pick an appropriate automatic thresholding method, see Figure 5. Depending on the structure of the histogram, one can choose from several automatic thresholding options: *Rosin's unimodal* [Ros01], *2-Gaussian mixture* [MP00], *2-Mean clustering* [Mac67], and *above one standard deviation*. Other options, such as RATS (robust automatic threshold selection) techniques [KIF85] [Wil98], will be added here in the future. What histogram information is shown in the UI depends on the segmentation algorithm used:

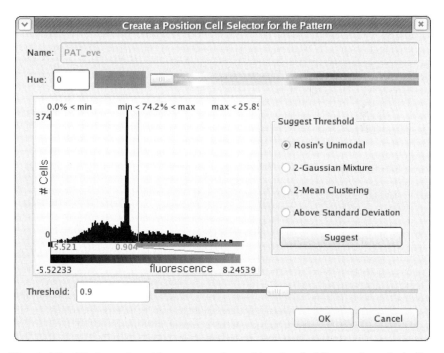

Fig. 5. The UI shows data histogram and provides thresholding options, including algorithms which can suggest a threshold.

When simple thresholding is used, the histogram of the original expression values is shown. For edge detection, the histogram of the gradient magnitude is used. For ridge detection, the divergence of the gradient is used.

4.2 Post-processing

Noise and outliers often exist in the data and can generate small false ridge regions. Since an objective quality measure is not easy to define in this application, we just provide two basic post-processing methods here to aid biologists more easily to edit the results to generate the final pattern. Pattern filtering and splitting are the two main pattern post-processing features supported to improve the quality of segmentation results. False pattern regions could be eliminated by keeping only the first few largest regions, or by filtering out regions smaller than a given number of nuclei. A pattern can also be split into its spatially independent components to allow detailed analysis of independent expression domains. An example of filtering and splitting is shown in Figure 6. In *PointCloudXplore*, segmentation results are in general stored in so-called *Cell-Selectors* (*brushes*) to make further analysis on the generated patterns possible [RWK+06].

4.3 Segment Editing and Comparison

Segments stored in *Cell-Selectors* can be shown and edited interactively in all physical views in *PointCloudXplore*. This allows the user to correct or fine-tune the segmentation results manually. The user can also combine *Cell-Selectors* with logical operations to create new data patterns [RWK+06]. An example is shown in Figure 7.

(a) Original expression (b) Before post-processing (c) After post-processing

Fig. 6. This example uses the ridge regions of the mRNA expression pattern for transcription factor gene *rhomboid* (*rho*). The left image shows original expressions of *rho*. The middle image shows segmentation results before post-processing. The right image shows the results after post-processing by requesting the largest three regions and splitting.

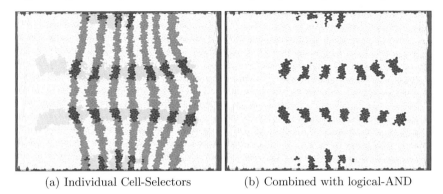

(a) Individual Cell-Selectors (b) Combined with logical-AND

Fig. 7. In the left image, *eve* ridge segments are shown in green while *rho* ridge segments are shown in red. Their logical-AND results are shown in yellow in both images highlighting those nuclei where both genes are expressed.

5 Results and Discussion

Simple thresholding is the easiest way to segment gene expression patterns. It is easy to understand and implement. However, using only one threshold value sometimes does not produce satisfactory segmentation results. For example, as shown in Figure 8(a), there are seven stripes in the *paired* (*prd*, a transcription factor) mRNA expression pattern. The first two stripes have high expression levels, but the remaining five have very low expression levels and can hardly be observed in the image. When using simple thresholding, we have to pick a very low threshold in order to see low-expression stripes. Unfortunately, such a threshold results in poor segmentation results as shown in the first two stripes to merge together, as can be seen in Figure 8(b). On the other hand, as shown in Figure 8(c) and 8(d), both edge detection and ridge detection can capture all seven stripes because these two approaches perform segmentation based on geometric properties.

The edge detection approach identifies those nuclei where gene expression changes most rapidly. One way to define a gene pattern is finding the regions enclosed by these edges. However, due to noise, detected edges are usually not continuous, as can be seen in Figure 8(c). Due to these discontinuities, no closed regions are defined and further processing is required before a binarized pattern can be produced. On the other hand, the ridge detection approach yields binarized patterns directly.

In some cases, ridge detection fails to segment gene expression patterns. For example, the "soft edges" or gradual slopes lack strong gradients. These low gradient values then get overpowered by the noise, invalidating both the edges detected and the estimated divergence used by the ridge region detection algorithm. In this case, the simple thresholding approach yields a segmentation result that is less fragmented and more consistent with human perception.

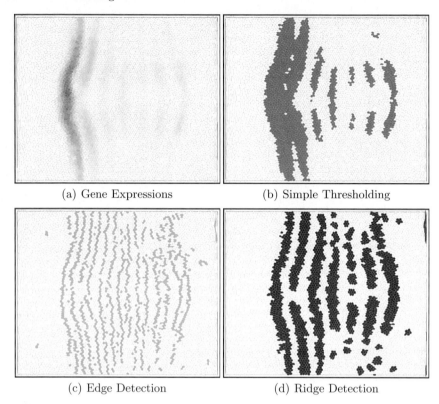

Fig. 8. mRNA expression of gene *prd*(a), segmented using simple threshold (b), edge detection (c), and ridge detection (d).

Figure 9 shows such an example. Hence, in *PointCloudXplore*, we provide all three approaches mentioned above to help users define gene patterns. They can choose the most suitable one and later edit the results interactively with our tool if necessary.

6 Conclusion and Future Work

Defining the pattern of the gene expression is a challenging task. Previous work in this field (e.g. [KJP+02] [PM04]) has concentrated on segmenting 2D images in order to extract the expression pattern of a gene. We have presented an interactive semi-automatic approach for 3D gene expression pattern segmentation based on ridge region detection. We have compared our method to standard thresholding and edge-detection-based segmentation techniques commonly used in 2D image analysis, which we have adapted to the problem.

Even though the data we show here in this paper mainly distributed on a 2D manifold, our algorithms do all the computations disregarding this

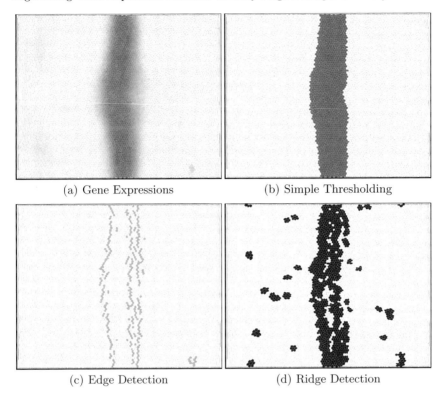

Fig. 9. mRNA expression of transcription factor gene *Krüppel* (*Kr*) (a), segmented using simple threshold (b), edge detection (c), and ridge detection (d).

manifold. In the future, we will apply these algorithms to later stage embryos, in which most organs are formed by 3D packing of cells.

Gene expression patterns are not static but show dynamic variation over time. One focus of our future work will therefore be development of analysis methods which take spatio-temporal variation of gene expression into account. Understanding how gene patterns evolve over time is essential in order to understand the complex relationships between genes. Current pattern segmentation methods take only the expression of one gene at a time into account. Development of new analysis techniques which incorporate the pattern of several genes at a time is likely to provide deeper insight into gene relationship.

Acknowledgements

This work was supported by the National Institutes of Health through grant GM70444, by the Director, Office of Science, Office of Advanced Scientific Computing Research, of the U.S. Department of Energy under Contract No.

DE-AC02-05CH11231, by the National Science Foundation through award ACI 9624034 (CAREER Award), through the Large Scientific and Software Data Set Visualization (LSSDSV) program under contract ACI 9982251, and a large Information Technology Research (ITR) grant; and by the LBNL Laboratory Directed Research Development (LDRD) program;

We thank the members of the Visualization and Computer Graphics Research Group at the Institute for Data Analysis and Visualization (IDAV) at the University of California, Davis; the members of the BDTNP at the Lawrence Berkeley National Laboratory (LBNL) and the members of the Visualization Group at LBNL.

References

[Can86] John F. Canny. A computational approach to edge detection. *IEEE Transactions on Pattern Analysis and Machine Intelligence (PAMI)*, 8(6):679–698, 1986.

[FSC04] Guoyi Fu, S.A.Hojjat, and A.C.F. Colchester. Integrating watersheds and critical point analysis for object detection in discrete 2D images. *Medical Image Analysis*, 8(3):177–185, Sep 2004.

[KFL+06] Soile V. E. Keränen, Charless C. Fowlkes, Cris L. Luengo Hendriks, Damir Sudar, David W. Knowles, Jitendra Malik, and Mark D. Biggin. Three-dimensional morphology and gene expression in the Drosophila blastoderm at cellular resolution II: Dynamics. *Genome Biology*, 7(12):R124, 2006.

[KIF85] Josef Kittler, John Illingworth, and J. Föglein. Threshold selection based on a simple image statistic. *Computer Vision, Graphics, and Image Processing*, 30(2):125–147, 1985.

[KJP+02] Sudhir Kumar, Karthik Jayaraman, Sethuraman Panchanathan, Rajalakshmi Gurunathan, Ana Marti-Subirana, and Stuart J. Newfeld. BEST: A novel computational approach for comparing gene expression patterns from early stages of Drosophila melanogaster development. *Genetics*, 162(4):2037–2047, Dec 2002.

[LKF+06] Cris L. Luengo Hendriks, Soile V. E. Keränen, Charless C. Fowlkes, Lisa Simirenko, Gunther H. Weber, Angela H. DePace, Clara Henriquez, David W. Kaszuba, Bernd Hamann, Michael B. Eisen, Jitendra Malik, Damir Sudar, Mark D. Biggin, and David W. Knowles. Three-dimensional morphology and gene expression in the Drosophila blastoderm at cellular resolution I: Data acquisition pipeline. *Genome Biology*, 7(12):R123, 2006.

[LLS97] Antonio M. López, Davis Lloret, and Joan Serrat. Multilocal creaseness based on the level-set extrinsic curvature. Technical Report 26, Centre de Visió per Computador, Dept. d'Informàtica, Universitat Autònoma de Barcelona, Spain, 1997.

[Mac67] James B. MacQueen. Some methods for classification and analysis of multivariate observations. In Lucien M. Le Cam and Jerzy Neyman, editors, *Proceedings of the Fifth Berkeley Symposium on Mathematical*

	Statistics and Probability, volume 1, pages 281–297, Berkely, CA, USA, 1967. University of California Press.
[MP00]	Geoffrey McLachlan and David Peel. *Finite Mixture Models*. Wiley, 2000.
[MvdEV96]	J.B. Antoine Maintzy, Petra A. van den Elsen, and Max A. Viergever. Evaluation of ridge seeking operators for multimodality medical image matching. *IEEE Transactions on Pattern Analysis and Machine Intelligence (PAMI)*, 18(4):353–365, Apr 1996.
[Nac84]	Lee R. Nackman. Two-dimensional critical point configuration graphs. *IEEE Transactions on Pattern Analysis and Machine Intelligence (PAMI)*, 6(4):442–449, 1984.
[PM04]	Hanchuan Peng and Eugene W. Myers. Comparing in situ mRNA expression patterns of Drosophila embryos. In *RECOMB '04: Proceedings of the eighth annual international conference on Resaerch in computational molecular biology*, pages 157–166, New York, NY, USA, 2004. ACM Press.
[RCH92]	Paul L. Rosin, Alan C. F. Colchester, and Davis J. Hawkes. Early image representation using regions defined by maximum gradient paths between singular points. *Pattern Recognition*, 25(7):695–711, 1992.
[Ros01]	Paul L. Rosin. Unimodal thresholding. *Pattern Recognition*, 34(11):2083–2096, 2001.
[RWK[+]06]	Oliver Rübel, Gunther H. Weber, Soile V. E. Keränen, Charless C. Fowlkes, Cris L. Luengo Hendriks, Nameeta Y. Shah, Mark D. Biggin, Hans Hagen, David W. Knowles, Jitendra Malik, Damir Sudar, and Bernd Hamann. PointCloudXplore: Visual analysis of 3D gene expression data using physical views and parallel coordinates. In Thomas Ertl, Ken Joy, and Beatriz Sousa Santos, editors, *Proceedings of the EUROGRAPHICS - IEEE VGTC Symposium on Visualization 2006*, pages 203–210, Lisbon, Portugal, May 2006.
[Wil98]	Michael H. F. Wilkinson. Optimizing edge detectors for robust automatic threshold selection: coping with edge curvature and noise. *Graph. Models Image Process.*, 60(5):385–401, 1998.

A

Color Plates

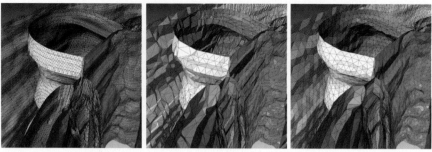

Fig. 8. Out-of-core surface reconstruction, simplification and remeshing. (a) Reconstruction of weighted label volume. (b) Simplified surface using our method (reduction 90% compared to (a)). (c) Simplified surface after remeshing step.

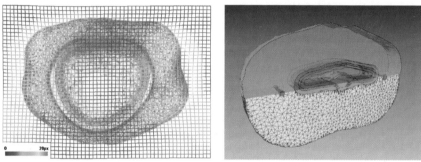

Fig. 6, 9. (a) Visualization of the elastic deformation of one section image. (b) Perspective view of a 3D model segment based on 10 automatically registered and segmented images of cross-sections of a barley grain (15-fold stretched in height).

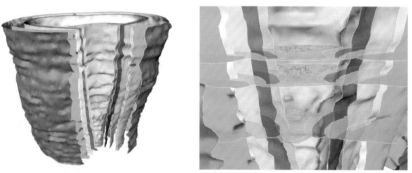

Fig. 10. Integration of experimental marker data (in red) into a part of a 3D model.

(DERCKSEN ET AL., PP. 3–25)

A Color Plates 331

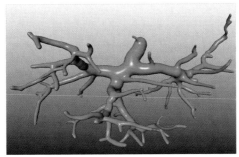

Fig. 10. Visualization of the portal vein derived from a clinical CT dataset with 136 edges.

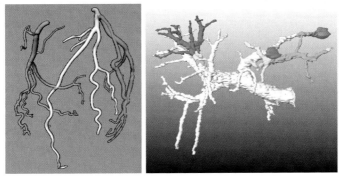

Fig. 12. Left: Colors encode to which major coronary artery the branches belong. The vascular model is visualized with convolution surfaces. Right: Colors encode the discretized distance to a tumor. Red indicates distances below 5 mm, yellow 5-10 mm and green 10-15 mm.

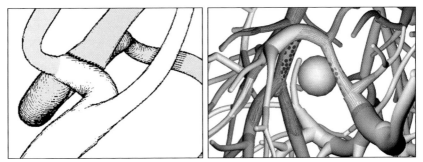

Fig. 13. Illustrative rendering techniques applied to visualizations with truncated cones. Left: Hatching lines support the shape perception. At a crossing, the depth perception is enhanced by interrupting hatching lines at the more distant vessel branch. Right: Two colors serve to distinguish two vascular trees. Different textures are mapped onto the surface to indicate distance intervals to a simulated tumor (yellow sphere). (Courtesy Christian Hansen, MeVis Research Bremen).

(PREIM AND OELTZE, PP. 39–59)

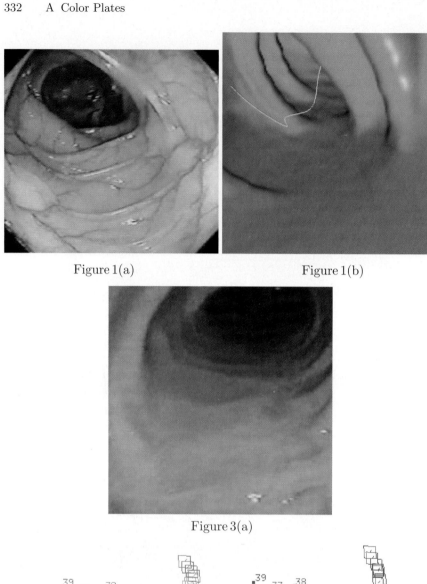

Figure 1(a) Figure 1(b)

Figure 3(a)

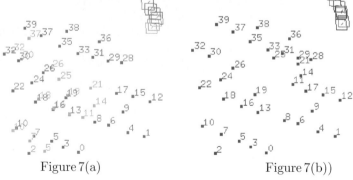

Figure 7(a) Figure 7(b))

(KAUFMAN AND WANG, PP. 61–74)

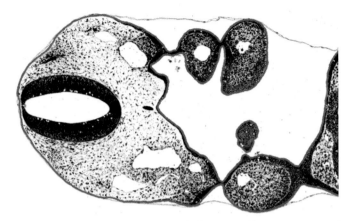

Fig. 11. 14-day-old embryo of *Tupaia belangeri*: Contour extraction result (surface ectoderm and endoderm shown in *blue*) of a quite complex embryonic surface.

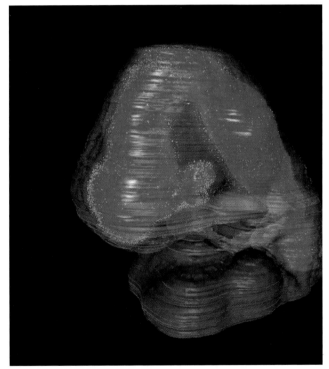

Fig. 12. 15-day-old embryo of *Tupaia belangeri*: Semitransparent surface ectoderm (*grey*, manually extracted), inner surface of the brain (*blue*, manually extracted), and cellular events (*red* and *yellow*; modeled as spheres) can be visualized in combination.

(KIENEL ET AL., PP. 77–97)

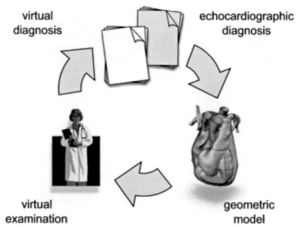

The standard tutoring cycle of the VES framework. Based upon a cardiac finding an instance of the pathologic heart is created and presented to the student. The diagnosis of the virtual examination are stored in a second instance of the finding module and automatically compared to the original finding.

Left: A trifurcation modeled with the vessel model. All vessels are composed of an inner an outer layer. Right: Multi resolution analysis of the vessel model. All vessels are composed of the same number of mesh control points, but the detail increases from left to right. The leftmost vessel can be described by a single value per layer and cross-section, respectively.

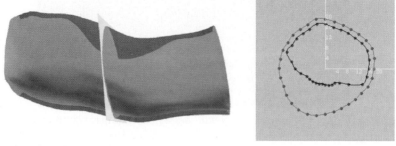

Stenosis of a blood vessel and the respective view of the cross-section in the 4D-blood vessel editor.

(REIS ET AL., PP. 99–119)

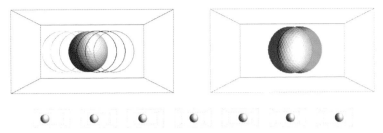

Fig. 1. Two visualization techniques to depict the motion of the golf ball volume dataset. The current 3D volume is rendered using isosurface shading. Edges of preceding and succeeding 3D volumes are shown (*left*); the combinations of the current and the succeeding 3D volumes are rendered with different colors (*right*). Renderings of seven time steps extracted from the dataset (*bottom*).

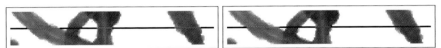

Fig. 2. Color coding of the depth structure. The detail images show the differences when using an absolute mapping (*left*) and a relative mapping (*right*). In the left image the right vessel seems to have the same depth value as the left ones. The right image show that actually the right vessel is further away from the observer.

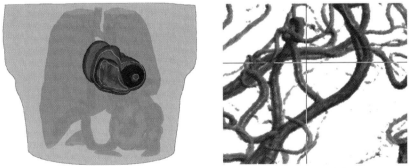

Fig. 5. Context information added to the visualization of heart motion.
Fig. 6. The cross-hair cursor visualizes depth information. Here a mapping to a rainbow gradient is used.

(MEYER-SPRADOW ET AL., PP. 121–133)

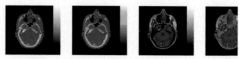

Fig. 9: Bone. (a) CT scan (green), (b) MRI scan (red), (c) superimposed (yellow).

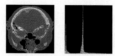

Fig. 10. (a) CT scan and (b) MRI scan with intensity histograms.

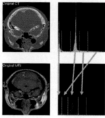

Fig. 11. Mapping of CT histogram (green) onto MRI histogram (red). Bins (purple).

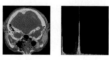

Fig. 12. CT (green) with MRI (a) original (b) transformed. Histogram in (b) shows both original (red) and transformed (yellow) MRI. Yellow is a better match.

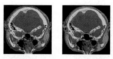

Fig. 13. (a), (c) unregistered; (b), (d) registered CT (cyan) and MRI (red). Left pair: Original MRI. Right pair: Transformed MRI. White: Matched bone.

(MEYER, PP. 137–151)

A Color Plates   337

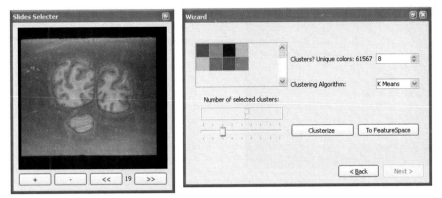

Fig. 3. The clusterization result is a sorted cluster list, which is drawn using cluster representatives (right). Clusters can be selected by clicking at the representatives and increased by dragging the slider. Any selection can be adaptively refined by reclusterization. The segmentation result is shown in a preview in object space by overlaying a 2D contour (pink) with any 2D slice (left)

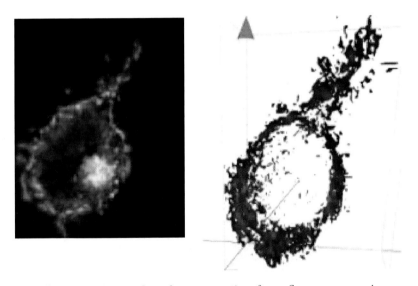

Fig. 12. Segmentation and surface extraction from fluorescence microscopy data representing a cancer cell. We extracted the regions that show up as red (right) and compare the result to one of the 2D slices (left).

(IVANOVSKA AND LINSEN, PP. 153–170)

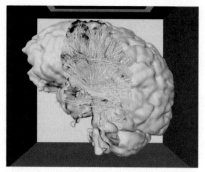 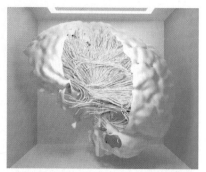

Fig. 4. White matter fibers within a cut-away approximation of the cortex. Left: local illumination. Right: global illumination. Brain dataset courtesy of Gordon Kindlmann at the Scientific Computing and Imaging Institute, University of Utah, and Andrew Alexander, W. M. Keck Laboratory for Functional Brain Imaging and Behavior, University of Wisconsin-Madison.

Fig. 6. Superquadric glyphs derived from DT-MRI data. The extent and orientation of the glyphs invite the visual system to integrate their small shapes into larger coherent structures. This visual integration process may be enhanced by global illumination. Left: local illumination. Right: global illumination.

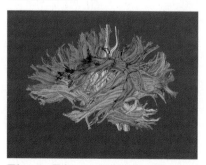 

Fig. 8. Fibers automatically clustered into fascicles. Left: local illumination. Right: global illumination.

(BANKS AND WESTIN, PP. 173–184)

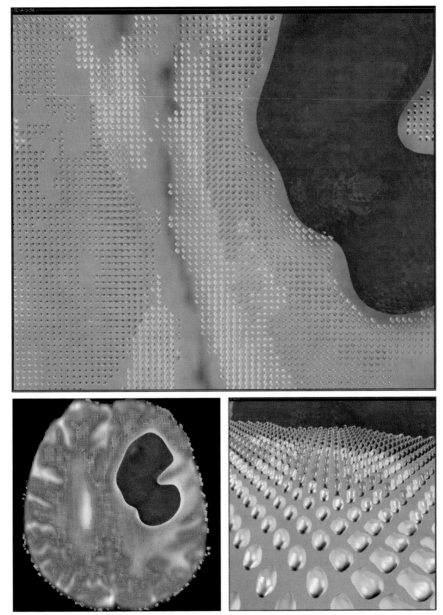

Fig. 13. Different views of a glyph-based visualization of a slice through a data set with a tumor using deformed spheres. The colors indicate the direction of maximum diffusion. Isotropic voxels with a fractional anisotropy of less than 0.2 are omitted.

(DOMIN ET AL., PP. 185–204)

Fig. 2. Topology of bioelectric field on epicardium combined with dense texture representation. Green points: potential minima, blue points: potentials maxima, red points: saddle points. Left and middle images: anterior view. Right image seen from the left of the torso.

Fig. 6. Textures computed on sagittal and coronal clipping planes reveal details of the dipolar source and its interaction with the surrounding anisotropic tissue.

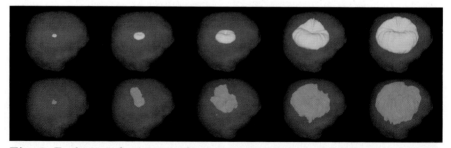

Fig. 8. Evolution of stream surface integrated along return current under the parameterization induced by the increasing radius of a seeding circle. *Top row:* isotropic white matter model. *Bottom row:* anisotropic white matter model.

(TRICOCHE ET AL., PP. 205–220)

A Color Plates 341

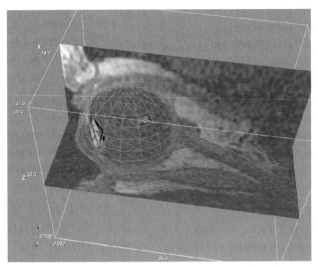

Fig. 3. Procedure for determining the actual directions of gaze. The lens is automatically segmented and its centroid is determined. The actual direction of gaze is determined by this centroid and the centre of the sphere that has been fitted to the eye.

Fig. 5. A visualisation with four advection volumes over 12 time-varying vector fields. The volumes have been chosen in the same plane as the optic nerve, on venous landmarks in order to verify the findings of a previous 2-D study. In each of the four cases, the green sphere in the middle is placed in the centre direction of gaze, and is advected both to the left and to the right.

(BOTHA ET AL., PP. 221–233)

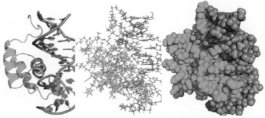

Fig. 1. Visualizing protein-DNA complexes. Complex of the antennapedia homeodomain of *drosophila melanogaster* (fruit fly) and its DNA binding site shown using three different types of visualization. The protein is shown in green, and the DNA fragments in orange. (a) The cartoon view highlights the position of the binding site where the DNA sits. (b) The skeletal model emphasizes the chemical nature of both molecules. (c) The space-filling diagram shows the tight binding between the protein and the ligand. Each representation is complementary to the others, and the biochemist uses all three of them when studying a protein.

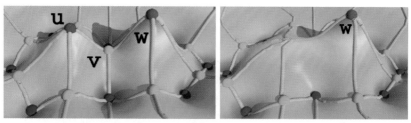

Fig. 5. (a) A simple height function with two maxima surrounded by multiple local minima and its Morse-Smale complex. (b) Smoother height functions are created by canceling pairs of critical points. Canceling a saddle-maximum pair removes a topological feature. The function is modified locally by rerouting integral lines to the remaining maximum.

Fig. 7. (a) The atomic density function computed for chain D of the Barnase-Barstar complex. Darker regions correspond to protrusions and lighter regions to cavities. (b) Peak-valley segmentation of the surface. (c) Coarse segmentation obtained by performing a series of critical pair cancellations.

(NATARAJAN ET AL., PP. 237–255)

A Color Plates 343

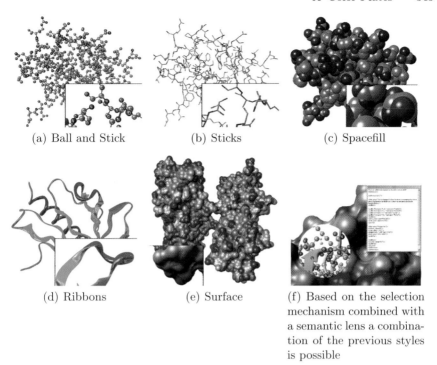

(a) Ball and Stick (b) Sticks (c) Spacefill

(d) Ribbons (e) Surface (f) Based on the selection mechanism combined with a semantic lens a combination of the previous styles is possible

Fig. 1. The images show the common visualization styles as screen shots from the *BioBrowser* showing the molecule 1I7A. In the lower corner of each style a close up view is shown. The last one shows part of the molecule 1A2J.

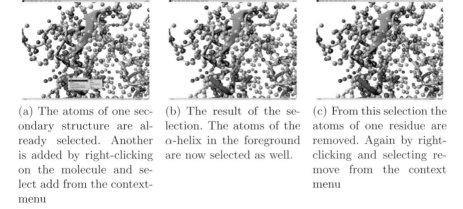

(a) The atoms of one secondary structure are already selected. Another is added by right-clicking on the molecule and select add from the context-menu

(b) The result of the selection. The atoms of the α-helix in the foreground are now selected as well.

(c) From this selection the atoms of one residue are removed. Again by right-clicking and selecting remove from the context menu

Fig. 6. Interactive selection process based on mouse input. The yellow atoms represent the current selection.

(OFFEN AND FELLNER, PP. 257–273)

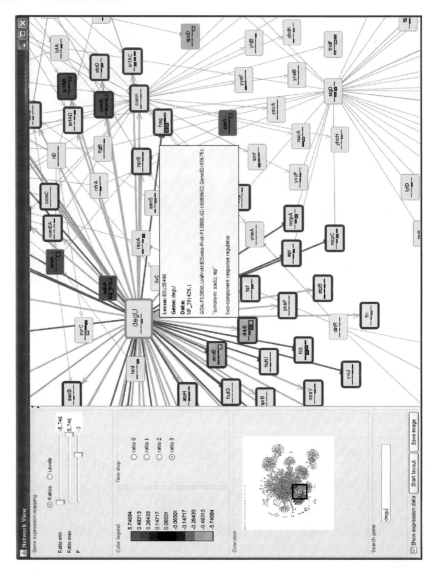

Fig. 6. Screen shot of GENeVis. The user can switch between expression level and expression ratio visualization with the radio buttons in the top left corner. The sliders together with the time point radio buttons can be used to select a subset of genes that have expression data within the ranges set by the sliders and buttons. The background of the selected genes is then colored according to their expression value. The color legend is shown to the left of the time point selection buttons. An overview display is shown to help the user in maintaining a view of the network context. The square indicates the area of the network visible in the main display, and it can be dragged to pan the view in the main display. The search box can be used to find a specific gene, and when found, the display automatically pans such that the gene found is centered in the view.

(WESTENBERG ET AL., PP. 293–311)

A Color Plates 345

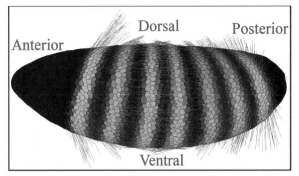
Figure 1

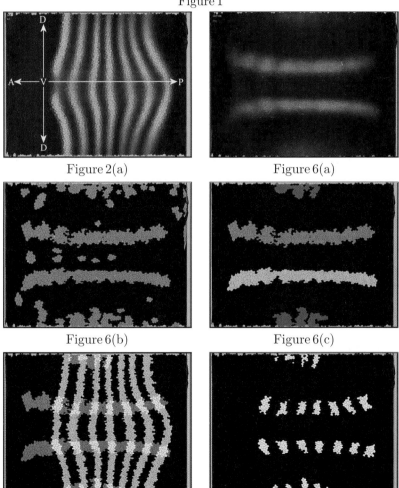

Figure 2(a)

Figure 6(a)

Figure 6(b)

Figure 6(c)

Figure 7(a)

Figure 7(b)

(HUANG ET AL., PP. 313–327)

Printing: Krips bv, Meppel, The Netherlands
Binding: Stürtz, Würzburg, Germany